THE BRAIN THAT CHANGES ITSELF

Norman Doidge, M.D., is a psychiatrist, a psychoanalyst, and the *New York Times* bestselling author of *The Brain's Way of Healing*, which was the winner of the 2015 Gold Nautilus Book Award in Science & Cosmology. He was on the Research Faculty of the Center for Psychoanalytic Training and Research at Columbia University's Department of Psychiatry in New York City, and on the faculty of the University of Toronto's Department of Psychiatry for thirty years. He lives in Toronto.

A *Slate* Pick for a Best Book of the Year
A *Globe & Mail* Best Book of the Year
A *Guardian* Best Book of the Year

Praise for *The Brain That Changes Itself*:

"Doidge's book is a remarkable and hopeful portrait of the endless adaptability of the human brain. . . . Only a few decades ago, scientists considered the brain to be fixed or 'hardwired,' and considered most forms of brain damage, therefore, to be incurable. Dr. Doidge, an eminent psychiatrist and researcher, was struck by how his patients' own transformations belied this, and set out to explore the new science of neuroplasticity by interviewing both scientific pioneers in neuroscience and patients who have benefited from neurorehabilitation. Here he describes in fascinating personal narratives how the brain, far from being fixed, has remarkable powers of changing its own structure and compensating for even the most challenging neurological conditions." —Oliver Sacks

"In bookstores, the science aisle generally lies well away from the self-help section, with hard reality on one set of shelves and wishful thinking on the other. But Norman Doidge's fascinating synopsis of the current revolution in neuroscience straddles this gap: the age-old distinction between the brain and the mind is crumbling fast as the power of positive thinking finally gains scientific credibility. Mind-bending, miracle-working, reality-busting stuff, with implications . . . not only for individual patients with neurologic disease but for all human beings, not to mention human culture, human learning and human history."

—*The New York Times*

"An utterly wonderful book—without question one of the most important books about the brain you will ever read; yet it is beautifully written, immensely approachable, and full of humanity. Its message is one of hope: it is not just our brains that shape our thinking, but our thinking that, very definitely, shapes our brains."

—Iain McGilchrist MA (Oxon), BM, FRCPsych, FRSA author of *The Master and His Emissary: The Divided Brain and the Making of the Western World*

"Lucid and absolutely fascinating . . . engaging, educational and riveting. It satisfies, in equal measure, the mind and the heart. Doidge is able to explain current research in neuroscience with clarity and thoroughness. He presents the ordeals of the patients about whom he writes—people born with parts of their brains missing, people with learning disabilities, people recovering from strokes—with grace and vividness. In the best medical narratives—and the works of Doidge . . . join that fraternity—the narrow bridge between body and soul is traversed with courage and eloquence." —*Chicago Tribune*

"Brilliant . . . Doidge has identified a tidal shift in basic science . . . The implications are monumental." —*London Times*

"Readers will want to read entire sections aloud and pass the book on to someone who can benefit from it. [Doidge] links scientific experimentation with personal triumph in a way that inspires awe for the brain, and for these scientists' faith in its capacity." —*The Washington Post*

"Doidge tells one spellbinding story after another as he travels the globe interviewing the scientists and their subjects who are on the cutting edge of a new age. It may be hard to imagine that a book so rich in science can also be a page-turner, but this one is hard to set down."

—Jeff Zimman, Posit Science, e-newsletter

"It takes a rare talent to explain science to the rest of us. Oliver Sacks is a master at this. So was the late Stephen Jay Gould. And now there is Norman Doidge. A terrific book. You don't have to be a brain surgeon to read it—just a person with a curious mind. Doidge is the best possible guide. He has a fluent and unassuming style, and is able to explain difficult concepts without talking down to his readers. The case study is the psychiatric literary genre par excellence, and Doidge does not disappoint. What makes neuroplasticity so exciting is that it completely upends how we look at the brain. It says that the brain, far from being a collection of specialized parts, each fixed in its location and function, is in fact a dynamic organ, one that can rewire and rearrange itself as the need arises. It is an insight from which all of us can benefit. People with severe afflictions—strokes, cerebral palsy, schizophrenia, learning disabilities, obsessive compulsive disorders and the like—are the most obvious candidates, but who among us would not like to tack on a few IQ points or improve our memories? Buy this book. Your brain will thank you." —*The Globe & Mail* (Toronto)

"The most readable and best general treatment of this subject to date."
—Michael M. Merzenich, Ph.D., Francis Sooy Professor, Keck Center for Integrative Neurosciences, University of California at San Francisco

"A masterfully guided tour through the burgeoning field of neuroplasticity research." —*Discover*

"Beautifully written and brings life and clarity to a variety of neuropsychiatric problems that affect children and adults. . . . It reads a bit like a science detective story—you do not need a Ph.D. to benefit from the wisdom imparted here." —Barbara Milrod, M.D. Psychiatry, Weill Medical College of Cornell University

"A riveting, essential book. Doidge covers an impressive amount of ground and is an expert guide, a sense of wonder always enriching his skill as an explicator of subject matter that in less able hands could be daunting or even impenetrable. These stories are most emotionally satisfying. . . . Doidge addresses how cultural influences literally 'shape' our brain. . . . It becomes clear that our response to the world around us is not only a social or psychological phenomenon, but often a lasting neurological process." —*The Gazette* (Montreal)

"Fascinating." —*Psychology Today*

"Brilliant . . . This book is a wonderful and engaging way of reimagining what kind of creatures we are."

—Jeanette Winterson, novelist, Order of the British Empire, *Guardian*, Best Book of 2008

"Superb. Brilliant. I devoured it."

—V.S. Ramachandran, MD, PHD, Director of the Center for Brain and Cognition, University of California, San Diego, Author of *Phantoms of the Brain*

"Astonishing. This book will inevitably draw comparisons to the work of Oliver Sacks. Doidge has a prodigious gift for rendering the highly technical highly readable. It's hard to imagine a more exciting topic—or a better introduction to it." —*The Kitchener Waterloo Record*

"We've long known that brain changes can affect our psychology and what we think. Norman Doidge has shown that what and how we think can change our brains. He has illuminated the foundations of psychological healing." —Charles Hanly, Ph.D., President, International Psychoanalytical Association

"A panoramic examination of plasticity's profound implications. Injured or dysfunctional cells and circuits can indeed be regenerated and rewired; the location of a given function can, astonishingly, move from one place to another. The body's lifespan may not have to outpace its mental lifespan. . . . Everything that you can see happen in a young brain can happen in an older brain. Deterioration can be reversed by twenty to thirty years." —*Toronto Daily Star*

"An eloquently written book about the boundless potential of the human brain. In addition to being a fascinating, informative and emotionally powerful read, it has the potential to enlighten parents

about the incredible learning-enhancing opportunities now available to them and their children. Addresses learning disabilities in a unique way and could revolutionize the way educational issues are addressed."
—*The Jewish Week*

"A rich banquet of brain-mind plasticity, communicated in a brilliantly clear writing style."
—Jaak Panksepp, Ph.D., Head, Affective Neuroscience Research, Northwestern University

"Why isn't this book on the top of the bestseller list of all time? In my mind the recognition that the brain is plastic and can actually change itself with exercise and understanding is a huge leap in the history or mankind—far greater than landing on the moon. Clear, fascinating, and gripping."
—Jane S. Hall, *International Psychoanalysis*

"A hymn to life."
—*Panorama* (Italy)

"An owner's manual for the brain, giving advice on how to maintain intellect and reasoning functions as we grow older, Doidge's book gives the reader hope for the future. I highly recommend this book to anyone who enjoys stories of triumph against all odds. Extremely engrossing, and always informative."
—Curled Up With a Good Book

"Doidge . . . turns everything we thought we knew about the brain upside down."
—*Publishers Weekly*

"Doidge leads the field in an extraordinary era for popular-science books about the brain. . . . This has to be explained in layperson's terms, no easy task, but there's a host of neurologists, psychologists, cognitive scientists and journalists explaining and interpreting the latest research. . . . Two years ago, when the journal *Cerebrum* at the Dana Foundation in the US updated its list of great books about the brain for the general reader, it found there were already 30,000 brain-related books in English (and the total would no doubt be higher now). Aided by scientific advisers and readers, it produced a new list—with *The Brain That Changes Itself* at No. 1."
—*The Age* (Australia)

The Brain That Changes Itself

Stories of Personal Triumph from the Frontiers of Brain Science

NORMAN DOIDGE, M.D.

A James H. Silberman Book

PENGUIN BOOKS
An imprint of Penguin Random House LLC
penguinrandomhouse.com

First published in the United States of America by Viking,
an imprint of Penguin Random House LLC, 2007
Published in Penguin Books 2007

A Penguin Life Book

ISBN 9780143113102 (paperback)

THE LIBRARY OF CONGRESS HAS CATALOGED THE HARDCOVER EDITION AS FOLLOWS:
Doidge, Norman.
The Brain that changes itself : stories of personal triumph from the frontiers of brain science / Norman Doidge.
p. cm.
ISBN 9780670038305 (hardcover)
1. Neuroplasticity. 2. Brain damage—Patients—Rehabilitation. I. Title.
QP363.3.D65 2007
612.8—dc22 2006049224

Printed in the United States of America
45th Printing

Set in Minion Pro
Designed by Spring Hoteling

For Eugene L. Goldberg, M.D.,
because you said you might like to read it

Contents

Note to the Reader

All the names of people who have undergone neuroplastic transformations are real, except in the few places indicated, and in the cases of children and their families.

The Notes and References section at the end of the book includes comments on both the chapters and the appendices.

Preface

This book is about the revolutionary discovery that the human brain can change itself, as told through the stories of the scientists, doctors, and patients who have together brought about these astonishing transformations. Without operations or medications, they have made use of the brain's hitherto unknown ability to change. Some were patients who had what were thought to be incurable brain problems; others were people without specific problems who simply wanted to improve the functioning of their brains or preserve them as they aged. For four hundred years this venture would have been inconceivable because mainstream medicine and science believed that brain anatomy was fixed. The common wisdom was that after childhood the brain changed only when it began the long process of decline; that when brain cells failed to develop properly, or were injured, or died, they could not be replaced. Nor could the brain ever alter its structure and find a new way to function if part of it was damaged. The theory of the unchanging brain decreed that people who were born with brain or mental limitations, or who sustained brain damage, would be limited or damaged for life. Scientists who wondered if the healthy brain might be improved or preserved through activity or mental exercise were told not to waste their time. A neurological nihilism—a sense that treatment for many brain problems was ineffective or even unwarranted—had taken hold, and it spread through our culture, even stunting our overall view of human nature. Since the brain could not change, human nature, which emerges from it, seemed necessarily fixed and unalterable as well.

The belief that the brain could not change had three major sources: the fact that brain-damaged patients could so rarely make full recoveries; our inability to observe the *living* brain's microscopic activities; and the idea—dating back to the beginnings of modern science—that the brain is like a glorious machine. And while machines do many extraordinary things, they don't change and grow.

I became interested in the idea of a changing brain because of my work as a research psychiatrist and psychoanalyst. When patients did not progress psychologically as much as hoped, often the conventional medical wisdom was that their problems were deeply "hardwired" into an unchangeable brain. "Hardwiring" was another machine metaphor coming from the idea of the brain as computer hardware, with permanently connected circuits, each designed to perform a specific, unchangeable function.

When I first heard news that the human brain might not be hardwired, I had to investigate and weigh the evidence for myself. These investigations took me far from my consulting room.

I began a series of travels, and in the process I met a band of brilliant scientists, at the frontiers of brain science, who had, in the late 1960s or early 1970s, made a series of unexpected discoveries. They showed that the brain changed its very structure with each different activity it performed, perfecting its circuits so it was better suited to the task at hand. If certain "parts" failed, then other parts could sometimes take over. The machine metaphor, of the brain as an organ with specialized parts, could not fully account for changes the scientists were seeing. They began to call this fundamental brain property "neuroplasticity."

Neuro is for "neuron," the nerve cells in our brains and nervous systems. *Plastic* is for "changeable, malleable, modifiable." At first many of the scientists didn't dare use the word "neuroplasticity" in their publications, and their peers belittled them for promoting a fanciful notion. Yet they persisted, slowly overturning the doctrine of the unchanging brain. They showed that children are not always stuck with the mental abilities they are born with; that the damaged brain can often reorganize itself so that when one part fails, another can often substitute; that if brain cells die, they can at times be replaced; that many "circuits" and even basic reflexes that we think are hardwired are not. One of these scientists even showed that thinking, learning, and acting can turn our genes on or off, thus shaping our brain anatomy and our behavior—surely one of the most extraordinary discoveries of the twentieth century.

In the course of my travels I met a scientist who enabled people who had been blind since birth to begin to see, another who enabled the deaf to hear; I spoke with people who had had strokes decades before and had been declared incurable, who were helped to recover with neuroplastic treatments; I met people whose learning disorders were cured and whose IQs were raised; I saw evidence that it is possible for eighty-year-olds to sharpen their memories to function the way they did when they were fifty-five. I saw people rewire their brains with their thoughts, to cure previously incurable obsessions and traumas. I spoke with Nobel laureates who were hotly debating how we must rethink our model of the brain now that we know it is ever changing.

The idea that the brain can change its own structure and function through thought and activity is, I believe, the most important alteration in our view of the brain since we first sketched out its basic anatomy and the workings of its basic component, the neuron. Like all revolutions, this one will have profound effects, and this book, I hope, will begin to show some of them. The neuroplastic revolution has implications for, among other things, our understanding of how love, sex, grief, relationships, learning, addictions, culture, technology, and psychotherapies change our brains. All of the humanities, social sciences, and physical sciences, insofar as they deal with human nature, are affected, as are all forms of training. All of these disciplines will have to come to terms with the fact of the self-changing brain and with the realization that the architecture of the brain differs from one person to the next and that it changes in the course of our individual lives.

While the human brain has apparently underestimated itself, neuroplasticity isn't all good news; it renders our brains not only more resourceful but also more vulnerable to outside influences. Neuroplasticity has the power to produce more flexible but also more rigid behaviors—a phenomenon I call "the plastic paradox." Ironically, some of our most stubborn habits and disorders are products of our plasticity. Once a particular plastic change occurs in the brain and becomes well established, it can prevent other changes from occurring. It is by understanding both the positive and negative effects of plasticity that we can truly understand the extent of human possibilities.

Because a new word is useful for those who do a new thing, I call the practitioners of this new science of changing brains "neuroplasticians."

What follows is the story of my encounters with them and the patients they have transformed.

1

A Woman Perpetually Falling . . .

Rescued by the Man Who Discovered the Plasticity of Our Senses

And they saw the voices.
Exodus 20:18

Cheryl Schiltz feels like she's perpetually falling. And because she feels like she's falling, she falls.

When she stands up without support, she looks, within moments, as if she were standing on a precipice, about to plummet. First her head wobbles and tilts to one side, and her arms reach out to try to stabilize her stance. Soon her whole body is moving chaotically back and forth, and she looks like a person walking a tightrope in that frantic seesaw moment before losing his balance—except that both her feet are firmly planted on the ground, wide apart. She doesn't look like she is only afraid of falling, more like she's afraid of being pushed.

"You look like a person teetering on a bridge," I say.

"Yeah, I feel I am going to jump, even though I don't want to."

Watching her more closely, I can see that as she tries to stand still, she jerks, as though an invisible gang of hoodlums were pushing and shoving her, first from one side, then from another, cruelly trying to knock her over. Only this gang is actually inside her and has been doing this to her for five years. When she tries to walk, she has to hold on to a wall, and still she staggers like a drunk.

For Cheryl there is no peace, even after she's fallen to the floor.

"What do you feel when you've fallen?" I ask her. "Does the sense of falling go away once you've landed?"

"There have been times," says Cheryl, "when I literally lose the sense of the feeling of the floor . . . and an imaginary trapdoor opens up and swallows me." Even when she has fallen, she feels she is still falling, perpetually, into an infinite abyss.

Cheryl's problem is that her vestibular apparatus, the sensory organ for the balance system, isn't working. She is very tired, and her sense that she is in free fall is driving her crazy because she can't think about anything else. She fears the future. Soon after her problem began, she lost her job as an international sales representative and now lives on a disability check of $1,000 a month. She has a newfound fear of growing old. And she has a rare form of anxiety that has no name.

An unspoken and yet profound aspect of our well-being is based on having a normally functioning sense of balance. In the 1930s the psychiatrist Paul Schilder studied how a healthy sense of being and a "stable" body image are related to the vestibular sense. When we talk of "feeling settled" or "unsettled," "balanced" or "unbalanced," "rooted" or "rootless," "grounded" or "ungrounded," we are speaking a vestibular language, the truth of which is fully apparent only in people like Cheryl. Not surprisingly, people with her disorder often fall to pieces psychologically, and many have committed suicide.

We have senses we don't know we have—until we lose them; balance is one that normally works so well, so seamlessly, that it is not listed among the five that Aristotle described and was overlooked for centuries afterward.

The balance system gives us our sense of orientation in space. Its sense organ, the vestibular apparatus, consists of three semicircular canals in the inner ear that tell us when we are upright and how gravity is affecting our bodies by detecting motion in three-dimensional space. One canal detects movement in the horizontal plane, another in the vertical plane, and another when we are moving forward or backward. The semicircular canals contain little hairs in a fluid bath. When we move our head, the fluid stirs the hairs, which send a signal to our brains telling us that we have increased our velocity in a particular direction. Each movement requires a corresponding adjustment of the rest of the body. If we move our heads forward, our brains tell an appropriate segment of our bodies to adjust, unconsciously, so that we can offset that change in our center of gravity and maintain our balance. The signals from the vestibular apparatus go along a nerve to a specialized clump of neurons in our brain, called the "vestibular nuclei," which process them, then send commands to our muscles to adjust themselves. A healthy vestibular apparatus also has a strong link to our visual system. When you run after a bus, with your head bouncing up and down as you race forward, you are able to keep that moving bus at the center of your gaze because your vestibular apparatus sends messages to your brain, telling it the speed and direction in which you are running. These signals allow your brain to rotate and adjust the position of your eyeballs to keep them directed at your target, the bus.

I am with Cheryl, and Paul Bach-y-Rita, one of the great pioneers in understanding brain plasticity, and his team, in one of his labs. Cheryl is hopeful about today's experiment and is stoical but open about her condition. Yuri Danilov, the team biophysicist, does the calculations on the data they are gathering on Cheryl's vestibular system. He is Russian, extremely smart, and has a deep accent. He says, "Cheryl is patient who has lost vestibular system—ninety-five to one hundred percent."

By any conventional standard, Cheryl's case is a hopeless one. The conventional view sees the brain as made up of a group of specialized processing modules, genetically hardwired to perform specific functions and those alone, each developed and refined over millions of years of evolution. Once one of them is this damaged, it can't be replaced. Now that her vestibular system is damaged, Cheryl has as much chance of regaining her balance as a person whose retina has been damaged has of seeing again.

But today all that is about to be challenged.

She is wearing a construction hat with holes in the side and a device inside it called an accelerometer. Licking a thin plastic strip with small electrodes on it, she places it on her tongue. The accelerometer in the hat sends signals to the strip, and both are attached to a nearby computer. She laughs at the way she looks in the hat, "because if I don't laugh I will cry."

This machine is one of Bach-y-Rita's bizarre-looking prototypes. It will replace her vestibular apparatus and send balance signals to her brain from her tongue. The hat may reverse Cheryl's current nightmare. In 1997 after a routine hysterectomy, Cheryl, then thirty-nine years old, got a postoperative infection and was given the antibiotic gentamicin. Excessive use of gentamicin is known to poison the inner ear structures and can be responsible for hearing loss (which Cheryl doesn't have), ringing in the ears (which she does), and devastation to the balance system. But because gentamicin is cheap and effective, it is still prescribed, though usually for only a brief period of time. Cheryl says she was given the drug way beyond the limit. And so she became one of a small tribe of gentamicin's casualties, known among themselves as Wobblers.

Suddenly one day she discovered she couldn't stand without falling. She'd turn her head, and the whole room would move. She couldn't figure out if she or the walls were causing the movement. Finally she got to her feet by hanging on to the wall and reached for the phone to call her doctor.

When she arrived at the hospital, the doctors gave her various tests to see if her vestibular function was working. They poured freezing-cold and warm water into her ears and tilted her on a table. When they asked her to stand with her eyes closed, she fell over. A doctor told her, "You have no vestibular function." The tests showed she had about 2 percent of the function left.

"He was," she says, "so nonchalant. 'It looks like a side effect of the gentamicin.' " Here Cheryl gets emotional. "Why in the world wasn't I told about that? 'It's permanent,' he said. I was alone. My mother had taken me to the doctor, but she went off to get the car and was waiting for me outside the hospital. My mother asked, 'Is it going to be okay?' And I looked at her and said, 'It's permanent . . . this is never going to go away.' "

Because the link between Cheryl's vestibular apparatus and her visual system is damaged, her eyes can't follow a moving target smoothly. "Everything I see bounces like a bad amateur video," she says. "It's as though everything I look at seems made of Jell-O, and with each step I take, everything wiggles."

Although she can't track moving objects with her eyes, her vision is all she has to tell her that she is upright. Our eyes help us know where we are in space by fixing on horizontal lines. Once when the lights went out, Cheryl immediately fell to the floor. But vision proves an unreliable crutch for her, because any kind of movement in front of her—even a person reaching out to her—exacerbates the falling feeling. Even zigzags on a carpet can topple her, by initiating a burst of false messages that make her think she's standing crookedly when she's not.

She suffers mental fatigue, as well, from being on constant high alert. It takes a lot of brain power to maintain an upright position—brain power that is taken away from such mental functions as memory and the ability to calculate and reason.

While Yuri is readying the computer for Cheryl, I ask to try the machine. I put on the construction worker's hat and slip into my mouth the plastic device with electrodes on it, called a tongue display. It is flat, no thicker than a stick of chewing gum.

The accelerometer, or sensor, in the hat detects movement in two planes. As I nod my head, the movement is translated onto a map on the computer screen that permits the team to monitor it. The same map is projected onto a small array of 144 electrodes implanted in the plastic strip on my tongue. As I tilt forward, electric shocks that feel like champagne bubbles go off on the front of my tongue, telling me that I am bending forward. On the computer screen I can see where my head is. As I tilt back, I feel the champagne swirl in a gentle wave to the back of my tongue. The same happens when I tilt to the sides. Then I close my eyes and experiment with finding my way in space with my tongue. I soon forget that the sensory information is coming from my tongue and can read where I am in space.

Cheryl takes the hat back; she keeps her balance by leaning against the table.

"Let's begin," says Yuri, adjusting the controls.

Cheryl puts on the hat and closes her eyes. She leans back from the table, keeping two fingers on it for contact. She doesn't fall, though she has no indication whatsoever of what is up and down except the swirling of the champagne bubbles over her tongue. She lifts her fingers from the table. She's not wobbling anymore. She starts to cry—the flood of tears that comes after a trauma; she can open up now that she has the hat on and feels safe. The first time she put on the hat, the sense of perpetual falling left her—for the first time in five years. Her goal today is to stand, free, for twenty minutes, with the hat on, trying to keep centered. For anyone—not to mention a Wobbler—to stand straight for twenty minutes requires the training and skill of a guard at Buckingham Palace.

She looks peaceful. She makes minor corrections. The jerking has stopped, and the mysterious demons that seemed to be inside her, pushing her, shoving her, have vanished. Her brain is decoding signals from her artificial vestibular apparatus. For her, these moments of peace are a miracle—a neuroplastic miracle, because somehow these tingling sensations on her tongue, which normally make their way to the part of the brain called the sensory cortex—the thin layer on the surface of the brain that processes the sense of touch—are making their way, through a novel pathway in the brain, to the brain area that processes balance.

"We are now working on getting this device small enough so that it is hidden in the mouth," says Bach-y-Rita, "like an orthodontist's mouth retainer. That's our goal. Then she, and anyone with this problem, will have a normal life restored. Someone like Cheryl should be able to wear the apparatus, talk, and eat without anyone knowing she has it.

"But this isn't just going to affect people damaged by gentamicin," he continues. "There was an article in *The New York Times* yesterday on falls in the elderly. Old people are more frightened of falling than of being mugged. A third of the elderly fall, and because they fear falling, they stay home, don't use their limbs, and become more physically frail. But I think part of the problem is that the vestibular sense—just like hearing, taste, eyesight, and our other senses—starts to weaken as we age. This device will help them."

"It's time," says Yuri, turning off the machine.

Now comes the second neuroplastic marvel. Cheryl removes the tongue device and takes off the hat. She gives a big grin, stands free with her eyes closed, and doesn't fall. Then she opens her eyes and, still not touching the table, lifts one foot off the ground, so she's balancing on the other.

"I love this guy," she says, and goes over and gives Bach-y-Rita a hug. She comes over to me. She's overflowing with emotion, overwhelmed by feeling the world under her feet again, and she gives me a hug too.

"I feel anchored and solid. I don't have to think where my muscles are. I can actually think of other things." She returns to Yuri and gives him a kiss.

"I have to emphasize why this is a miracle," says Yuri, who considers himself a data-driven skeptic. "She has almost no natural sensors. For the past twenty minutes we provided her with an artificial sensor. But the real miracle is what is happening *now* that we have removed the device, and she doesn't have either an artificial or a natural vestibular apparatus. We are awakening some kind of force inside her."

The first time they tried the hat, Cheryl wore it for only a minute. They noticed that after she took it off, there was a "residual effect" that lasted about twenty seconds, a third of the time she wore the device. Then Cheryl wore the hat for two minutes and the residual effect lasted about forty seconds. Then they went up to about twenty minutes, expecting a residual effect of just under seven minutes. But instead of lasting a third of the time, it lasted triple the time, a full hour. Today, Bach-y-Rita says, they are experimenting to see if twenty more minutes on the device will lead to some kind of training effect, so that the residual effect will last even longer.

Cheryl starts clowning and showing off. "I can walk like a woman again. That's probably not important to most people, but it means a lot that I don't have to walk with my feet wide apart now."

She gets up on a chair and jumps off. She bends down to pick things up off the floor, to show she can right herself. "Last time I did this I was able to jump rope in the residual time."

"What is amazing," says Yuri, "is that she doesn't just keep her posture. After some time on the device, she behaves almost normally. Balancing on a beam. Driving a car. It is the recovery of the vestibular function. When she moves her head, she can keep her focus on her target—the link between the visual and vestibular systems is also recovered."

I look up, and Cheryl is dancing with Bach-y-Rita.

She leads.

How is it that Cheryl can dance and has returned to normal functioning without the machine? Bach-y-Rita thinks there are several reasons. For one, her damaged vestibular system is disorganized and "noisy," sending off random signals. Thus, noise from the damaged tissue blocks any signals sent by healthy tissue. The machine helps to reinforce the signals from her healthy tissues. He thinks the machine also helps recruit other pathways, which is where plasticity comes in. A brain system is made of many neuronal pathways, or neurons that are connected to one another and working together. If certain key pathways are blocked, then the brain uses older pathways to go around them. "I look at it this way," says Bach-y-Rita. "If you are driving from here to Milwaukee, and the main bridge goes out, first you are paralyzed. Then you take old secondary roads through the farmland. Then, as you use these roads more, you find shorter paths to use to get where you want to go, and you start to get there faster." These "secondary" neural pathways are "unmasked," or exposed, and, with use, strengthened. This "unmasking" is generally thought to be one of the main ways the plastic brain reorganizes itself.

The fact that Cheryl is gradually lengthening the residual effect suggests that the unmasked pathway is getting stronger. Bach-y-Rita hopes that Cheryl, with training, will be able to continue extending the length of the residual effect.

A few days later an e-mail for Bach-y-Rita arrives from Cheryl, her report from home about how long the residual time lasted. "Total residual time was: 3 hours, 20 minutes . . . The wobbling begins in my head—just like usual . . . I am having trouble finding words . . . Swimming feeling in my head. Tired, exhausted . . . Depressed."

A painful Cinderella story. Coming down from normalcy is very hard. When it happens, she feels she has died, come to life, and then died again. On the other hand, three hours and twenty minutes after only twenty minutes on the machine is residual time ten times greater than the time on the device. She is the first Wobbler ever to have been treated, and even if the residual time never grows longer, she could now wear the device briefly four times a day and have a normal life. But there is good reason to expect more, since each session seems to be training her brain to extend the residual time. If this keeps up . . .

. . . It did keep up. Over the next year Cheryl wore the device more frequently to get relief and build up her residual effect. Her residual effect progressed to multiple hours, to days, and then to four months. Now she does not use the device at all and no longer considers herself a Wobbler.

In 1969, *Nature,* Europe's premier science journal, published a short article that had a distinctly sci-fi feel about it. Its lead author, Paul Bach-y-Rita, was both a basic scientist and a rehabilitation physician—a rare combination. The article described a device that enabled people who had been blind from birth to see. All had damaged retinas and had been considered completely untreatable.

The *Nature* article was reported in *The New York Times, Newsweek,* and *Life,* but perhaps because the claim seemed so implausible, the device and its inventor soon slipped into relative obscurity.

Accompanying the article was a picture of a bizarre-looking machine—a large old dentist's chair with a vibrating back, a tangle of wires, and bulky computers. The whole contraption, made of castaway parts combined with 1960s electronics, weighed four hundred pounds.

A congenitally blind person—someone who had never had any experience of sight—sat in the chair, behind a large camera the size of those used in television studios at the time. He "scanned" a scene in front of him by turning hand cranks to move the camera, which sent electrical signals of the image to a computer that processed them. Then the electrical signals were conveyed to four hundred vibrating stimulators, arranged in rows on a metal plate attached to the inside of the chair back, so the stimulators rested against the blind subject's skin. The stimulators functioned like pixels vibrating for the dark part of a scene and holding still for the brighter shades. This "tactile-vision device," as it was called, enabled blind subjects to read, make out faces and shadows, and distinguish which objects were closer and which farther away. It allowed them to discover perspective and observe how objects seem to change shape depending upon the angle from which they were viewed. The six subjects of the experiment learned to recognize such objects as a telephone, even when it was partially obscured by a vase. This being the 1960s, they even learned to recognize a picture of the anorexic supermodel Twiggy.

Everyone who used the relatively clunky tactile-vision device had a remarkable perceptual experience, as they went from having tactile sensations to "seeing" people and objects.

With a little practice, the blind subjects began to experience the space in front of them as three-dimensional, even though the information entered from the two-dimensional array on their backs. If someone threw a ball toward the camera, the subject would automatically jump back to duck it. If the plate of vibrating stimulators was moved from their backs to their abdomens, subjects still accurately perceived the scene as happening in front of the camera. If tickled near the stimulators, they didn't confuse the tickle with a visual stimulus. Their mental perceptual experience took place not on the skin surface but in the world. And their perceptions were complex. With practice, subjects could move the camera around and say things like "That is Betty; she is wearing her hair down today and does not have her glasses on; her mouth is open, and she is moving her right hand from her left side to the back of her head." True, the resolution was often poor, but as Bach-y-Rita would explain, vision doesn't have to be perfect to be vision. "When we walk down a foggy street and see the outline of a building," he would ask, "are we seeing it any less for the lack of resolution? When we see something in black and white, are we not seeing it for lack of color?"

This now-forgotten machine was one of the first and boldest applications of neuroplasticity—an attempt to use one sense to replace another—and it worked. Yet it was thought implausible and ignored because the scientific mind-set at the time assumed that the brain's structure is fixed, and that our senses, the avenues by which experience gets into our minds, are hardwired. This idea, which still has many adherents, is called "localizationism." It's closely related to the idea that the brain is like a complex machine, made up of parts, each of which performs a specific mental function and exists in a genetically predetermined or hardwired *location*—hence the name. A brain that is hardwired, and in which each mental function has a strict location, leaves little room for plasticity.

The idea of the machinelike brain has inspired and guided neuroscience since it was first proposed in the seventeenth century, replacing more mystical notions about the soul and the body. Scientists, impressed by the discoveries of Galileo (1564–1642), who showed that the planets could be understood as inanimate bodies moved by mechanical forces, came to believe that all nature functioned as a large cosmic clock, subject to the laws of physics, and they began to explain individual living things, including our bodily organs, mechanistically, as though they too were machines. This idea that all nature was like a vast mechanism, and that our organs were machinelike, replaced the two-thousand-year-old Greek idea that viewed all nature as a vast living organism, and our bodily organs as anything but inanimate mechanisms. But the first great accomplishment of this new "mechanistic biology" was a brilliant and original achievement. William Harvey (1578–1657), who studied anatomy in Padua, Italy, where Galileo lectured, discovered how our blood circulates through our bodies and demonstrated that the heart functions like a pump, which is, of course, a simple machine. It soon seemed to many scientists that for an explanation to be scientific it had to be mechanistic—that is, subject to the mechanical laws of motion. Following Harvey, the French philosopher René Descartes (1596–1650) argued that the brain and nervous system also functioned like a pump. Our nerves were really tubes, he argued, that went from our limbs to the brain and back. He was the first person to theorize how reflexes work, proposing that when a person is touched on the skin, a fluidlike substance in the nerve tubes flows to the brain and is mechanically "reflected" back down the nerves to move the muscles. As crude as it sounds, he wasn't so far off. Scientists soon refined his primitive picture, arguing that not some fluid but an electric current moved through the nerves. Descartes's idea of the brain as a complex machine culminated in our current idea of the brain as a computer and in localizationism. Like a machine, the brain came to be seen as made of parts, each one in a preassigned location, each performing a single function, so that if one of those parts was damaged, nothing could be done to replace it; after all, machines don't grow new parts.

Localizationism was applied to the senses as well, theorizing that each of our senses—sight, hearing, taste, touch, smell, balance—has a receptor cell that specializes in detecting one of the various forms of energy around us. When stimulated, these receptor cells send an electric signal along their nerve to a specific brain area that processes that sense. Most scientists believed that these brain areas were so specialized that one area could never do the work of another.

Almost in isolation from his colleagues, Paul Bach-y-Rita rejected these localizationist claims. Our senses have an unexpectedly plastic nature, he discovered, and if one is damaged, another can sometimes take over for it, a process he calls "sensory substitution." He developed ways of triggering sensory substitution and devices that give us "supersenses." By discovering that the nervous system can adapt to seeing with cameras instead of retinas, Bach-y-Rita laid the groundwork for the greatest hope for the blind: retinal implants, which can be surgically inserted into the eye.

Unlike most scientists, who stick to one field, Bach-y-Rita has become an expert in many—medicine, psychopharmacology, ocular neurophysiology (the study of eye muscles), visual neurophysiology (the study of sight and the nervous system), and biomedical engineering. He follows ideas wherever they take him. He speaks five languages and has lived for extended periods in Italy, Germany, France, Mexico, Sweden, and throughout the United States. He has worked in the labs of major scientists and Nobel Prize winners, but he has never much cared what others thought and doesn't play the political games that many researchers do in order to get ahead. After becoming a physician, he gave up medicine and switched to basic research. He asked questions that seemed to defy common sense, such as, "Are eyes necessary for vision, or ears for hearing, tongues for tasting, noses for smelling?" And then, when he was forty-four years old, his mind ever restless, he switched back to medicine and began a medical residency, with its endless days and sleepless nights, in one of the dreariest specialties of all: rehabilitation medicine. His ambition was to turn an intellectual backwater into a science by applying to it what he had learned about plasticity.

Bach-y-Rita is a completely unassuming man. He is partial to five-dollar suits and wears Salvation Army clothes whenever his wife lets him get away with it. He drives a rusty twelve-year-old car, his wife a new model Passat.

He has a full head of thick, wavy gray hair, speaks softly and rapidly, has the darkish skin of a Mediterranean man of Spanish and Jewish ancestry, and appears a lot younger than his sixty-nine years.

He's obviously cerebral but radiates a boyish warmth toward his wife, Esther, a Mexican of Mayan descent.

He is used to being an outsider. He grew up in the Bronx, was four foot ten when he entered high school because of a mysterious disease that stunted his growth for eight years, and was twice given a preliminary diagnosis of leukemia. He was beaten up by the larger students every day and during those years developed an extraordinarily high pain threshold. When he was twelve, his appendix burst, and the mysterious disease, a rare form of chronic appendicitis, was properly diagnosed. He grew eight inches and won his first fight.

We are driving through Madison, Wisconsin, his home when he's not in Mexico. He is devoid of pretension, and after many hours of our talking together, he lets only one even remotely self-congratulatory remark leave his lips.

"I can connect anything to anything." He smiles.

"WE SEE WITH OUR BRAINS, not with our eyes," he says.

This claim runs counter to the commonsensical notion that we see with our eyes, hear with our ears, taste with our tongues, smell with our noses, and feel with our skin. Who would challenge such facts? But for Bach-y-Rita, our eyes merely sense changes in light energy; it is our brains that perceive and hence see.

How a sensation enters the brain is not important to Bach-y-Rita. "When a blind man uses a cane, he sweeps it back and forth, and has only one point, the tip, feeding him information through the skin receptors in the hand. Yet this sweeping allows him to sort out where the doorjamb is, or the chair, or distinguish a foot when he hits it, because it will give a little. Then he uses this information to guide himself to the chair to sit down. Though his hand sensors are where he gets the information and where the cane 'interfaces' with him, what he *subjectively* perceives is not the cane's pressure on his hand but the layout of the room: chairs, walls, feet, the three-dimensional space. The actual receptor surface in the hand becomes merely a relay for information, a data port. The receptor surface loses its identity in the process."

Bach-y-Rita determined that skin and its touch receptors could substitute for a retina, because both the skin and the retina are two-dimensional sheets, covered with sensory receptors, that allow a "picture" to form on them.

It's one thing to find a new data port, or way of getting sensations to the brain. But it's another for the brain to decode these skin sensations and turn them into pictures. To do that, the brain has to learn something new, and the part of the brain devoted to processing touch has to adapt to the new signals. This adaptability implies that the brain is plastic in the sense that it can reorganize its sensory-perceptual system.

If the brain can reorganize itself, simple localizationism cannot be a correct image of the brain. At first even Bach-y-Rita was a localizationist, moved by its brilliant accomplishments. Serious localizationism was first proposed in 1861, when Paul Broca, a surgeon, had a stroke patient who lost the ability to speak and could utter only one word. No matter what he was asked, the poor man responded, "Tan, tan." When he died, Broca dissected his brain and found damaged tissue in the left frontal lobe. Skeptics doubted that speech could be localized to a single part of the brain until Broca showed them the injured tissue, then reported on other patients who had lost the ability to speak and had damage in the same location. That place came to be called "Broca's area" and was presumed to coordinate the movements of the muscles of the lips and tongue. Soon afterward another physician, Carl Wernicke, connected damage in another brain area farther back to a different problem: the inability to understand language. Wernicke proposed that the damaged area was responsible for the mental representations of words and comprehension. It came to be known as "Wernicke's area." Over the next hundred years localizationism became more specific as new research refined the brain map.

Unfortunately, though, the case for localizationism was soon exaggerated. It went from being a series of intriguing correlations (observations that damage to specific brain areas led to the loss of specific mental functions) to a general theory that declared that every brain function had only one hardwired location—an idea summarized by the phrase "one function, one location," meaning that if a part was damaged, the brain could not reorganize itself or recover that lost function.

A dark age for plasticity began, and any exceptions to the idea of "one function, one location" were ignored. In 1868 Jules Cotard studied children who had early massive brain disease, in which the left hemisphere (including Broca's area) wasted away. Yet these children could still speak normally. This meant that even if speech tended to be processed in the left hemisphere, as Broca claimed, the brain might be plastic enough to reorganize itself, if necessary. In 1876 Otto Soltmann removed the motor cortex from infant dogs and rabbits—the part of the brain thought to be responsible for movement—yet found they were still able to move. These findings were submerged in the wave of localizationist enthusiasm.

Bach-y-Rita came to doubt localizationism while in Germany in the early 1960s. He had joined a team that was studying how vision worked by measuring with electrodes electrical discharge from the visual processing area of a cat's brain. The team fully expected that when they showed the cat an image, the electrode in its visual processing area would send off an electric spike, showing it was processing that image. And it did. But when the cat's paw was accidentally stroked, the visual area also fired, indicating that it was processing touch as well. And they found that the visual area was also active when the cat heard sounds.

Bach-y-Rita began to think that the localizationist idea of "one function, one location" couldn't be right. The "visual" part of the cat's brain was processing at least two other functions, touch and sound. He began to conceive of much of the brain as "polysensory"—that its sensory areas were able to process signals from more than one sense.

This can happen because all our sense receptors translate different kinds of energy from the external world, no matter what the source, into electrical patterns that are sent down our nerves. These electrical patterns are the universal language "spoken" inside the brain—there are no visual images, sounds, smells, or feelings moving inside our neurons. Bach-y-Rita realized that the areas that process these electrical impulses are far more homogeneous than neuroscientists appreciated, a belief that was reinforced when the neuroscientist Vernon Mountcastle discovered that the visual, auditory, and sensory cortices all have a similar six-layer processing structure. To Bach-y-Rita, this meant that any part of the cortex should be able to process whatever electrical signals were sent to it, and that our brain modules were not so specialized after all.

Over the next few years Bach-y-Rita began to study all the exceptions to localizationism. With his knowledge of languages, he delved into the untranslated, older scientific literature and rediscovered scientific work done before the more rigid versions of localizationism had taken hold. He discovered the work of Marie-Jean-Pierre Flourens, who in the 1820s showed that the brain could reorganize itself. And he read the oft-quoted but seldom translated work of Broca in French and found that even Broca had not closed the door to plasticity as his followers had.

The success of his tactile-vision machine further inspired Bach-y-Rita to reinvent his picture of the human brain. After all, it was not his machine that was the miracle, but the brain that was alive, changing, and adapting to new kinds of artificial signals. As part of the reorganization, he guessed that signals from the sense of touch (processed initially in the sensory cortex, near the top of the brain) were rerouted to the visual cortex at the back of the brain for further processing, which meant that any neuronal paths that ran from the skin to the visual cortex were undergoing development.

Forty years ago, just when localization's empire had extended to its farthest reaches, Bach-y-Rita began his protest. He praised localization's accomplishments but argued that "a large body of evidence indicates that the brain demonstrates both motor and sensory plasticity." One of his papers was rejected for publication six times by journals, not because the evidence was disputed but because he dared to put the word "plasticity" in the title. After his *Nature* article came out, his beloved mentor, Ragnar Granit, who had received the Nobel Prize in physiology in 1965 for his work on the retina, and who had arranged for the publication of Bach-y-Rita's medical school thesis, invited him over for tea. Granit asked his wife to leave the room and, after praising Bach-y-Rita's work on the eye muscles, asked him—for his own good—why he was wasting his time with "that adult toy." Yet Bach-y-Rita persisted and began to lay out, in a series of books and several hundred articles, the evidence for brain plasticity and to develop a theory to explain how it might work.

Bach-y-Rita's deepest interest became explaining plasticity, but he continued to invent sensory-substitution devices. He worked with engineers to shrink the dentist-chair–computer-camera device for the blind. The clumsy, heavy plate of vibrating stimulators that had been attached to the back has now been replaced by a paper-thin strip of plastic covered with electrodes, the diameter of a silver dollar, that is slipped onto the tongue. The tongue is what he calls the ideal "brain-machine interface," an excellent entry point to the brain because it has no insensitive layer of dead skin on it. The computer too has shrunk radically, and the camera that was once the size of a suitcase now can be worn strapped to the frame of eyeglasses.

He has been working on other sensory-substitution inventions as well. He received NASA funding to develop an electronic "feeling" glove for astronauts in space. Existing space gloves were so thick that it was hard for the astronauts to feel small objects or perform delicate movements. So on the outside of the glove he put electric sensors that relayed electrical signals to the hand. Then he took what he learned making the glove and invented one to help people with leprosy, whose illness mutilates the skin and destroys peripheral nerves so that the lepers lose sensation in their hands. This glove, like the astronaut's glove, had sensors on the outside, and it sent its signals to a healthy part of the skin—away from the diseased hands—where the nerves were unaffected. That healthy skin became the portal of entry for hand sensations. He then began work on a glove that would allow blind people to read computer screens, and he even has a project for a condom that he hopes will allow spinal cord injury victims who have no feeling in their penises to have orgasms. It is based on the premise that sexual excitement, like other sensory experiences, is "in the brain," so the sensations of sexual movement, picked up by sensors on the condom, can be translated into electrical impulses that can then be transmitted to the part of the brain that processes sexual excitement. Other potential uses of his work include giving people "supersenses," such as infrared or night vision. He has developed a device for the Navy SEALs that helps them sense how their bodies are oriented underwater, and another, successfully tested in France, that tells surgeons the exact position of a scalpel by sending signals from an electronic sensor attached to the scalpel to a small device attached to their tongues and to their brains.

The origin of Bach-y-Rita's understanding of brain rehabilitation lies in the dramatic recovery of his own father, the Catalan poet and scholar Pedro Bach-y-Rita, after a disabling stroke. In 1959 Pedro, then a sixty-five-year-old widower, had a stroke that paralyzed his face and half of his body and left him unable to speak.

George, Paul's brother, now a psychiatrist in California, was told that his father had no hope of recovery and would have to go into an institution. Instead, George, then a medical student in Mexico, brought his paralyzed father from New York, where he lived, back to

Mexico to live with him. At first he tried to arrange rehabilitation for his father at the American British Hospital, which offered only a typical four-week rehab, as nobody believed the brain could benefit from extended treatment. After four weeks his father was nowhere near better. He was still helpless and needed to be lifted onto and off the toilet and showered, which George did with the help of the gardener.

"Fortunately, he was a little man, a hundred and eighteen pounds, and we could manage him," says George.

George knew nothing about rehabilitation, and his ignorance turned out to be a godsend, because he succeeded by breaking all its current rules, unencumbered by pessimistic theories.

"I decided that instead of teaching my father to walk, I was going to teach him first to crawl. I said, 'You started off crawling, you are going to have to crawl again for a while.' We got kneepads for him. At first we held him on all fours, but his arms and legs didn't hold him very well, so it was a struggle." As soon as Pedro could support himself somewhat, George then got him to crawl with his weak shoulder and arm supported by a wall. "That crawling beside the wall went on for months. After that I even had him practicing in the garden, which led to problems with the neighbors, who were saying it wasn't nice, it was unseemly, to be making the professor crawl like a dog. The only model I had was how babies learn. So we played games on the floor, with me rolling marbles, and him having to catch them. Or we'd throw coins on the floor, and he'd have to try and pick them up with his weak right hand. Everything we tried involved turning normal life experiences into exercises. We turned washing pots into an exercise. He'd hold the pot with his good hand and make his weak hand—it had little control and made spastic jerking movements—go round and round, fifteen minutes clockwise, fifteen minutes counterclockwise. The circumference of the pot kept his hand contained. There were steps, each one overlapping with the one before, and little by little he got better. After a while he helped to design the steps. He wanted to get to the point where he could sit down and eat with me and the other medical students." The regime took many hours every day, but gradually Pedro went from crawling, to moving on his knees, to standing, to walking.

Pedro struggled with his speech on his own, and after about three months there were signs it too was coming back. After a number of months he wanted to resume his writing. He would sit in front of the typewriter, his middle finger over the desired key, then drop his whole arm to strike it. When he had mastered that, he would drop just the wrist, and finally the fingers, one at a time. Eventually he learned to type normally again.

At the end of a year his recovery was complete enough for Pedro, now sixty-eight, to start full-time teaching again at City College in New York. He loved it and worked until he retired at seventy. Then he got another teaching job at San Francisco State, remarried, and kept working, hiking, and traveling. He was active for seven more years after his stroke. On a visit to friends in Bogotá, Colombia, he went climbing high in the mountains. At nine thousand feet he had a heart attack and died shortly thereafter. He was seventy-two.

I asked George if he understood how unusual this recovery was so long after his father's stroke and whether he thought at the time that the recovery might have been the result of brain plasticity.

"I just saw it in terms of taking care of Papa. But Paul, in subsequent years, talked about it in terms of neuroplasticity. Not right away, though. It wasn't until after our father died."

Pedro's body was brought to San Francisco, where Paul was working. It was 1965, and in those days, before brain scans, autopsies were routine because they were one way doctors could learn about brain diseases, and about why a patient died. Paul asked Dr. Mary Jane Aguilar to perform the autopsy.

"A few days later Mary Jane called me and said, 'Paul, come down. I've got something to show you.' When I got to the old Stanford Hospital, there, spread out on the table, were slices of my father's brain on slides."

He was speechless.

"I was feeling revulsion, but I could also see Mary Jane's excitement, because what the slides showed was that my father had had a huge lesion from his stroke and that it had never healed, even though he recovered all those functions. I freaked out. I got numb. I was thinking, 'Look at all this damage he has.' And she said, 'How can you recover with all this damage?' "

When he looked closely, Paul saw that his father's seven-year-old lesion was mainly in the brain stem—the part of the brain closest to the spinal cord—and that other major brain centers in the cortex that control movement had been destroyed by the stroke as well. Ninety-seven percent of the nerves that run from the cerebral cortex to the spine were destroyed—catastrophic damage that had caused his paralysis.

"I knew that meant that somehow his brain had totally reorganized itself with the work he did with George. We didn't know how remarkable his recovery was until that moment, because we had no idea of the extent of his lesion, since there were no brain scans in those days. When people did recover, we tended to assume that there really hadn't been much damage in the first place. She wanted me to be a coauthor on the paper she wrote about his case. I couldn't."

His father's story was firsthand evidence that a "late" recovery could occur even with a massive lesion in an elderly person. But after examining that lesion and reviewing the literature, Paul found more evidence that the brain can reorganize itself to recover functions after devastating strokes, discovering that in 1915 an American psychologist, Shepherd Ivory Franz, had shown that patients who had been paralyzed for twenty years were capable of making late recoveries with brain-stimulating exercises.

His father's "late recovery" triggered a career change for Bach-y-Rita. At forty-four, he went back to practicing medicine and did residencies in neurology and rehabilitation medicine. He understood that for patients to recover they needed to be motivated, as his father had been, with exercises that closely approximated real-life activities.

He turned his attention to treating strokes, focusing on "late rehabilitation," helping people overcome major neurological problems years after they'd begun, and developing computer video games to train stroke patients to move their arms again. And he began to integrate what he knew about plasticity into exercise design. Traditional rehabilitation exercises typically ended after a few weeks, when a patient stopped improving, or "plateaued," and doctors lost the motivation to continue. But Bach-y-Rita, based on his knowledge of nerve growth, began to argue that these learning plateaus were temporary—part of a plasticity-based learning cycle—in which stages of learning are followed by periods of consolidation. Though there was no *apparent* progress in the consolidation stage, biological changes were happening internally, as new skills became more automatic and refined.

Bach-y-Rita developed a program for people with damaged facial motor nerves, who could not move their facial muscles and so couldn't close their eyes, speak properly, or express emotion, making them look like monstrous automatons. Bach-y-Rita had one of the "extra" nerves that normally goes to the tongue surgically attached to a patient's facial muscles. Then he developed a program of brain exercises to train the "tongue nerve" (and particularly the part of the brain that controls it) to act like a facial nerve. These patients learned to express normal facial emotions, speak, and close their eyes—one more instance of Bach-y-Rita's ability to "connect anything to anything."

Thirty-three years after Bach-y-Rita's *Nature* article, scientists using the small modern version of his tactile-vision machine have put patients under brain scans and confirmed that the tactile images that enter patients through their tongues are indeed processed in their brains' visual cortex.

All reasonable doubt that the senses can be rewired was recently put to rest in one of the most amazing plasticity experiments of our time. It involved rewiring not touch and vision pathways, as Bach-y-Rita had done, but those for hearing and vision—literally. Mriganka Sur, a neuroscientist, surgically rewired the brain of a very young ferret. Normally the optic nerves run from the eyes to the visual cortex, but Sur surgically redirected the optic nerves from the ferret's visual to its auditory (hearing) cortex and discovered that the ferret learned to see. Using electrodes inserted into the ferret's brain, Sur proved that when the ferret was seeing, the neurons in its auditory cortex were firing and doing the visual processing. The auditory cortex, as plastic as Bach-y-Rita had always imagined, had reorganized itself, so that it had the structure of the visual cortex. Though the ferrets that had this surgery did not have 20/20 vision, they had about a third of that, or 20/60—no worse than some people who wear eyeglasses.

Till recently, such transformations would have seemed utterly inexplicable. But Bach-y-Rita, by showing that our brains are more flexible than localizationism admits, has helped to invent a more accurate view of the brain that allows for such changes. Before he did this work, it was acceptable to say, as most neuroscientists do, that we have a "visual cortex" in our occipital lobe that processes vision, and an "auditory cortex" in our temporal lobe that processes hearing. From Bach-y-Rita we have learned that the matter is more complicated and that these areas of the brain are plastic processors, connected to each other and capable of processing an unexpected variety of input.

Cheryl has not been the only one to benefit from Bach-y-Rita's strange hat. The team has since used the device to train fifty more patients to improve their balance and walking. Some had the same damage Cheryl had; others have had brain trauma, stroke, or Parkinson's disease.

Paul Bach-y-Rita's importance lies in his being the first of his generation of neuroscientists both to understand that the brain is plastic and to apply this knowledge in a practical way to ease human suffering. Implicit in all his work is the idea that we are all born with a far more adaptable, all-purpose, opportunistic brain than we have understood.

When Cheryl's brain developed a renewed vestibular sense—or blind subjects' brains developed new paths as they learned to recognize objects, perspective, or movement—these changes were not the mysterious exception to the rule but the rule: the sensory cortex is plastic and adaptable. When Cheryl's brain learned to respond to the artificial receptor that replaced her damaged one, it was not doing anything out of the ordinary. Recently Bach-y-Rita's work has inspired cognitive scientist Andy Clark to wittily argue that we are "natural-born cyborgs," meaning that brain plasticity allows us to attach ourselves to machines, such as computers and electronic tools, quite naturally. But our brains also restructure themselves in response to input from the simplest tools too, such as a blind man's cane. Plasticity has been, after all, a property inherent in the brain since prehistoric times. The brain is a far more open system than we ever imagined, and nature has gone very far to help us perceive and take in the world around us. It has given us a brain that survives in a changing world by changing itself.

2

Building Herself a Better Brain

A Woman Labeled "Retarded" Discovers How to Heal Herself

The scientists who make important discoveries about the brain are often those whose own brains are extraordinary, working on those whose brains are damaged. It is rare that the person who makes an important discovery is the one with the defect, but there are some exceptions. Barbara Arrowsmith Young is one of these.

"Asymmetry" is the word that best describes her mind when she was a schoolgirl. Born in Toronto in 1951 and raised in Peterborough, Ontario, Barbara had areas of brilliance as a child—her auditory and visual memory both tested in the ninety-ninth percentile. Her frontal lobes were remarkably developed, giving her a driven, dogged quality. But her brain was "asymmetrical," meaning that these exceptional abilities coexisted with areas of retardation.

This asymmetry left its chaotic handwriting on her body as well. Her mother made a joke of it. "The obstetrician must have yanked you out by your right leg," which was longer than her left, causing her pelvis to shift. Her right arm never straightened, her right side was larger than her left, her left eye less alert. Her spine was asymmetrical and twisted with scoliosis.

She had a confusing assortment of serious learning disabilities. The area of her brain devoted to speech, Broca's area, was not working properly, so she had trouble pronouncing words. She also lacked the capacity for spatial reasoning. When we wish to move our bodies in space, we use spatial reasoning to construct an imaginary pathway in our heads before executing our movements. Spatial reasoning is important for a baby crawling, a dentist drilling a tooth, a hockey player planning his moves. One day when Barbara was three she decided to play matador and bull. She was the bull, and the car in the driveway was the matador's cape. She charged, thinking she would swerve and avoid it, but she misjudged the space and ran into the car, ripping her head open. Her mother declared she would be surprised if Barbara lived another year.

Spatial reasoning is also necessary for forming a mental map of where things are. We use this kind of reasoning to organize our desks or remember where we have left our keys. Barbara lost everything all the time. With no mental map of things in space, out of sight was literally out of mind, so she became a "pile person" and had to keep everything she was playing with or working on in front of her in piles, and her closets and dressers open. Outdoors she was always getting lost.

She also had a "kinesthetic" problem. Kinesthetic perception allows us to be aware of where our body or limbs are in space, enabling us to control and coordinate our movements. It also helps us recognize objects by touch. But Barbara could never tell how far her arms or legs had moved on her left side. Though a tomboy in spirit, she was clumsy. She couldn't hold a cup of juice in her left hand without spilling it. She frequently tripped or stumbled. Stairs were treacherous. She also had a decreased sense of touch on her left and was always bruising herself on that side. When she eventually learned to drive, she kept denting the left side of the car.

She had a visual disability as well. Her span of vision was so narrow that when she looked at a page of writing, she could take in only a few letters at a time.

But these were not her most debilitating problems. Because the part of her brain that helps to understand the relationships between symbols wasn't functioning normally, she had trouble understanding grammar, math concepts, logic, and cause and effect. She couldn't distinguish between "the father's brother" and "the brother's father." The double negative was impossible for her to decipher. She couldn't read a clock because she couldn't understand the relationship between the hands. She literally couldn't tell her left hand from her right, not only because she lacked a spatial map but because she couldn't understand the relationship between "left" and "right." Only with extraordinary mental effort and constant repetition could she learn to relate symbols to one another.

She reversed *b, d, q,* and *p,* read "was" as "saw," and read and wrote from right to left, a disability called mirror writing. She was right-handed, but because she wrote from right to left, she smeared all her work. Her teachers thought she was being obstreperous. Because she was dyslexic, she made reading errors that cost her dearly. Her brothers kept sulfuric acid for experiments in her old nose-drops bottle. Once when she decided to treat herself for sniffles, Barbara misread the new label they had written. Lying in bed with acid running into her sinuses, she was too ashamed to tell her mother of yet another mishap.

Unable to understand cause and effect, she did odd things socially because she couldn't connect behavior with its consequences. In kindergarten she couldn't understand why, if her brothers were in the same school, she couldn't leave her class and visit them in theirs whenever she wanted. She could memorize math procedures but couldn't understand math concepts. She could recall that five times five equals twenty-five but couldn't understand why. Her teachers responded by giving her extra drills, and her father spent hours tutoring her, to no avail. Her mother held up flash cards with simple math problems on them. Because Barbara couldn't figure them out, she found a place to sit where the sun made the paper translucent, so she could read the answers on the back. But the attempts at remediation didn't get at the root of the problem; they just made it more agonizing.

Wanting desperately to do well, she got through elementary school by memorizing during lunch hours and after school. In high school her performance was extremely erratic. She learned to use her memory to cover her deficits and with practice could remember pages of facts. Before tests she prayed they would be fact-based, knowing she could score 100; if they were based on understanding relationships, she would probably score in the low teens.

Barbara understood nothing in real time, only after the fact, in lag time. Because she did not understand what was happening around her while it was occurring, she spent hours reviewing the past, to make its confusing fragments come together and become comprehensible. She had to replay simple conversations, movie dialogue, and song lyrics twenty times over in her head because by the time she got to the end of a sentence, she could not recall what the beginning meant.

Her emotional development suffered. Because she had trouble with logic, she could not pick up inconsistencies when listening to smooth talkers and so she was never sure whom to trust. Friendships were difficult, and she could not have more than one relationship at a time.

But what plagued her most was the chronic doubt and uncertainty that she felt about everything. She sensed meaning everywhere but could never verify it. Her motto was "I don't get it." She told herself, "I live in a fog, and the world is no more solid than cotton candy." Like many children with serious learning disabilities, she began to think she might be crazy.

Barbara grew up in a time when little help was available.

"In the 1950s, in a small town like Peterborough, you didn't talk about these things," she says. "The attitude was, you either make it or you don't. There were no special-ed teachers, no visits to medical specialists or psychologists. The term 'learning disabilities' wouldn't be widely used for another two decades. My grade-one teacher told my parents I had 'a mental block' and I wouldn't ever learn the way others did. That was as specific as it got. You were either bright, average, slow, or mentally retarded."

If you were mentally retarded, you were placed in "opportunity classes." But that was not the place for a girl with a brilliant memory who could ace vocabulary tests. Barbara's childhood friend Donald Frost, now a sculptor, says, "She was under incredible academic pressure. The whole Young family were high achievers. Her father, Jack, was an electrical engineer and inventor with thirty-four patents for Canadian General Electric. If you could pull Jack from a book for dinner, it was a miracle. Her mother, Mary, had the attitude: 'You will succeed; there is no doubt,' and 'If you have a problem, fix it.' Barbara was always incredibly sensitive, warm, and caring," Frost continues, "but she hid her problems well. It was hush-hush. In the postwar years there was a sense of integrity that meant you didn't draw attention to your disabilities any more than you would to your pimples."

Barbara gravitated toward the study of child development, hoping somehow to sort things out for herself. As an undergraduate at the University of Guelph, her great mental disparities were again apparent. But fortunately her teachers saw that she had a remarkable ability to pick up nonverbal cues in the child-observation laboratory, and she was asked to teach the course. She felt there must have been some mistake. Then she was accepted into graduate school at the Ontario Institute for Studies in Education (OISE). Most students read a research paper once or twice, but typically Barbara had to read one twenty times as well as many of its sources to get even a fleeting sense of its meaning. She survived on four hours of sleep a night.

Because Barbara was brilliant in so many ways, and so adept at child observation, her teachers in graduate school had trouble believing she was disabled. It was Joshua Cohen, another gifted but learning-disabled student at OISE, who first understood. He ran a small clinic for learning-disabled kids that used the standard treatment, "compensations," based on the accepted theory of the time: once brain cells die or fail to develop, they cannot be restored. Compensations work around the problem. People with trouble reading listen to audiotapes. Those who are "slow" are given more time on tests. Those who have trouble following an argument are told to color-code the main points. Joshua designed a compensation program for Barbara, but she found it too time-consuming. Moreover, her thesis, a study of learning-disabled children treated with compensations at the OISE clinic, showed that most of them were not really improving. And she herself had so many deficits that it was sometimes hard to find healthy functions that could work around her deficits. Because she had had such success developing her memory, she told Joshua she thought there must be a better way.

One day Joshua suggested she look into some books by Aleksandr Luria that he'd been reading. She tackled them, going over the difficult passages countless times, especially a section in Luria's *Basic Problems of Neurolinguistics* about people with strokes or wounds who had trouble with grammar, logic, and reading clocks. Luria, born in 1902, came of age in revolutionary Russia. He was deeply interested in psychoanalysis, corresponded with Freud, and wrote papers on the psychoanalytic technique of "free association," in which patients say everything that comes to mind. His goal was to develop objective methods to assess Freudian ideas. While still in his twenties, he invented the prototype of the lie detector. When the Great Purges of the Stalin era began, psychoanalysis became *scientia non grata,* and Luria was denounced. He delivered a public recantation, admitting to having made certain "ideological mistakes." Then, to remove himself from view, he went to medical school.

But he had not totally finished with psychoanalysis. Without calling attention to his work, he integrated aspects of the psychoanalytic method and of psychology into neurology, becoming the founder of neuropsychology. His case histories, instead of being brief vignettes focused on symptoms, described his patients at length. As Oliver Sacks wrote, "Luria's case histories, indeed, can only be compared to Freud's in their precision, their vitality, their wealth and depth of detail." One of Luria's books, *The Man with a Shattered World,* was the summary of, and commentary on, the diary of a patient with a very peculiar condition.

At the end of May 1943 Comrade Lyova Zazetsky, a man who seemed like a boy, came to Luria's office in the rehabilitation hospital where he was working. Zazetsky was a young Russian lieutenant who had just been injured in the battle of Smolensk, where poorly equipped Russians had been thrown against the invading Nazi war machine. He had sustained a bullet wound to the head, with massive damage on the left side, deep inside his brain. For a long time he lay in a coma. When Zazetsky awoke, his symptoms were very odd. The shrapnel had lodged in the part of the brain that helped him understand relationships between symbols. He could no longer understand logic, cause and effect, or spatial relationships. He couldn't distinguish his left from his right. He couldn't understand the elements of grammar dealing with relationships. Prepositions such as "in," "out," "before," "after," "with," and "without" had become meaningless to him. He couldn't comprehend a whole word, understand a whole sentence, or recall a complete memory because doing any of those things would require relating symbols. He could grasp only fleeting fragments. Yet his frontal lobes—which allowed him to seek out what is relevant and to plan, strategize, form intentions, and pursue them—were spared, so he had the capacity to recognize his defects, and the wish to overcome them. Though he could not read, which is largely a perceptual activity, he could write, because it is an intentional one. He began a fragmentary diary he called *I'll Fight On* that swelled to three thousand pages. "I was killed March 2, 1943," he wrote, "but because of some vital power of my organism, I miraculously remained alive."

Over thirty years Luria observed him and reflected on the way Zazetsky's wound affected his mental activities. He would witness Zazetsky's relentless fight "to live, not merely exist."

Reading Zazetsky's diary, Barbara thought, "He is describing my life."

"I knew what the words 'mother' and 'daughter' meant but not the expression 'mother's daughter,'" Zazetsky wrote. "The expressions 'mother's daughter' and 'daughter's mother' sounded just the same to me. I also had trouble with expressions like 'Is an elephant bigger than a fly?' All I could figure out was that a fly was small and an elephant is big, but I didn't understand the words 'bigger' and 'smaller.'"

While watching a film, Zazetsky wrote, "before I've had a chance to figure out what the actors are saying, a new scene begins."

Luria began to make sense of the problem. Zazetsky's bullet had lodged in the left hemisphere, at the junction of three major perceptual areas where the temporal lobe (which normally processes sound and language), the occipital lobe (which normally processes visual images), and the parietal lobe (which normally processes spatial relationships and integrates information from different senses) meet. At this junction perceptual input from those three areas is brought together and associated. While Zazetsky could perceive properly, Luria realized he could not relate his different perceptions, or parts of things to wholes. Most important, he had great difficulty relating a number of symbols to one another, as we normally do when we think with words. Thus Zazetsky often spoke in malapropisms. It was as though he didn't have a large enough net to catch and hold words and their meanings, and he often could not relate words to their meanings or definitions. He lived with fragments and wrote, "I'm in a fog all the time . . . All that flashes through my mind are images . . . hazy visions that suddenly appear and just as suddenly disappear . . . I simply can't understand or remember what these mean."

For the first time, Barbara understood that her main brain deficit had an address. But Luria did not provide the one thing she needed: a treatment. When she realized how impaired she really was, she found herself more exhausted and depressed and thought she could not go on this way. On subway platforms she looked for a spot from which to jump for maximum impact.

It was at this point in her life, while she was twenty-eight and still in graduate school, that a paper came across her desk. Mark Rosenzweig of the University of California at Berkeley had studied rats in stimulating and nonstimulating environments, and in postmortem exams he found that the brains of the stimulated rats had more neurotransmitters, were heavier, and had better blood supply than those from the less stimulating environments. He was one of the first scientists to demonstrate neuroplasticity by showing that activity could produce changes in the structure of the brain.

For Barbara, lightning struck. Rosenzweig had shown that the brain could be modified. Though many doubted it, to her this meant that compensation might not be the only answer. Her own breakthrough would be to link Rosenzweig's and Luria's research.

She isolated herself and began toiling to the point of exhaustion, week after week—with only brief breaks for sleep—at mental exercises she designed, though she had no guarantee they would lead anywhere. Instead of practicing compensation, she exercised her most weakened function—relating a number of symbols to each other. One exercise involved reading hundreds of cards picturing clock faces showing different times. She had Joshua Cohen write the correct time on the backs. She shuffled the cards so she couldn't memorize the answers. She turned up a card, attempted to tell the time, checked the answer, then moved on to the next card as fast as she could. When she couldn't get the time right, she'd spend hours with a real clock, turning the hands slowly, trying to understand why, at 2:45, the hour hand was three-quarters of the way toward the three.

When she finally started to get the answers, she added hands for seconds and sixtieths of a second. At the end of many exhausting weeks, not only could she read clocks faster than normal people, but she noticed improvements in her other difficulties relating to symbols and began for the first time to grasp grammar, math, and logic. Most important, she could understand what people were saying as they said it. For the first time in her life, she began to live in real time.

Spurred on by her initial success, she designed exercises for her other disabilities—her difficulties with space, her trouble with knowing where her limbs were, and her visual disabilities—and brought them up to average level.

Barbara and Joshua Cohen married, and in 1980 they opened the Arrowsmith School in Toronto. They did research together, and Barbara continued to develop brain exercises and to run the school from day to day. Eventually they parted, and Joshua died in 2000.

Because so few others knew about or accepted neuroplasticity or believed that the brain might be exercised as though it were a muscle, there was seldom any context in which to understand her work. She was viewed by some critics as making claims—that learning disabilities were treatable—that couldn't be substantiated. But far from being plagued by uncertainty, she continued to design exercises for the brain areas and functions most commonly weakened in those with learning disabilities. In these years before high-tech brain scans were available, she relied on Luria's work to understand which areas of the brain commonly processed which mental functions. Luria had formed his own map of the brain by working with patients like Zazetsky. He observed where a soldier's wound had occurred and related this location to the mental functions lost. Barbara found that learning disorders were often milder versions of the thinking deficits seen in Luria's patients.

Applicants to the Arrowsmith School—children and adults alike—undergo up to forty hours of assessments, designed to determine precisely which brain functions are weak and whether they might be helped. Accepted students, many of whom were distracted in regular schools, sit quietly working at their computers. Some, diagnosed with attention-deficit as well as learning disorders, were on Ritalin when they entered the school. As their exercises progress, some can come off medication, because their attention problems are secondary to their underlying learning disorders.

At the school, children who, like Barbara, had been unable to read a clock now work at computer exercises reading mind-numbingly complex ten-handed clocks (with hands not only for minutes, hours, and seconds but also for other time divisions, such as days, months, years) in mere seconds. They sit quietly, with intense concentration, until they get enough answers right to progress to the next level, when they shriek out a loud "Yes!" and their computer screen lights up to congratulate them. By the time they finish, they can read clocks far more complex than those any "normal" person can read.

At other tables children are studying Urdu and Persian letters to strengthen their visual memories. The shapes of these letters are unfamiliar, and the brain exercise requires the students to learn to recognize these alien shapes quickly.

Other children, like little pirates, wear eye patches on their left eyes and diligently trace intricate lines, squiggles, and Chinese letters with pens. The eye patch forces visual input into the right eye, then to the side of the brain where they have a problem. These children are not simply learning to write better. Most of them come with three related problems: trouble speaking in a smooth, flowing way, writing neatly, and reading. Barbara, following Luria, believes that all three difficulties are caused by a weakness in the brain function that normally helps us to coordinate and string together a number of movements when we perform these tasks.

When we speak, our brain converts a sequence of symbols—the letters and words of the thought—into a sequence of movements made by our tongue and lip muscles. Barbara believes, again following Luria, that the part of the brain that strings these movements together is the left premotor cortex of the brain. I referred several people with a weakness in this brain function to the school. One boy with this problem was always frustrated, because his thoughts came faster than he could turn them into speech, and he would often leave out chunks of information, have trouble finding words, and ramble. He was a very social person yet could not express himself and so remained silent much of the time. When he was asked a question in class, he often knew the answer but took such a painfully long time to get it out that he appeared much less intelligent than he was, and he began to doubt himself.

When we write a thought, our brain converts the words—which are symbols—into movements of the fingers and hands. The same boy had very jerky writing because his processing capacity for converting symbols into movements was easily overloaded, so he had to write with many separate, small movements instead of long, flowing ones. Even though he had been taught cursive writing, he preferred to print. (As adults, people with this problem can often be identified because they prefer to print or type. When we print, we make each letter separately, with just a few pen movements, which is less demanding on the brain. In cursive we write several letters at a time, and the brain must process more complex movements.) Writing was especially painful for the boy, since he often knew the right answers on tests but wrote so slowly that he couldn't get them all down. Or he would think of one word, letter, or number but write another. These children are often accused of being careless, but actually their overloaded brains fire the wrong motor movements.

Students with this disability also have reading problems. Normally when we read, the brain reads part of a sentence, then directs the eyes to move the right distance across the page to take in the next part of the sentence, requiring an ongoing sequence of precise eye movements.

The boy's reading was very slow because he skipped words, lost his place, and then lost his concentration. Reading was overwhelming and exhausting. On exams he would often misread the question, and when he tried to proofread his answers, he'd skip whole sections.

At the Arrowsmith School this boy's brain exercises involved tracing complex lines to stimulate his neurons in the weakened premotor area. Barbara has found that tracing exercises improve children in all three areas—speaking, writing, and reading. By the time the boy graduated, he read above grade level and could read for pleasure for the first time. He spoke more spontaneously in longer, fuller sentences, and his writing improved.

At the school some students listen to CDs and memorize poems to improve their weak auditory memories. Such children often forget instructions and are thought to be irresponsible or lazy, when in fact they have a brain difficulty. Whereas the average person can remember seven unrelated items (such as a seven-digit phone number), these people can remember only two or three. Some take notes compulsively, so they won't forget. In severe cases, they can't follow a song lyric from beginning to end, and they get so overloaded they just tune out. Some have difficulty remembering not only spoken language but even their own thoughts, because thinking with language is slow. This deficit can be treated with exercises in rote memorizing.

Barbara has also developed brain exercises for children who are socially clumsy because they have a weakness in the brain function that would allow them to read nonverbal cues. Other exercises are for those who have frontal lobe deficits and who are impulsive or have problems planning, developing strategies, sorting out what is relevant, forming goals, and sticking to them. They often appear disorganized, flighty, and unable to learn from their mistakes. Barbara believes that many people labeled "hysterical" or "antisocial" have weaknesses in this area.

The brain exercises are life-transforming. One American graduate told me that when he came to the school at thirteen, his math and reading skills were still at a third-grade level. He had been told after neuropsychological testing at Tufts University that he would never improve. His mother had tried him in ten different schools for students with learning disabilities, but none had helped. After three years at Arrowsmith, he was reading and doing math at a tenth-grade level. Now he has graduated from college and works in venture capital. Another student came to Arrowsmith at sixteen reading at a first-grade level. His parents, both teachers, had tried all the standard compensation techniques. After fourteen months at Arrowsmith he is reading at a seventh-grade level.

We all have some weak brain functions, and such neuroplasticity-based techniques have great potential to help almost everyone. Our weak spots can have a profound effect on our professional success, since most careers require the use of multiple brain functions. Barbara used brain exercises to rescue a talented artist who had a first-rate drawing ability and sense of color but a weak ability to recognize the shape of objects. (The ability to recognize shapes depends on a brain function quite different from those functions required for drawing or seeing color; it is the same skill that allows some people to excel at games like Where's Waldo? Women are often better at it at than men, which is why men seem to have more difficulty finding things in the refrigerator.)

Barbara also helped a lawyer, a promising litigator who, because of a Broca's area pronunciation deficit, spoke poorly in court. Since expending the extra mental effort to support a weak area seems to divert resources from strong areas, a person with a Broca's problem may also find it harder to think while talking. After practicing brain exercises focused on Broca's area, the lawyer went on to a successful courtroom career.

The Arrowsmith approach, and the use of brain exercises generally, has major implications for education. Clearly many children would benefit from a brain-area-based assessment to identify their weakened functions and a program to strengthen them—a far more productive approach than tutoring that simply repeats a lesson and leads to endless frustration. When "weak links in the chain" are strengthened, people gain access to skills whose development was formerly blocked, and they feel enormously liberated. A patient of mine, before he did the brain exercises, had a sense that he was very bright but could not make full use of his intelligence. For a long time I mistakenly thought his problems were based primarily on psychological conflicts, such as a fear of competition, and buried conflicts about surpassing his parents and siblings. Such conflicts did exist and did hold him back. But I came to see that his conflict about learning—his wish to avoid it—was based mostly on years of frustration and on a very legitimate fear of failure based on his brain's limits. Once he was liberated from his difficulties by Arrowsmith's exercises, his innate love of learning emerged full force.

The irony of this new discovery is that for hundreds of years educators did seem to sense that children's brains had to be built up through exercises of increasing difficulty that strengthened brain functions. Up through the nineteenth and early twentieth centuries a classical education often included rote memorization of long poems in foreign languages, which strengthened the auditory memory (hence thinking in language) and an almost fanatical attention to handwriting, which probably helped strengthen motor capacities and thus not only helped handwriting but added speed and fluency to reading and speaking. Often a great deal of attention was paid to exact elocution and to perfecting the pronunciation of words. Then in the 1960s educators dropped such traditional exercises from the curriculum, because they were too rigid, boring, and "not relevant." But the loss of these drills has been costly; they may have been the only opportunity that many students had to systematically exercise the brain function that gives us fluency and grace with symbols. For the rest of us, their disappearance may have contributed to the general decline of eloquence, which requires memory and a level of auditory brainpower unfamiliar to us now. In the Lincoln-Douglas debates of 1858 the debaters would comfortably speak for an hour or more without notes, in extended memorized paragraphs; today many of the most learned among us, raised in our most elite schools since the 1960s, prefer the omnipresent PowerPoint presentation—the ultimate compensation for a weak premotor cortex.

Barbara Arrowsmith Young's work compels us to imagine how much good might be accomplished if every child had a brain-based assessment and, if problems were found, a tailor-made program created to strengthen essential areas in the early years, when neuroplasticity is greatest. It is far better to nip brain problems in the bud than to allow the child to wire into his brain the idea that he is "stupid," begin to hate school and learning, and stop work in the weakened area, losing whatever strength he may have. Younger children often progress more quickly through brain exercises than do adolescents, perhaps because in an immature brain the number of connections among neurons, or synapses, is 50 percent greater than in the adult brain. When we reach adolescence, a massive "pruning back" operation begins in the brain, and synaptic connections and neurons that have not been used extensively suddenly die off—a classic case of "use it or lose it." It is probably best to strengthen weakened areas while all this extra cortical real estate is available. Still, brain-based assessments can be helpful all through school and even in college and university, when many students who did well in high school fail because their weak brain functions are overloaded by the increased demand. Even apart from these crises, every adult could benefit from a brain-based cognitive assessment, a cognitive fitness test, to help them better understand their own brain.

It's been years since Mark Rosenzweig first did the rat experiments that inspired Barbara and showed her that enriched environments and stimulation lead the brain to grow. Over the years his labs and others have shown that stimulating the brain makes it grow in almost every conceivable way. Animals raised in enriched environments—surrounded by other animals, objects to explore, toys to roll, ladders to climb, and running wheels—learn better than genetically identical animals that have been reared in impoverished environments. Acetylcholine, a brain chemical essential for learning, is higher in rats trained on difficult spatial problems than in rats trained on simpler problems. Mental training or life in enriched environments increases brain weight by 5 percent in the cerebral cortex of animals and up to 9 percent in areas that the training directly stimulates. Trained or stimulated neurons develop 25 percent more branches and increase their size, the number of connections per neuron, and their blood supply. These changes can occur late in life, though they do not develop as rapidly in older animals as in younger ones. Similar effects of training and enrichment on brain anatomy have been seen in all types of animals tested to date.

For people, postmortem examinations have shown that education increases the number of branches among neurons. An increased number of branches drives the neurons farther apart, leading to an increase in the volume and thickness of the brain. The idea that the brain is like a muscle that grows with exercise is not just a metaphor.

Some things can never be put together again. Lyova Zazetsky's diaries remained mostly a series of fragmented thoughts till the end. Aleksandr Luria, who figured out the meaning of those fragments, could not really help him. But Zazetsky's life story made it possible for Barbara Arrowsmith Young to heal herself and now others.

Today Barbara Arrowsmith Young is sharp and funny, with no noticeable bottlenecks in her mental processes. She flows from one activity to the next, from one child to the next, a master of many skills.

She has shown that children with learning disabilities can often go beyond compensations and correct their underlying problem. Like all brain exercise programs, hers work best and most quickly for people with only a few areas of difficulty. But because she has developed exercises for so many brain dysfunctions, she is often able to help children with multiple learning disabilities—children like herself, before she built herself a better brain.

3

Redesigning the Brain

A Scientist Changes Brains to Sharpen Perception and Memory, Increase Speed of Thought, and Heal Learning Problems

Michael Merzenich is a driving force behind scores of neuroplastic innovations and practical inventions, and I am on the road to Santa Rosa, California, to find him. His is the name most frequently praised by other neuroplasticians, and he's by far the hardest to track down. Only when I found out that he would be at a conference in Texas, went there, and sat myself down beside him, was I finally able to set up a meeting in San Francisco.

"Use *this* e-mail address," he says.

"And if you don't respond again?"

"Be persistent."

At the last minute, he switches our meeting to his villa in Santa Rosa.

Merzenich is worth the search.

The Irish neuroscientist Ian Robertson has described him as

"the world's leading researcher on brain plasticity." Merzenich's specialty is improving people's ability to think and perceive by redesigning the brain by training specific processing areas, called brain maps, so that they do more mental work. He has also, perhaps more than any other scientist, shown in rich scientific detail *how* our brain-processing areas change.

This villa in the Santa Rosa hills is where Merzenich slows down and regenerates himself. This air, these trees, these vineyards, seem like a piece of Tuscany transplanted into North America. I spend the night here with him and his family, and then in the morning we are off to his lab in San Francisco.

Those who work with him call him "Merz," to rhyme with "whirs" and "stirs." As he drives his small convertible to meetings—he's been double-booked much of the afternoon—his gray hair flies in the wind, and he tells me that many of his most vivid memories, in this, the second half of his life—he's sixty-one—are of conversations about scientific ideas. I hear him pour them into his cell phone, in his crackling voice. As we pass over one of San Francisco's glorious bridges, he pays a toll he doesn't have to because he's so involved with the concepts we are discussing. He has dozens of collaborations and experiments all going on at once and has started several companies. He describes himself as "just this side of crazy." He is not, but he is an interesting mix of intensity and informality. He was born in Lebanon, Oregon, of German stock, and though his name is Teutonic and his work ethic unrelenting, his speech is West Coast, easygoing, down-to-earth.

Of neuroplasticians with solid hard-science credentials, it is Merzenich who has made the most ambitious claims for the field: that brain exercises may be as useful as drugs to treat diseases as severe as schizophrenia; that plasticity exists from the cradle to the grave; and that radical improvements in cognitive functioning—how we learn, think, perceive, and remember—are possible even in the elderly. His latest patents are for techniques that show promise in allowing adults to learn language skills, without effortful memorization. Merzenich argues that practicing a new skill, under the right conditions, can change hundreds of millions and possibly billions of the connections between the nerve cells in our brain maps.

If you are skeptical of such spectacular claims, keep in mind that they come from a man who has already helped cure some disorders that were once thought intractable. Early in his career Merzenich developed, along with his group, the most commonly used design for the cochlear implant, which allows congenitally deaf children to hear. His current plasticity work helps learning-disabled students improve their cognition and perception. These techniques—his series of plasticity-based computer programs, *Fast ForWord*—have already helped hundreds of thousands. *Fast ForWord* is disguised as a children's game. What is amazing about it is how quickly the change occurs. In some cases people who have had a lifetime of cognitive difficulties get better after only thirty to sixty hours of treatment. Unexpectedly, the program has also helped a number of autistic children.

Merzenich claims that when learning occurs in a way consistent with the laws that govern brain plasticity, the mental "machinery" of the brain can be improved so that we learn and perceive with greater precision, speed, and retention.

Clearly when we learn, we increase what we know. But Merzenich's claim is that we can also change the very structure of the brain itself and increase its capacity to learn. Unlike a computer, the brain is constantly adapting itself.

"The cerebral cortex," he says of the thin outer layer of the brain, "is actually selectively refining its processing capacities to fit each task at hand." It doesn't simply learn; it is always "learning how to learn." The brain Merzenich describes is not an inanimate vessel that we fill; rather it is more like a living creature with an appetite, one that can grow and change itself with proper nourishment and exercise. Before Merzenich's work, the brain was seen as a complex machine, having unalterable limits on memory, processing speed, and intelligence. Merzenich has shown that each of these assumptions is wrong.

Merzenich did not set out to understand how the brain changes. He only stumbled on the realization that the brain could reorganize its maps. And though he was not the first scientist to demonstrate neuroplasticity, it was through experiments he conducted early in his career that mainstream neuroscientists came to accept the plasticity of the brain.

To understand how brain maps can be changed, we need first to have a picture of them. They were first made vivid in human beings by the neurosurgeon Dr. Wilder Penfield at the Montreal Neurological Institute in the 1930s. For Penfield, "mapping" a patient's brain meant finding where in the brain different parts of the body were represented and their activities processed—a solid localizationist project. Localizationists had discovered that the frontal lobes were the seat of the brain's *motor* system, which initiates and coordinates the movement of our muscles. The three lobes behind the frontal lobe, the temporal, parietal, and occipital lobes, comprise the brain's *sensory* system, processing the signals sent to the brain from our sense receptors—eyes, ears, touch receptors, and so on.

Penfield spent years mapping the sensory and motor parts of the brain, while performing brain surgery on cancer and epilepsy patients who could be conscious during the operation, because there are no pain receptors in the brain. Both the sensory and motor maps are part of the cerebral cortex, which lies on the brain's surface and so is easily accessible with a probe. Penfield discovered that when he touched a patient's sensory brain map with an electric probe, it triggered sensations that the patient felt in his body. He used the electric probe to help him distinguish the healthy tissue he wanted to preserve from the unhealthy tumors or pathological tissue he needed to remove.

Normally, when one's hand is touched, an electrical signal passes to the spinal cord and up to the brain, where it turns on cells in the map that make the hand feel touched. Penfield found he could also make the patient feel his hand was touched by turning on the hand area of the brain map electrically. When he stimulated another part of the map, the patient might feel his arm being touched; another part, his face. Each time he stimulated an area, he asked his patients what they'd felt, to make sure he didn't cut away healthy tissue. After many such operations he was able to show where on the brain's sensory map all parts of the body's surface were represented.

He did the same for the motor map, the part of the brain that controls movement. By touching different parts of this map, he could trigger movements in a patient's leg, arm, face, and other muscles.

One of the great discoveries Penfield made was that sensory and motor brain maps, like geographical maps, are topographical, meaning that areas adjacent to each other on the body's surface are generally adjacent to each other on the brain maps. He also discovered that when he touched certain parts of the brain, he triggered long-lost childhood memories or dreamlike scenes—which implied that higher mental activities were also mapped in the brain.

The Penfield maps shaped several generations' view of the brain. But because scientists believed that the brain couldn't change, they assumed, and taught, that the maps were fixed, immutable, and universal—the same in each of us—though Penfield himself never made either claim.

Merzenich discovered that these maps are neither immutable within a single brain nor universal but vary in their borders and size from person to person. In a series of brilliant experiments he showed that the shape of our brain maps changes depending upon what we do over the course of our lives. But in order to prove this point he needed a tool far finer than Penfield's electrodes, one that would be able to detect changes in just a few neurons at a time.

While an undergraduate at the University of Portland, Merzenich and a friend used electronic lab equipment to demonstrate the storm of electrical activity in insects' neurons. These experiments came to the attention of a professor who admired Merzenich's talent and curiosity and recommended him for graduate school at both Harvard and Johns Hopkins. Both accepted him. Merzenich opted for Hopkins to do his Ph.D. in physiology under one of the great neuroscientists of the time, Vernon Mountcastle, who in the 1950s was demonstrating that the subtleties of brain architecture could be discovered by studying the electrical activity of neurons using a new technique: micromapping with pin-shaped microelectrodes.

Microelectrodes are so small and sensitive that they can be inserted inside or beside a single neuron and can detect when an *individual* neuron fires off its electrical signal to other neurons. The neuron's signal passes from the microelectrode to an amplifier and then to an oscilloscope screen, where it appears as a sharp spike. Merzenich would make most of his major discoveries with microelectrodes.

This momentous invention allowed neuroscientists to decode the communication of neurons, of which the adult human brain has approximately 100 billion. Using large electrodes as Penfield did, scientists could observe thousands of neurons firing at once. With microelectrodes, scientists could "listen in on" one or several neurons at a time as they communicated with one another. Micromapping is still about a thousand times more precise than the current generation of brain scans, which detect bursts of activity that last one second in thousands of neurons. But a neuron's electrical signal often lasts a thousandth of a second, so brain scans miss an extraordinary amount of information. Yet micromapping hasn't replaced brain scans because it requires an extremely tedious kind of surgery, conducted under a microscope with microsurgical instruments.

Merzenich took to this technology right away. To map the area of the brain that processes feeling from the hand, Merzenich would cut away a piece of a monkey's skull over the sensory cortex, exposing a 1- to 2-millimeter strip of brain, then insert a microelectrode beside a sensory neuron. Next, he would tap the monkey's hand until he touched a part—say, the tip of a finger—that caused that neuron to fire an electrical signal into the microelectrode. He would record the location of the neuron that represented the fingertip, establishing the first point on the map. Then he would remove the microelectrode, reinsert it near another neuron, and tap different parts of the hand, until he located the part that turned on that neuron. He did this until he'd mapped the entire hand. A single mapping might require five hundred insertions and take several days, and Merzenich and his colleagues did thousands of these laborious surgeries to make their discoveries.

At about this time, a crucial discovery was made that would forever affect Merzenich's work. In the 1960s, just as Merzenich was beginning to use microelectrodes on the brain, two other scientists, who had also worked at Johns Hopkins with Mountcastle, discovered that the brain in very young animals is plastic. David Hubel and Torsten Wiesel were micromapping the visual cortex to learn how vision is processed. They'd inserted microelectrodes into the visual cortex of kittens and discovered that different parts of the cortex processed the lines, orientations, and movements of visually perceived objects. They also discovered that there was a "critical period," from the third to the eighth week of life, when the newborn kitten's brain *had to* receive visual stimulation in order to develop normally. In the crucial experiment Hubel and Wiesel sewed shut one eyelid of a kitten during its critical period, so the eye got no visual stimulation. When they opened this shut eye, they found that the visual areas in the brain map that normally processed input from the shut eye had failed to develop, leaving the kitten blind in that eye for life. Clearly the brains of kittens during the critical period were plastic, their structure literally shaped by experience.

When Hubel and Wiesel examined the brain map for that blind eye, they made one more unexpected discovery about plasticity. The part of the kitten's brain that had been deprived of input from the shut eye did not remain idle. It had begun to process visual input from the open eye, as though the brain didn't want to waste any "cortical real estate" and had found a way to rewire itself—another indication that the brain is plastic in the critical period. For this work Hubel and Wiesel received the Nobel Prize. Yet even though they had discovered plasticity in infancy, they remained localizationists, defending the idea that the adult brain is hardwired by the end of infancy to perform functions in fixed locations.

The discovery of the critical period became one of the most famous in biology in the second half of the twentieth century. Scientists soon showed that other brain systems required environmental stimuli to develop. It also seemed that each neural system had a different critical period, or window of time, during which it was especially plastic and sensitive to the environment, and during which it had rapid, formative growth. Language development, for instance, has a critical period that begins in infancy and ends between eight years and puberty. After this critical period closes, a person's ability to learn a second language without an accent is limited. In fact, second languages learned after the critical period are not processed in the same part of the brain as is the native tongue.

The notion of critical periods also lent support to ethologist Konrad Lorenz's observation that goslings, if exposed to a human being for a brief period of time, between fifteen hours and three days after birth, bonded with that person, instead of with their mother, for life. To prove it, he got goslings to bond to him and follow him around. He called this process "imprinting." In fact, the psychological version of the critical period went back to Freud, who argued that we go through developmental stages that are brief windows of time, during which we must have certain experiences to be healthy; these periods are formative, he said, and shape us for the rest of our lives.

Critical-period plasticity changed medical practice. Because of Hubel and Wiesel's discovery, children born with cataracts no longer faced blindness. They were now sent for corrective surgery as infants, during their critical period, so their brains could get the light required to form crucial connections. Microelectrodes had shown that plasticity is an indisputable fact of childhood. And they also seemed to show that, like childhood, this period of cerebral suppleness is short-lived.

Merzenich's first glimpse of adult plasticity was accidental. In 1968, after completing his doctorate, he went to do a postdoc with Clinton Woolsey, a researcher in Madison, Wisconsin, and peer of Penfield's. Woolsey asked Merzenich to supervise two neurosurgeons, Drs. Ron Paul and Herbert Goodman. The three decided to observe what happens in the brain when one of the peripheral nerves in the hand is cut and then starts to regenerate.

It is important to understand that the nervous system is divided into two parts. The first part is the central nervous system (the brain and spinal cord), which is the command-and-control center of the system; it was thought to lack plasticity. The second part is the peripheral nervous system, which brings messages from the sense receptors to the spinal cord and brain and carries messages from the brain and spinal cord to the muscles and glands. The peripheral nervous system was long known to be plastic; if you cut a nerve in your hand, it can "regenerate" or heal itself.

Each neuron has three parts. The *dendrites* are treelike branches that receive input from other neurons. These dendrites lead into the *cell body,* which sustains the life of the cell and contains its DNA. Finally the *axon* is a living cable of varying lengths (from microscopic lengths in the brain, to some that can run down to the legs and reach up to six feet long). Axons are often compared to wires because they carry electrical impulses at very high speeds (from 2 to 200 miles per hour) toward the dendrites of neighboring neurons.

A neuron can receive two kinds of signals: those that excite it and those that inhibit it. If a neuron receives enough *excitatory* signals from other neurons, it will fire off its own signal. When it receives enough *inhibitory* signals, it becomes less likely to fire. Axons don't quite touch the neighboring dendrites. They are separated by a microscopic space called a *synapse.* Once an electrical signal gets to the end of the axon, it triggers the release of a chemical messenger, called a neurotransmitter, into the synapse. The chemical messenger floats over to the dendrite of the adjacent neuron, exciting or inhibiting it. When we say that neurons "rewire" themselves, we mean that alterations occur at the synapse, strengthening and increasing, or weakening and decreasing, the number of connections between the neurons.

Merzenich, Paul, and Goodman wanted to investigate a well-known but mysterious interaction between the peripheral and central nervous systems. When a *large* peripheral nerve (which consists of many axons) is cut, sometimes in the process of regeneration the "wires get crossed." When axons reattach to the axons of the wrong nerve, the person may experience "false localization," so that a touch on the index finger is felt in the thumb. Scientists assumed that this false localization occurred because the regeneration process "shuffled" the nerves, sending the signal from the index finger to the brain map for the thumb.

The model scientists had of the brain and the nervous system was that each point on the body surface had a nerve that passed signals directly to a specific point on the brain map, anatomically hardwired at birth. Thus a nerve branch for the thumb always passed its signals directly to the spot on the sensory brain map for the thumb. Merzenich and the group accepted this "point-to-point" model of the brain map and innocently set out to document what was happening *in the brain* during this shuffling of nerves.

They micromapped the hand maps in the brains of several adolescent monkeys, cut a peripheral nerve to the hand, and immediately sewed the two severed ends close together but not quite touching, hoping the many axonal wires in the nerve would get crossed as the nerve regenerated itself. After seven months they remapped the brain. Merzenich assumed they would see a very disturbed, chaotic brain map. Thus, if the nerves for the thumb and the index finger had been crossed, he expected that touching the index finger would generate activity in the map area for the thumb. But he saw nothing of the kind. The map was almost normal.

"What we saw," says Merzenich, "was absolutely astounding. I couldn't understand it." It was *topographically* arranged as though the brain had unshuffled the signals from the crossed nerves.

This breakthrough week changed Merzenich's life. He realized that he, and mainstream neuroscience, had fundamentally misinterpreted how the human brain forms maps to represent the body and the world. If the brain map could normalize its structure in response to abnormal input, the prevailing view that we are born with a *hardwired* system had to be wrong. The brain had to be plastic.

How could the brain do it? Moreover, Merzenich also observed that the new topographical maps were forming in slightly different places than before. The localizationist view, that each mental function was always processed in the same location in the brain, had to be either wrong or radically incomplete. What was Merzenich to make of it?

He went back to the library to look for evidence that contradicted localizationism. He found that in 1912 Graham Brown and Charles Sherrington had shown that stimulating *one point* in the motor cortex might cause an animal to bend its leg at one time and straighten it at another. This experiment, lost in the scientific literature, implied that there was no point-to-point relationship between the brain's motor map and a given movement. In 1923 Karl Lashley, using equipment far cruder than microelectrodes, exposed a monkey's motor cortex, stimulated it in a particular place, and observed the resulting movement. He then sewed the monkey back up. After some time he repeated the experiment, stimulating the monkey in that same spot, only to find that the movement produced often changed. As Harvard's great historian of psychology of the time,

Edwin G. Boring, put it, "One day's mapping would no longer be valid on the morrow."

Maps were dynamic.

Merzenich immediately saw the revolutionary implications of these experiments. He discussed the Lashley experiment with Vernon Mountcastle, a localizationist, who, Merzenich told me, "had actually been bothered by the Lashley experiment. Mountcastle did not instinctively want to believe in plasticity. He wanted things to be in their place, forever. And Mountcastle knew that this experiment represented an important challenge to how you think about the brain. Mountcastle thought that Lashley was an extravagant exaggerator."

Neuroscientists were willing to accept Hubel and Wiesel's discovery that plasticity exists in infancy, because they accepted that the infant brain was in the midst of development. But they rejected Merzenich's discovery that plasticity continues into adulthood.

Merzenich leans back with an almost mournful expression and remembers, "I had all of these reasons why I wanted to believe that the brain wasn't plastic in this way, and they were thrown over in a week."

Merzenich now had to find his mentors among the ghosts of dead scientists, like Sherrington and Lashley. He wrote a paper on the shuffled nerve experiment, and in the discussion section he argued for several pages that the adult brain is plastic—though he didn't use the word.

But the discussion was never published. Clinton Woolsey, his supervisor, wrote a big X across it, saying that it was too conjectural and that Merzenich was going way beyond the data. When the paper was published, no mention was made of plasticity, and only minimal emphasis was given to explaining the new topographic organization. Merzenich backed down from the opposition, at least in print. He was still, after all, a postdoc working in another man's lab.

But he was angry, and his mind was churning. He was beginning to think that plasticity might be a basic property of the brain that had evolved to give humans a competitive edge and that it might be "a fabulous thing."

In 1971 Merzenich became a professor at the University of California at San Francisco, in the department of otolaryngology and physiology, which did research on diseases of the ear. Now his own boss, he began the series of experiments that would prove the existence of plasticity beyond a doubt. Because the area was still so controversial, he did his plasticity experiments in the guise of more acceptable research. Thus he spent much of the early 1970s mapping the auditory cortex of different species of animals, and he helped others invent and perfect the cochlear implant.

The cochlea is the microphone inside our ears. It sits beside the vestibular apparatus that deals with position sense and that was damaged in Cheryl, Bach-y-Rita's patient. When the external world produces sound, different frequencies vibrate different little hair cells within the cochlea. There are three thousand such hair cells, which convert the sound into patterns of electrical signals that travel down the auditory nerve into the auditory cortex. The micromappers discovered that in the auditory cortex, sound frequencies are mapped "tonotopically." That is, they are organized like a piano: the lower sound frequencies are at one end, the higher ones at the other.

A cochlear implant is not a hearing aid. A hearing aid amplifies sound for those who have partial hearing loss due to a partially functioning cochlea that works well enough to detect some sound. Cochlear implants are for those who are deaf because of a profoundly damaged cochlea. The implant replaces the cochlea, transforming speech sounds into bursts of electrical impulses, which it sends to the brain. Because Merzenich and his colleagues could not hope to match the complexity of a natural organ with three thousand hair cells, the question was, could the brain, which had evolved to decode complex signals coming from so many hair cells, decode impulses from a far simpler device? If it could, it would mean that the auditory cortex was plastic, capable of modifying itself and responding to artificial inputs. The implant consists of a sound receiver, a converter that translates sound into electrical impulses, and an electrode inserted by surgeons into the nerves that run from the ear to the brain.

In the mid-1960s some scientists were hostile to the very idea of cochlear implants. Some said the project was impossible. Others argued that they would put deaf patients at risk of further damage. Despite the risks, patients volunteered for implants. At first some heard only noise; others heard just a few tones, hisses, and sounds starting and stopping.

Merzenich's contribution was to use what he had learned from mapping the auditory cortex to determine the kind of input patients needed from the implant to be able to decode speech, and where to implant the electrode. He worked with communication engineers to design a device that could transmit complex speech on a small number of bandwidth channels and still be intelligible. They developed a highly accurate, multichannel implant that allowed deaf people to hear, and the design became the basis for one of the two primary cochlear implant devices available today.

What Merzenich most wanted, of course, was to investigate plasticity directly. Finally, he decided to do a simple, radical experiment in which he would cut off all sensory input to a brain map and see how it responded. He went to his friend and fellow neuroscientist Jon Kaas, of Vanderbilt University in Nashville, who worked with adult monkeys. A monkey's hand, like a human's, has three main nerves: the radial, the median, and the ulnar. The *median* nerve conveys sensation mostly from the *middle* of the hand, the other two from either side of the hand. Merzenich cut the median nerve in one of the monkeys to see how the median nerve brain map would respond when *all* input was cut off. He went back to San Francisco and waited.

Two months later he returned to Nashville. When he mapped the monkey, he saw, as he expected, that the portion of the brain map that serves the median nerve showed no activity when he touched the middle part of the hand. But he was shocked by something else.

When he stroked the *outsides* of the monkey's hand—the areas that send their signals through the radial and ulnar nerves—the median nerve map lit up! The brain maps for the radial and ulnar nerves had almost doubled in size and *invaded* what used to be the median nerve map. And these new maps were topographical. This time he and Kaas, writing up the findings, called the changes "spectacular" and used the word "plasticity" to explain the change, though they put it in quotes.

The experiment demonstrated that if the median nerve was cut, other nerves, still brimming with electrical input, would take over the unused map space to process their input. When it came to allocating brain-processing power, brain maps were governed by competition for precious resources and the principle of *use it or lose it.*

The competitive nature of plasticity affects us all. There is an endless war of nerves going on inside each of our brains. If we stop exercising our mental skills, we do not just forget them: the brain map space for those skills is turned over to the skills we practice instead. If you ever ask yourself, "How often must I practice French, or guitar, or math to keep on top of it?" you are asking a question about competitive plasticity. You are asking how frequently you must practice one activity to make sure its brain map space is not lost to another.

Competitive plasticity in adults even explains some of our limitations. Think of the difficulty most adults have in learning a second language. The conventional view now is that the difficulty arises because the critical period for language learning has ended, leaving us with a brain too *rigid* to change its structure on a large scale. But the discovery of competitive plasticity suggests there is more to it. As we age, the more we use our native language, the more it comes to dominate our linguistic map space. Thus it is also because our brain is *plastic*—and because plasticity is competitive—that it is so hard to learn a new language and end the tyranny of the mother tongue.

But why, if this is true, is it easier to learn a second language when we are young? Is there not competition then too? Not really. If two languages are learned at the same time, during the critical period, both get a foothold. Brain scans, says Merzenich, show that in a bilingual child all the sounds of its two languages share a single large map, a library of sounds from both languages.

Competitive plasticity also explains why our bad habits are so difficult to break or "unlearn." Most of us think of the brain as a container and learning as putting something in it. When we try to break a bad habit, we think the solution is to put something new into the container. But when we learn a bad habit, it takes over a brain map, and each time we repeat it, it claims more control of that map and prevents the use of that space for "good" habits. That is why "unlearning" is often a lot harder than learning, and why early childhood education is so important—it's best to get it right early, before the "bad habit" gets a competitive advantage.

Merzenich's next experiment, ingeniously simple, made plasticity famous among neuroscientists and eventually did more to win over skeptics than any plasticity experiment before or since.

He mapped a monkey's hand map in the brain. Then he amputated the monkey's middle finger. After a number of months he remapped the monkey and found that the brain map for the amputated finger had disappeared and that the maps for the adjacent fingers had grown into the space that had originally mapped for the middle finger. Here was the clearest possible demonstration that brain maps are dynamic, that there is a competition for cortical real estate, and that brain resources are allocated according to the principle of use it or lose it.

Merzenich also noticed that animals of a particular species may have similar maps, but they are *never* identical. Micromapping allowed him to see differences that Penfield, with larger electrodes, could not. He also found that the maps of normal body parts change every few weeks. Every time he mapped a normal monkey's face, it was unequivocally different. Plasticity doesn't require the provocation of cut nerves or amputations. Plasticity is a normal phenomenon, and brain maps are constantly changing. When he wrote up this new experiment, Merzenich finally took the word "plasticity" out of quotes. Yet despite the elegance of his experiment, opposition to Merzenich's ideas did not melt away overnight.

He laughs when he says it. "Let me tell you what happened when I began to declare that the brain was plastic. I received hostile treatment. I don't know how else to put it. I got people saying things in reviews such as, 'This would be really interesting if it could possibly be true, but it could not be.' It was as if I just made it up."

Because Merzenich was arguing that brain maps could alter their borders and location and change their functions well into adulthood, localizationists opposed him. "Almost everybody I knew in the mainstream of neuroscience," he says, "thought that this was sort of *semi*-serious stuff—that the experiments were sloppy, that the effects described were uncertain. But actually the experiment had been done enough times that I realized that the position of the majority was arrogant and indefensible."

One of the major figures who voiced doubts was Torsten Wiesel. Despite the fact that Wiesel had shown that plasticity exists in the critical period, he still opposed the idea that it existed in adults, and wrote that he and Hubel "firmly believed that once cortical connections were established in their mature form, they stayed in place permanently." He had indeed won the Nobel Prize for establishing where visual processing occurs, a finding considered one of localizationism's greatest triumphs. Wiesel now accepts adult plasticity and has gracefully acknowledged in print that for a long time he was wrong and that Merzenich's pioneering experiments ultimately led him and his colleagues to change their minds. Hardcore localizationists took notice when a man of Wiesel's stature changed his mind.

"The most frustrating thing," says Merzenich, "was that I saw that neuroplasticity had all kinds of potential implications for medical therapeutics—for the interpretation of human neuropathology and psychiatry. And nobody paid any attention."

Since plastic change is a process, Merzenich realized he would only really be able to understand it if he could see it unfolding in the brain over time. He cut a monkey's median nerve and then did multiple mappings over a number of months.

The first mapping, immediately after he cut the nerve, showed, as he expected, that the brain map for the median nerve was completely silent when the middle of the hand was stroked. But when he stroked the part of the hand served by the outside nerves, the silent median nerve portion of the map lit up immediately. Maps for the outside nerves, the radial and ulnar nerves, now appeared in the median map space. These maps sprang up so quickly, it was as though they had been hidden there all along, since early development, and now they were "unmasked."

On the twenty-second day Merzenich mapped the monkey again. The radial and ulnar maps, which had been lacking in detail when they first appeared, had grown more refined and detailed and had now expanded to occupy almost the entire median nerve map. (A primitive map lacks detail; a refined map has a lot and thus conveys more information.)

By the 144th day the whole map was every bit as detailed as a normal map.

By doing multiple mappings over time, Merzenich observed that the new maps were changing their borders, becoming more detailed, and even moving around the brain. In one case he even saw a map disappear altogether, like Atlantis.

It seemed reasonable to assume that if totally new maps were forming, then new connections must have been forming among neurons. To help understand this process, Merzenich invoked the ideas of Donald O. Hebb, a Canadian behavioral psychologist who had worked with Penfield. In 1949 Hebb proposed that learning linked neurons in new ways. He proposed that when two neurons fire at the same time repeatedly (or when one fires, causing another to fire), chemical changes occur in both, so that the two tend to connect more strongly. Hebb's concept—actually proposed by Freud sixty years before—was neatly summarized by neuroscientist Carla Shatz: *Neurons that fire together wire together.*

Hebb's theory thus argued that neuronal structure can be altered by experience. Following Hebb, Merzenich's new theory was that neurons in brain maps develop strong connections to one another when they are activated at the same moment in time. And if maps could change, thought Merzenich, then there was reason to hope that people born with problems in brain map–processing areas—people with learning problems, psychological problems, strokes, or brain injuries—might be able to form new maps if he could help them form new neuronal connections, by getting their healthy neurons to fire together and wire together.

Starting in the late 1980s, Merzenich designed or participated in brilliant studies to test whether brain maps are time based and whether their borders and functioning can be manipulated by "playing" with the timing of input to them.

In one ingenious experiment, Merzenich mapped a normal monkey's hand, then sewed together two of the monkey's fingers, so that both fingers moved as one. After several months of allowing the monkey to use its sewn fingers, the monkey was remapped. The two maps of the originally separate fingers had now merged into a single map. If the experimenters touched any point on either finger, this new single map would light up. Because all the movements and sensations in those fingers always occurred simultaneously, they'd formed the same map. The experiment showed that timing of the input to the neurons in the map was the key to forming it—neurons that fired together *in time* wired together to make one map.

Other scientists tested Merzenich's findings on human beings. Some people are born with their fingers fused, a condition called syndactyly or "webbed-finger syndrome." When two such people were mapped, the brain scan found that they each had one large map for their fused fingers instead of two separate ones.

After surgeons separated the webbed fingers, the subjects' brains were remapped, and two distinct maps emerged for the two separated digits. Because the fingers could move independently, the neurons no longer fired simultaneously, illustrating another principle of plasticity: if you separate the signals to neurons in time, you create separate brain maps. In neuroscience this finding is now summarized as *Neurons that fire apart wire apart*—or *Neurons out of sync fail to link.*

In the next experiment in the sequence, Merzenich created a map for what might be called a nonexistent finger that ran perpendicular to the other fingers. The team stimulated all five fingertips of a monkey simultaneously, five hundred times a day for over a month, preventing the monkey from using its fingers one at a time. Soon the monkey's brain map had a new, elongated finger map, in which the five fingertips were merged. This new map ran perpendicular to the other fingers, and all the fingertips were part of it, instead of part of their individual finger maps, which had started to melt away from disuse.

In the final and most brilliant demonstration, Merzenich and his team proved that maps cannot be anatomically based. They took a small patch of skin from one finger, and—this is the key point—with the nerve to its brain map still attached, surgically grafted the skin onto an adjacent finger. Now that piece of skin and its nerve were stimulated whenever the finger it was attached to was moved or touched in the course of daily use. According to the anatomical-hardwiring model, the signals should *still* have been sent from the skin along its nerve to the brain map for the finger that the skin and nerve originally came from. Instead, when the team stimulated the patch of skin, the map of its *new* finger responded. The map for the patch of skin migrated from the brain map of the original finger to its new one, because both the patch and the new finger were stimulated simultaneously.

In a few short years Merzenich had discovered that adult brains are plastic, persuaded skeptics in the scientific community this was the case, and shown that experience changes the brain. But he still hadn't explained a crucial enigma: how the maps organize themselves to become topographical and function in a way that is useful to us.

When we say a brain map is organized topographically, we mean that the map is ordered as the body itself is ordered. For instance, our middle finger sits between our index finger and our ring finger. The same is true for our brain map: the map for the middle finger sits between the map for our index finger and that of our ring finger. Topographical organization is efficient, because it means that parts of the brain that often work together are close together in the brain map, so signals don't have to travel far in the brain itself.

The question for Merzenich was, how does this topographic order emerge in the brain map? The answer he and his group came to was ingenious. A topographic order emerges because many of our everyday activities involve repeating sequences in a fixed order. When we pick up an object the size of an apple or baseball, we usually grip it first with our thumb and index finger, then wrap the rest of our fingers around it one by one. Since the thumb and index finger often touch at almost the same time, sending their signals to the brain almost simultaneously, the thumb map and the index finger map tend to form close together in the brain. (Neurons that fire together wire together.) As we continue to wrap our hand around the object, our middle finger will touch it next, so its brain map will tend to be beside the index finger and farther away from the thumb. As this common grasping sequence—thumb first, index finger second, middle finger third—is repeated thousands of times, it leads to a brain map where the thumb map is next to the index finger map, which is next to the middle finger map, and so on. Signals that tend to arrive at separate times, like thumbs and pinkies, have more distant brain maps, because neurons that fire apart wire apart.

Many if not all brain maps work by spatially grouping together events that happen together. As we have seen, the auditory map is arranged like a piano, with mapping regions for low notes at one end and for high notes at the other. Why is it so orderly? Because the low frequencies of sounds tend to come together with one another in nature. When we hear a person with a low voice, most of the frequencies are low, so they get grouped together.

The arrival of Bill Jenkins at Merzenich's lab ushered in a new phase of research that would help Merzenich develop practical applications of his discoveries. Jenkins, trained as a behavioral psychologist, was especially interested in understanding how we learn. He suggested they teach animals to learn new skills, to observe how learning affected their neurons and maps.

In one basic experiment they mapped a monkey's sensory cortex. Then they trained it to touch a spinning disk with its fingertip, with just the right amount of pressure for ten seconds to get a banana-pellet reward. This required the monkey to pay close attention, learning to touch the disk very lightly and judge time accurately. After thousands of trials, Merzenich and Jenkins remapped the monkey's brain and saw that the area mapping the monkey's fingertip had enlarged as the monkey had learned how to touch the disk with the right amount of pressure. The experiment showed that when an animal is motivated to learn, the brain responds plastically.

The experiment also showed that as brain maps get bigger, the individual neurons get more efficient in two stages. At first, as the monkey trained, the map for the fingertip grew to take up more space. But after a while individual neurons within the map became more efficient, and eventually fewer neurons were required to perform the task.

When a child learns to play piano scales for the first time, he tends to use his whole upper body—wrist, arm, shoulder—to play each note. Even the facial muscles tighten into a grimace. With practice the budding pianist stops using irrelevant muscles and soon uses only the correct finger to play the note. He develops a "lighter touch," and if he becomes skillful, he develops "grace" and relaxes when he plays. This is because the child goes from using a massive number of neurons to an appropriate few, well matched to the task. This more efficient use of neurons occurs whenever we become proficient at a skill, and it explains why we don't quickly run out of map space as we practice or add skills to our repertoire.

Merzenich and Jenkins also showed that individual neurons got more selective with training. Each neuron in a brain map for the sense of touch has a "receptive field," a segment on the skin's surface that "reports" to it. As the monkeys were trained to feel the disk, the receptive fields of individual neurons got smaller, firing only when small parts of the fingertip touched the disk. Thus, despite the fact that the size of the brain map increased, each neuron in the map became responsible for a smaller part of the skin surface, allowing the animal to have finer touch discrimination. Overall, the map became more precise.

Merzenich and Jenkins also found that as neurons are trained and become more efficient, they can process *faster*. This means that the speed at which we think is itself plastic. Speed of thought is essential to our survival. Events often happen quickly, and if the brain is slow, it can miss important information. In one experiment Merzenich and Jenkins successfully trained monkeys to distinguish sounds in shorter and shorter spans of time. The trained neurons fired more quickly in response to the sounds, processed them in a shorter time, and needed less time to "rest" between firings. Faster neurons ultimately lead to faster thought—no minor matter—because speed of thought is a crucial component of intelligence. IQ tests, like life, measure not only whether you can get the right answer but how long it takes you to get it.

They also discovered that as they trained an animal at a skill, not only did its neurons fire faster, but because they were faster their signals were clearer. Faster neurons were more likely to fire in sync with each other—becoming better team players—wiring together more and forming groups of neurons that gave off clearer and more powerful signals. This is a crucial point, because a powerful signal has greater impact on the brain. When we want to remember something we have heard we must hear it clearly, because a memory can be only as clear as its original signal.

Finally, Merzenich discovered that paying close attention is essential to long-term plastic change. In numerous experiments he found that lasting changes occurred *only* when his monkeys paid close attention. When the animals performed tasks automatically, without paying attention, they changed their brain maps, but the changes did not last. We often praise "the ability to multitask." While you can learn when you divide your attention, divided attention doesn't lead to abiding change in your brain maps.

When Merzenich was a boy, his mother's first cousin, a grade-school teacher in Wisconsin, was chosen teacher of the year for the entire United States. After the ceremony at the White House, she visited the Merzenich family in Oregon.

"My mother," he recalls, "asked the inane question that you'd ask in conversation: 'What are your most important principles in teaching?' And her cousin answered, 'Well, you test them when they come into school, and you figure out whether they are worthwhile. And if they are worthwhile, you really pay attention to them, and you don't waste time on the ones that aren't.' That's what she said. And you know, in one way or another, that's reflected in how people have treated children who are different, forever. It's just so destructive to imagine that your neurological resources are permanent and enduring and cannot be substantially improved and altered."

Merzenich now became aware of the work of Paula Tallal at Rutgers, who had begun to analyze why children have trouble learning to read. Somewhere between 5 and 10 percent of preschool children have a language disability that makes it difficult for them to read, write, or even follow instructions. Sometimes these children are called dyslexic.

Babies begin talking by practicing consonant-vowel combinations, cooing "da, da, da" and "ba, ba, ba." In many languages their first words consist of such combinations. In English their first words are often "mama" and "dada," "pee pee," and so on. Tallal's research showed that children with language disabilities have auditory processing problems with common consonant-vowel combinations that are spoken quickly and are called "the fast parts of speech." The children have trouble hearing them accurately and, as a result, reproducing them accurately.

Merzenich believed that these children's auditory cortex neurons were firing too slowly, so they couldn't distinguish between two very similar sounds or be certain, if two sounds occurred close together, which was first and which was second. Often they didn't hear the beginnings of syllables or the sound changes within syllables. Normally neurons, after they have processed a sound, are ready to fire again after about a 30-millisecond rest. Eighty percent of language-impaired children took at least three times that long, so that they lost large amounts of language information. When their neuron-firing patterns were examined, the signals weren't clear.

"They were muddy in, muddy out," says Merzenich. Improper hearing led to weaknesses in *all* the language tasks, so they were weak in vocabulary, comprehension, speech, reading, and writing. Because they spent so much energy decoding words, they tended to use shorter sentences and failed to exercise their memory for longer sentences.

Their language processing was more childlike, or "delayed," and they still needed practice distinguishing "da, da, da" and "ba, ba, ba."

When Tallal originally discovered their problems, she feared that "these kids were 'broken' and there was nothing you could do" to fix their basic brain defect. But that was before she and Merzenich combined forces.

In 1996 Merzenich, Paula Tallal, Bill Jenkins, and one of Tallal's colleagues, psychologist Steve Miller, formed the nucleus of a company, Scientific Learning, that is wholly devoted to using neuroplastic research to help people rewire their brains.

Their head office is in the Rotunda, a Beaux Arts masterpiece with an elliptical glass dome, 120 feet high, its edges painted in 24-karat gold leaf, in the middle of downtown Oakland, California. When you enter, you enter another world. The Scientific Learning staff includes child psychologists, plasticity researchers, experts in human motivation, speech pathologists, engineers, programmers, and animators. From their desks these researchers, bathed in natural light, can look up into the gorgeous dome.

Fast ForWord is the name of the training program they developed for language-impaired and learning-disabled children. The program exercises every basic brain function involved in language from decoding sounds up to comprehension—a kind of cerebral cross-training.

The program offers seven brain exercises. One teaches the children to improve their ability to distinguish short sounds from long. A cow flies across the computer screen, making a series of mooing sounds. The child has to catch the cow with the computer cursor and hold it by depressing the mouse button. Then suddenly the length of the moo sound changes subtly. At this point the child must release the cow and let it fly away. A child who releases it just after the sound changes scores points. In another game children learn to identify easily confused consonant-vowel combinations, such as "ba" and "da," first at slower speeds than they occur in normal language, and then at increasingly faster speeds. Another game teaches the children to hear faster and faster frequency glides (sounds like "whooooop" that sweep up). Another teaches them to remember and match sounds. The "fast parts of speech" are used throughout the exercises but have been slowed down with the help of computers, so the language-disabled children can hear them and develop clear maps for them; then gradually, over the course of the exercises, they are sped up. Whenever a goal is achieved, something funny happens: the character in the animation eats the answer, gets indigestion, gets a funny look on its face, or makes some slapstick move that is unexpected enough to keep the child attentive. This "reward" is a crucial feature of the program, because each time the child is rewarded, his brain secretes such neurotransmitters as dopamine and acetylcholine, which help consolidate the map changes he has just made. (Dopamine reinforces the reward, and acetylcholine helps the brain "tune in" and sharpen memories.)

Children with milder difficulties typically work at *Fast ForWord* for an hour and forty minutes a day, five days a week for several weeks, and those with more severe difficulties work for eight to twelve weeks.

The first study results, reported in the journal *Science* in January 1996, were remarkable. Children with language impairments were divided into two groups, one that did *Fast ForWord* and a control group that did a computer game that was similar but didn't train temporal processing or use modified speech. The two groups were matched for age, IQ, and language-processing skills. The children who did *Fast ForWord* made significant progress on standard speech, language, and auditory-processing tests, ended up with normal or better-than-normal language scores, and kept their gains when retested six weeks after training. They improved far more than children in the control group.

Further study followed five hundred children at thirty-five sites—hospitals, homes, and clinics. All were given standardized language tests before and after *Fast ForWord* training. The study showed that most children's ability to understand language normalized after *Fast ForWord*. In many cases, their comprehension rose above normal. The average child who took the program moved ahead 1.8 years of language development in six weeks, remarkably fast progress. A Stanford group did brain scans of twenty dyslexic children, before and after *Fast ForWord*. The opening scans showed that the children used different parts of their brains for reading than normal children do. After *Fast ForWord* new scans showed that their brains had begun to normalize. (For instance, they developed increased activity, on average, in the left temporo-parietal cortex, and their scans began to show patterns that were similar to those of children who have no reading problems.)

Willy Arbor is a seven-year-old from West Virginia. He's got red hair and freckles, belongs to Cub Scouts, likes going to the mall, and, though barely over four feet tall, loves wrestling. He's just gone through *Fast ForWord* and has been transformed.

"Willy's main problem was hearing the speech of others clearly," his mother explains. "I might say the word 'copy,' and he would think I said 'coffee.' If there was any background noise, it was especially hard for him to hear. Kindergarten was depressing. You could *see* his insecurity. He got into nervous habits like chewing on his clothes, or his sleeve, because everybody else was getting the answer right, and he wasn't. The teacher had actually talked about holding him back in first grade." Willy had trouble reading, both to himself and aloud.

"Willy," his mother continues, "couldn't hear change in pitch properly. So he couldn't tell when a person was making an exclamation or just a general statement, and he didn't grasp inflections in speech, which made it hard for him to read people's emotions. Without the high and low pitch he wasn't hearing that *wow* when people are excited. It was like everything was the same."

Willy was taken to a hearing specialist, who diagnosed his "hearing problem" as caused by an auditory-processing disorder that originated in his brain. He had difficulty remembering strings of words because his auditory system was so easily overloaded. "If you gave him more than three instructions, such as 'please put your shoes upstairs—put them in the closet—then come down for dinner,' he'd forget them. He'd take his shoes off, go up the steps, and ask 'Mom what did you want me to do?' Teachers had to repeat instructions all the time." Though he appeared to be a gifted child—he was good at math—his problems held him back in that area too.

His mother protested making Willy repeat first grade and over the summer sent him to *Fast ForWord* for eight weeks.

"Before he did *Fast ForWord*," his mother recalls, "you'd put him at the computer, and he got very stressed out. With this program, though, he spent a hundred minutes a day for a solid eight weeks at the computer. He loved doing it and loved the scoring system because he could see himself going up, up, up," says his mother. As he improved, he became able to perceive inflections in speech, got better at reading the emotions of others, and became a less anxious child. "So much changed for him. When he brought his midterms home, he said, 'It is better than last year, Mommy.' He began bringing home A and B marks on his papers most of the time—a noticeable difference . . . Now it's 'I can do this. This is my grade. I can make it better.' I feel like I had my prayer answered, it's done so much for him. It's amazing." A year later he continues to improve.

Merzenich's team started hearing that *Fast ForWord* was having a number of spillover effects. Children's handwriting improved. Parents reported that many of the students were starting to show sustained attention and focus. Merzenich thought these surprising benefits were occurring because *Fast ForWord* led to some general improvements in mental processing.

One of the most important brain activities—one we don't often think about—is the determination of how long things go on, or temporal processing. You can't move properly, perceive properly, or predict properly if you can't determine how long events last. Merzenich discovered that when you train people to distinguish very fast vibrations on their skin, lasting only 75 milliseconds, these same people could detect 75-millisecond *sounds* as well. It seemed that *Fast ForWord* was improving the brain's general ability to keep time. Sometimes these improvements spilled over into visual processing as well. Before *Fast ForWord,* when Willy was given a game that asked which items are out of place—a boot up in the tree, or a tin can on the roof—his eyes jumped all over the page. He was trying to see the whole page instead of taking in a little section at a time. At school he skipped lines when he read. After *Fast ForWord* his eyes no longer jumped around the page, and he was able to focus his visual attention.

A number of children who took standardized tests shortly after completing *Fast ForWord* showed improvements not only in language, speaking, and reading, but in math, science, and social studies as well. Perhaps these children were hearing what was going on in class better or were better able to read—but Merzenich thought it might be more complicated.

"You know," he says, "IQ goes up. We used the matrix test, which is a *visual*-based measurement of IQ—and IQ goes up."

The fact that a *visual* component of the IQ went up meant that the IQ improvements were not caused simply because *Fast ForWord* improved the children's ability to read verbal test questions. Their mental processing was being improved in a general way, possibly because their temporal processing was improving. And there were other unexpected benefits. Some children with autism began to make some general progress.

The mystery of autism—a human mind that cannot conceive of other minds—is one of the most baffling and poignant in psychiatry and one of the most severe developmental disorders of childhood. It is called a "pervasive developmental disorder," because so many aspects of development are disturbed: intelligence, perception, socializing skills, language, and emotion.

Most autistic children have an IQ of less than 70. They have major problems connecting socially to others and may, in severe cases, treat people like inanimate objects, neither greeting them nor acknowledging them as human beings. At times it seems that autistics don't have a sense that "other minds" exist in the world. They also have perceptual processing difficulties and are thus often hypersensitive to sound and touch, easily overloaded by stimulation. (That may be one reason autistic children often avoid eye contact: the stimulation from people, especially when coming from many senses at once, is too intense.) Their neural networks appear to be overactive, and many of these children have epilepsy.

Because so many autistic children have language impairments, clinicians began to suggest the *Fast ForWord* program for them. They never anticipated what might happen. Parents of autistic children who did *Fast ForWord* told Merzenich that their children became more connected socially. He began asking, were the children simply being trained to be more attentive listeners? And he was fascinated by the fact that with *Fast ForWord* both the language symptoms and the autistic symptoms seemed to be fading together. Could this mean that the language and autistic problems were different expressions of a common problem?

Two studies of autistic children confirmed what Merzenich had been hearing. One, a language study, showed that *Fast ForWord* quickly moved autistic children from severe language impairment to the normal range. But another pilot study of one hundred autistic children showed that *Fast ForWord* had a significant impact on their autistic symptoms as well. Their attention spans improved. Their sense of humor improved. They became more connected to people. They developed better eye contact, began greeting people and addressing them by name, spoke with them, and said good-bye at the end of their encounters. It seemed the children were beginning to experience the world as filled with other human minds.

Lauralee, an eight-year-old autistic girl, was diagnosed with moderate autism when she was three. Even as an eight-year-old she rarely used language. She didn't answer to her name, and to her parents, it seemed she was not hearing it. Sometimes she would speak, but when she did, "she had her own language," says her mother, "which was often unintelligible." If she wanted juice, she didn't ask for it. She would make gestures and pull her parents over to the cabinets to get things for her.

She had other autistic symptoms, among them the repetitive movements that autistic children use to try to contain their sense of being overwhelmed. According to her mother, Lauralee had "the whole works—the flapping of the hands, toe-walking, a lot of energy, biting. And she couldn't tell me what she was feeling."

She was very attached to trees. When her parents took her walking in the evening to burn off energy, she'd often stop, touch a tree, hug it, and speak to it.

Lauralee was unusually sensitive to sounds. "She had bionic ears," says her mother. "When she was little, she would often cover her ears. She couldn't tolerate certain music on the radio, like classical and slow music." At her pediatrician's office she heard sounds from the floor upstairs that others didn't. At home she would go over to the sinks, fill them with water, then wrap herself around the pipes, hugging them, listening to the water drain through them.

Lauralee's father is in the navy and served in the Iraq war in 2003. When the family was transferred to California, Lauralee was enrolled in a public school with a special-ed class that used *Fast ForWord*. The program took her about two hours a day for eight weeks to complete.

When she finished it, "she had an explosion in language," says her mother, "and began to speak more and use complete sentences. She could tell me about her days at school. Before I would just say, 'Did you have a good day or a bad day?' Now she was able to say what she did, and she remembered details. If she got into a bad situation, she would be able to tell me, and I wouldn't have to prompt her to get it out of her. She also found it easier to remember things." Lauralee has always loved to read, but now she is reading longer books, nonfiction and the encyclopedia. "She is listening to quieter sounds now and can tolerate different sounds from the radio," says her mother. "It was an awakening for her. And with the better communication, there was an awakening for all of us. It was a big blessing."

Merzenich decided that to deepen his understanding of autism and its many developmental delays, he would have to go back to the lab. He thought the best way to go about it was first to produce an "autistic animal"—one that had multiple developmental delays, as autistic children do. Then he could study it and try to treat it.

As Merzenich began to think through what he calls the "infantile catastrophe" of autism, he had a hunch that something might be going wrong in infancy, when most critical periods occur, plasticity is at its height, and a massive amount of development should be occurring. But autism is largely an inherited condition. If one identical twin is autistic, there is an 80 to 90 percent chance the other twin will be as well. In cases of *non*identical twins, where one is autistic, the nonautistic twin will often have some language and social problems.

Yet the incidence of autism has been climbing at a staggering rate that can't be explained by genetics alone. When the condition was first recognized over forty years ago, about one in 5,000 people had it. Now Merzenich believes it is at least fifteen in 5,000. That number has risen partly because autism is more often diagnosed, and because some children are labeled mildly autistic to get public funding for treatment. "But," says Merzenich, "even when all of the corrections are made by very hard-assed epidemiologists, it looks like it's about a threefold increase over the last fifteen years. There is a world emergency that relates to risk factors for autism."

He has come to think it likely that an environmental factor affects the neural circuits in these children, forcing the critical periods to shut down early, before the brain maps are fully differentiated. When we are born, our brain maps are often "rough drafts," or sketches, lacking detail, *undifferentiated.* In the critical period, when the structure of our brain maps is literally getting shaped by our first worldly experiences, the rough draft normally becomes detailed and differentiated.

Merzenich and his team used micromapping to show how maps in newborn rats are formed in the critical period. Right after birth, at the beginning of the critical period, auditory maps were undifferentiated, with only two broad regions in the cortex. Half of the map responded to *any* high-frequency sound. The other half responded to *any* low-frequency sound.

When the animal was exposed to a particular frequency during the critical period, that simple organization changed. If the animal was repeatedly exposed to a high C, after a while only a few neurons would turn on, becoming *selective* for high C. The same would happen when the animal was exposed to a D, E, F, and so on. Now the map, instead of having two broad areas, had many different areas, each responding to different notes. It was now differentiated.

What is remarkable about the cortex in the critical period is that it is so plastic that its structure can be changed just by exposing it to new stimuli. That sensitivity allows babies and very young children in the critical period of language development to pick up new sounds and words effortlessly, simply by hearing their parents speak; mere exposure causes their brain maps to wire in the changes. After the critical period older children and adults can, of course, learn languages, but they really have to *work* to pay attention. For Merzenich, the difference between critical-period plasticity and adult plasticity is that in the critical period the brain maps can be changed just by being exposed to the world because "the learning machinery is continuously on."

It makes good biological sense for this "machinery" always to be on because babies can't possibly know what will be important in life, so they pay attention to everything. Only a brain that is already somewhat organized can sort out what is worth paying attention to.

The next clue Merzenich needed in order to understand autism came from a line of research that was originated during the Second World War, in Fascist Italy, by a young Jewish woman, Rita Levi-Montalcini, while in hiding. Levi-Montalcini was born in Turin in 1909 and attended medical school there. In 1938, when Mussolini barred Jews from practicing medicine and doing scientific research, she fled to Brussels to continue her studies; when the Nazis threatened Belgium, she went back to Turin and built a secret laboratory in her bedroom, to study how nerves form, forging microsurgical equipment from sewing needles. When the Allies bombed Turin in 1940, she fled to Piedmont. One day in 1940, traveling to a small northern Italian village in a cattle car that had been converted into a passenger train, she sat down on the floor and read a scientific paper by Viktor Hamburger, who had been doing pioneering work on the development of neurons by studying chick embryos. She decided to repeat and extend his experiments, working on a table in a mountain house with eggs from a local farmer. When she finished each experiment, she ate the eggs. After the war Hamburger invited Levi-Montalcini to join him and his researchers in St. Louis to work on their discovery that the nerve fibers of chicks grew faster in the presence of tumors from mice. Levi-Montalcini speculated that the tumor might be releasing a substance to promote nerve growth. With biochemist Stanley Cohen she isolated the protein responsible and called it nerve growth factor, or NGF. Levi-Montalcini and Cohen were awarded the Nobel Prize in 1986.

Levi-Montalcini's work led to the discovery of a number of such nerve growth factors, one of which, brain-derived neurotrophic factor, or BDNF, caught Merzenich's attention.

BDNF plays a crucial role in reinforcing plastic changes made in the brain in the critical period. According to Merzenich, it does this in four different ways.

When we perform an activity that requires specific neurons to fire together, they release BDNF. This growth factor consolidates the connections between those neurons and helps to wire them together so they fire together reliably in the future. BDNF also promotes the growth of the thin fatty coat around every neuron that speeds up the transmission of electrical signals.

During the critical period BDNF turns on the nucleus basalis, the part of our brain that allows us to focus our attention—*and keeps it on, throughout the entire critical period.* Once turned on, the nucleus basalis helps us not only pay attention but remember what we are experiencing. It allows map differentiation and change to take place effortlessly. Merzenich told me, "It is like a teacher in the brain saying, 'Now *this* is really important—this you have to know for the exam of life.' " Merzenich calls the nucleus basalis and the attention system the "modulatory control system of plasticity"—the neurochemical system that, when turned on, puts the brain in an extremely plastic state.

The fourth and final service that BDNF performs—when it has completed strengthening key connections—is to help close down the critical period. Once the main neuronal connections are laid down, there is a need for stability and hence less plasticity in the system. When BDNF is released in sufficient quantities, it turns off the nucleus basalis and ends that magical epoch of effortless learning. Henceforth the nucleus can be activated only when something important, surprising, or novel occurs, or if we make the effort to pay close attention.

Merzenich's work on the critical period and BDNF helped him develop a theory that explains how so many different problems could be part of a single autistic whole. During the critical period, he argues, some situations overexcite the neurons in children who have genes that predispose them to autism, leading to *the massive, premature release of BDNF.* Instead of *important* connections being reinforced, *all* connections are. So much BDNF is released that it turns off the critical period prematurely, sealing all these connections in place, and the child is left with scores of undifferentiated brain maps and hence pervasive developmental disorders. Their brains are hyperexcitable and hypersensitive. If they hear one frequency, the whole auditory cortex starts firing. This is what seemed to be happening in Lauralee, who had to cover her "bionic" ears when she heard music. Other autistic children are hypersensitive to touch and feel tormented when the labels in their clothes touch their skin. Merzenich's theory also explains the high rates of epilepsy in autism: because of BDNF release, the brain maps are poorly differentiated, and because so many connections in the brain have been indiscriminately reinforced, once a few neurons start firing, the whole brain can be set off. It also explains why autistic children have bigger brains—the substance increases the fatty coating around the neurons.

If BDNF release was contributing to autism and language problems, Merzenich needed to understand what might cause young neurons to get "overexcited" and release massive amounts of the chemical.

Several studies alerted him to how an environmental factor might contribute. One disturbing study showed that the closer children lived to the noisy airport in Frankfurt, Germany, the lower their intelligence was. A similar study, on children in public housing high-rises above the Dan Ryan Expressway in Chicago, found that the closer their floor was to the highway, the lower their intelligence. So Merzenich began wondering about the role of a new environmental risk factor that might affect everyone but have a more damaging effect on genetically predisposed children: the continuous background noise from machines, sometimes called white noise. White noise consists of many frequencies and is very stimulating to the auditory cortex.

"Infants are reared in continuously more noisy environments. There is always a din," he says. White noise is everywhere now, coming from fans in our electronics, air conditioners, heaters, and car engines. How would such noise affect the developing brain? Merzenich wondered.

To test this hypothesis, his group exposed rat pups to pulses of white noise throughout their critical period and found that the pups' cortices were devastated.

"Every time you have a pulse," Merzenich says, "you are exciting everything in the auditory cortex—every neuron." So many neurons firing results in a massive BDNF release. And as his model predicted, this exposure brings the critical period to a premature close. The animals are left with undifferentiated brain maps and utterly indiscriminate neurons that get turned on by any frequency.

Merzenich found that these rat pups, like autistic children, were predisposed to epilepsy, and exposing them to normal speech caused them to have epileptic fits. (Human epileptics find that strobe lights at rock concerts set off their seizures. Strobes are pulsed emissions of white light and consist of many frequencies as well.) Merzenich now had his animal model for autism.

Recent brain scan studies now confirm that autistic children do indeed process sound in an abnormal way. Merzenich thinks that the undifferentiated cortex helps to explain why they have trouble learning, because a child with an undifferentiated cortex has a very difficult time paying attention. When asked to focus on one thing, these children experience booming, buzzing confusion—one reason autistic children often withdraw from the world and develop a shell. Merzenich thinks this same problem, in a milder form, may contribute to more common attention disorders.

Now the question for Merzenich was, could anything be done to normalize undifferentiated brain maps after the critical period? If he and his team could do so, they could offer hope for autistic children.

Using white noise, they first dedifferentiated the auditory maps of rats. Then, after the damage was done, they normalized and redifferentiated the maps using very simple tones, one at a time. With training, in fact, they brought the maps to an above-normal range. "And that," says Merzenich, "is exactly what we are trying to do in these autistic children." He is currently developing a modification of *Fast ForWord* that is designed for autism, a refinement of the program that helped Lauralee.

What if it were possible to reopen critical-period plasticity, so that adults could pick up languages the way children do, just by being exposed to them? Merzenich had already shown that plasticity extends into adulthood, and that with work—by paying close attention—we can rewire our brains. But now he was asking, could the critical period of effortless learning be extended?

Learning in the critical period is effortless because during that period the nucleus basalis is always on. So Merzenich and his young colleague Michael Kilgard set up an experiment in which they artificially turned on the nucleus basalis in adult rats and gave them learning tasks where they wouldn't have to pay attention and wouldn't receive a reward for learning.

They inserted microelectrodes into the nucleus basalis and used an electric current to keep it turned on. Then they exposed the rats to a 9 Hz sound frequency to see if they could effortlessly develop a brain map location for it, the way pups do during the critical period. After a week Kilgard and Merzenich found they could *massively* expand the brain map for that particular sound frequency. They had found an artificial way to reopen the critical period in adults.

They then used the same technique to get the brain to speed up its processing time. Normally an adult rat's auditory neurons can only respond to tones at a maximum of 12 pulses per second. By stimulating the nucleus basalis, it was possible to "educate" the neurons to respond to ever more rapid inputs.

This work opens up the possibility of high-speed learning later in life. The nucleus basalis could be turned on by an electrode, by microinjections of certain chemicals, or by drugs. It is hard to imagine that people will not—for better or for worse—be drawn to a technology that would make it relatively effortless to master the facts of science, history, or a profession, merely by being exposed to them briefly. Imagine immigrants coming to a new country, now able to pick up their new language, with ease and without an accent, in a matter of months. Imagine how the lives of older people who have been laid off from a job might be transformed, if they were able to learn a new skill with the alacrity they had in early childhood. Such techniques would no doubt be used by high school and university students in their studies and in competitive entrance exams. (Already many students who do not have attention deficit disorder use stimulants to study.) Of course, such aggressive interventions might have unanticipated, adverse effects on the brain—not to mention our ability to discipline ourselves—but they would likely be pioneered in cases of dire medical need, where people are willing to take the risk. Turning on the nucleus basalis might help brain-injured patients, so many of whom cannot relearn the lost functions of reading, writing, speaking, or walking because they can't pay close enough attention.

Merzenich has started a new company, Posit Science, devoted to helping people preserve the plasticity of their brains as they age and extend their mental lifespans. He's sixty-one but is not reluctant about calling himself old. "I love old people. I've always loved old people. Probably my favorite person was my paternal grandfather, one of the three or four most intelligent and interesting people

I've met in life." Grandpa Merzenich came from Germany at nine on one of the last clipper ships. He was self-educated, an architect and a building contractor. He lived to be seventy-nine, at a time when life expectancy was closer to forty.

"It's estimated that by the time someone who is sixty-five now dies, the life expectancy will be in the late eighties. Well, when you are eighty-five, there is a forty-seven percent chance that you will have Alzheimer's disease." He laughs. "So we've created this bizarre situation in which we are keeping people alive long enough so that on the average, half of them get the black rock before they die. We've got to do something about the mental lifespan, to extend it out and into the body's lifespan."

Merzenich thinks our neglect of intensive learning as we age leads the systems in the brain that modulate, regulate, and control plasticity to waste away. In response he has developed brain exercises for age-related cognitive decline—the common decline of memory, thinking, and processing speed.

Merzenich's way of attacking mental decline is at odds with mainstream neuroscience. Tens of thousands of papers, written about the physical and chemical changes that occur in the aging brain, describe processes that occur as neurons die. There are many drugs on the market and scores of drugs in the pipeline designed to block these processes and raise levels of falling chemicals in the brain. Yet, Merzenich believes that such drugs, worth billions in sales, provide only about four to six months of improvement.

"And there is something really wrong about all this," he says. "It all neglects the role of what is required to *sustain* normal skills and abilities . . . It is as if your skills and abilities, acquired in the brain at some young age, are just destined to deteriorate as the physical brain deteriorates." The mainstream approach, he argues, is based on no real understanding of what it takes to develop a new skill in the brain, never mind to sustain it. "It is imagined," he says, "that if you manipulate the levels of the right neurotransmitter . . . that memory will be recovered, and cognition will be useful, and that you will start moving like a gazelle again."

The mainstream approach doesn't take into account what is required to maintain a sharp memory. A major reason memory loss occurs as we age is that we have trouble *registering* new events in our nervous systems, because processing speed slows down, so that the accuracy, strength, and sharpness with which we perceive declines. If you can't register something clearly, you won't be able to remember it well.

Take one of the most common problems of aging, trouble finding words. Merzenich thinks this problem often occurs because of the gradual neglect and atrophy of the brain's attentional system and nucleus basalis, which have to be engaged for plastic change to occur. This atrophy leads to our representing oral speech with "fuzzy engrams," meaning that the representation of sounds or words is not sharp because the neurons that encode these fuzzy engrams are not firing in the coordinated, quick way needed to send a powerful sharp signal. Because the neurons that represent speech pass on fuzzy signals to all the neurons downstream from them ("muddy in, muddy out") we also have trouble remembering, finding, and using words. It is similar to the problem we saw occurring in the brains of language-impaired children, who also have "noisy brains."

When our brains are "noisy," the signal for a new memory can't compete against the background electrical activity of the brain, causing a "signal-noise problem."

Merzenich says the system gets noisier for two reasons. First because as everyone knows, "everything is progressively going to hell." But "the main reason it is getting noisier is that it is not being appropriately exercised." The nucleus basalis, which works by secreting acetylcholine—which, as we said, helps the brain "tune in" and form sharp memories—has been totally neglected. In a person with mild cognitive impairment the acetylcholine produced in the nucleus basalis is not even measurable.

"We have an intense period of learning in childhood. Every day is a day of new stuff. And then, in our early employment, we are intensely engaged in learning and acquiring new skills and abilities. And more and more as we progress in life we are operating as users of mastered skills and abilities."

Psychologically, middle age is often an appealing time because, all else being equal, it can be a relatively placid period compared with what has come before. Our bodies aren't changing as they did in adolescence; we're more likely to have a solid sense of who we are and be skilled at a career. We still regard ourselves as active, but we have a tendency to deceive ourselves into thinking that we are learning as we were before. We rarely engage in tasks in which we must focus our attention as closely as we did when we were younger, trying to learn a new vocabulary or master new skills. Such activities as reading the newspaper, practicing a profession of many years, and speaking our own language are mostly the replay of mastered skills, not learning. By the time we hit our seventies, we may not have systematically engaged the systems in the brain that regulate plasticity for fifty years.

That's why learning a new language in old age is so good for improving and maintaining the memory generally. Because it requires intense focus, studying a new language turns on the control system for plasticity and keeps it in good shape for laying down sharp memories of all kinds. No doubt *Fast ForWord* is responsible for so many general improvements in thinking, in part because it stimulates the control system for plasticity to keep up its production of acetylcholine and dopamine. Anything that requires highly focused attention will help that system—learning new physical activities that require concentration, solving challenging puzzles, or making a career change that requires that you master new skills and material. Merzenich himself is an advocate of learning a new language in old age. "You will gradually sharpen *everything* up again, and that will be very highly beneficial to you."

The same applies to mobility. Just doing the dances you learned years ago won't help your brain's motor cortex stay in shape. To keep the mind alive requires learning something truly *new* with intense focus. That is what will allow you to both lay down new memories and have a system that can easily access and preserve the older ones.

The thirty-six scientists at Posit Science are working on five areas that tend to fall apart as we age. The key in developing exercises is to give the brain the right stimuli, in the right order, with the right timing to drive plastic change. Part of the scientific challenge is to find the most efficient way to train the brain, by finding mental functions to train that apply to real life.

Merzenich told me, "Everything that you can see happen in a young brain can happen in an older brain." The only requirement is that the person must have enough of a reward, or punishment, to keep paying attention through what might otherwise be a boring training session. If so, he says, "the changes can be every bit as great as the changes in a newborn."

Posit Science has exercises for memory of words and language, using *Fast ForWord*–like listening exercises and computer games for auditory memory designed for adults. Instead of giving people with fading memories lists of words to memorize, as many self-help books recommend, these exercises rebuild the brain's basic ability to process sound, by getting people to listen to slowed, refined speech sounds. Merzenich doesn't believe you can improve a fading memory by asking people to do what they can't. "We don't want to kick a dead horse with training," he says. Adults do exercises that refine their ability to hear in a way they haven't since they were in the crib trying to separate out Mother's voice from background noise. The exercises increase processing speed and make basic signals stronger, sharper, and more accurate, while stimulating the brain to produce the dopamine and acetylcholine.

Various universities are now testing the memory exercises, using standardized tests of memory, and Posit Science has published its first control study in the *Proceedings of the National Academy of Sciences, USA*. Adults between the ages of sixty and eighty-seven trained on the auditory memory program an hour a day, five days a week, for eight to ten weeks—a total of forty to fifty hours of exercises. Before the training, the subjects functioned on average like typical seventy-year-olds on standard memory tests. After, they functioned like people in the broad forty-to-sixty-year-old range. Thus, many turned back their memory clock ten or more years, and some individuals turned it back about twenty-five years. These improvements held at a three-month follow-up. A group at the University of California at Berkeley, led by William Jagust, did "before" and "after" PET (positron emission tomography) scans of people who underwent the training, and found that their brains did not show the signs of "metabolic decline"—neurons gradually becoming less active—typically seen in people of their age. The study also compared seventy-one-year-old subjects who used the auditory memory program with those of the same age who spent the same amount of time reading newspapers, listening to audiobooks, or playing computer games. Those who didn't use the program showed signs of continuing metabolic decline in their frontal lobes, while those who used it didn't. Rather, program users showed increased metabolic activity in their right parietal lobes and in a number of other brain areas, which correlated with their better performance on memory and attention tests. These studies show that brain exercises not only slow age-related cognitive decline but can lead to improved functioning. And keep in mind that these changes were seen with only forty to fifty hours of brain exercise; it may be that with more work, greater change is possible.

Merzenich says they have been able to turn back the clock on people's cognitive functioning so that their memories, problem-solving abilities, and language skills are more youthful again. "We've driven people to abilities that apply to a much more youthful person—twenty or thirty years of reversal. An eighty-year-old is acting, operationally, like they are fifty or sixty years old." These exercises are now available in thirty independent-living communities and for individuals through the Posit Science Web site.

Posit Science is also working on visual processing. As we age, we stop seeing clearly, not just because our eyes fail but because the vision processors in the brain weaken. The elderly are more easily distracted and more prone to lose control of their "visual attention." Posit Science is developing computer exercises to keep people on task and speed up visual processing by asking subjects to search for various objects on a computer screen.

There are exercises for the frontal lobes that support our "executive functions" such as focusing on goals, extracting themes from what we perceive, and making decisions. These exercises are also designed to help people categorize things, follow complex instructions, and strengthen associative memory, which helps put people, places, and things into context.

Posit Science is also working on fine motor control. As we age, many of us give up on tasks such as drawing, knitting, playing musical instruments, or woodworking because we can't control the fine movements in our hands. These exercises, now being developed, will make fading hand maps in the brain more precise.

Finally, they are working on "gross motor control," a function that declines as we age, leading to loss of balance, the tendency to fall, and difficulties with mobility. Aside from the failure of vestibular processing, this decline is caused by the decrease in sensory feedback from our feet. According to Merzenich, shoes, worn for decades, limit the sensory feedback from our feet to our brain. If we went barefoot, our brains would receive many different kinds of input as we went over uneven surfaces. Shoes are a relatively flat platform that spreads out the stimuli, and the surfaces we walk on are increasingly artificial and perfectly flat. This leads us to dedifferentiate the maps for the soles of our feet and limit how touch guides our foot control. Then we may start to use canes, walkers, or crutches or rely on other senses to steady ourselves. By resorting to these compensations instead of exercising our failing brain systems, we hasten their decline.

As we age, we want to look down at our feet while walking down stairs or on slightly challenging terrain, because we're not getting much information from our feet. As Merzenich escorted his mother-in-law down the stairs of the villa, he urged her to stop looking down and start feeling her way, so that she would maintain, and develop, the sensory map for her foot, rather than letting it waste away.

Having devoted years to enlarging brain maps, Merzenich now believes there are times you want to shrink them. He has been working on developing a mental eraser that can eliminate a problematic brain map. This technique could be of great use for people who have post-traumatic flashbacks, recurring obsessional thoughts, phobias, or problematic mental associations. Of course, its potential for abuse is chilling.

Merzenich continues to challenge the view that we are stuck with the brain we have at birth. The Merzenich brain is structured by its constant collaboration with the world, and it is not only the parts of the brain most exposed to the world, such as our senses, that are shaped by experience. Plastic change, caused by our experience, travels deep into the brain and ultimately even into our genes, molding them as well—a topic to which we shall return.

This Mediterranean-style villa where he spends so much time sits among low mountains. He has just planted his own vineyard, and we walk through it. At night we talk about his early years studying philosophy, while four generations of his spirited family tease each other, breaking into peals of laughter. On the couch sits Merzenich's latest grandchild, just a few months old and in the midst of many critical periods. She makes everyone around her happy because she is such a good audience. You can coo at her, and she listens, thrilled. You tickle her toes, and she is completely attentive. As she looks around the room she takes in everything.

4

Acquiring Tastes and Loves

What Neuroplasticity Teaches Us About Sexual Attraction and Love

A. was a single, handsome young man who came to me because he was depressed. He had just gotten involved with a beautiful woman who had a boyfriend, and she had begun to encourage him to abuse her. She tried to draw A. into acting out sexual fantasies in which she dressed up as a prostitute, and he was to "take charge" of her and become violent in some way. When A. began to feel an alarming wish to oblige her, he got very upset, broke it off, and sought treatment. He had a history of involvement with women who were already attached to other men and emotionally out of control. His girlfriends had either been demanding and possessive or castratingly cruel. Yet these were the women who thrilled him. "Nice" girls, thoughtful, kind women, bored him, and he felt that any woman who fell in love with him in a tender, uncomplicated way was defective.

His own mother was a severe alcoholic, frequently needy, seductive, and given to emotional storms and violent rages throughout his childhood. A. recalled her banging his sister's head against the radiator and burning his stepbrother's fingers as a punishment for playing with matches. She was frequently depressed, often threatening suicide, and his role was to be on the alert, calm her, and prevent her. His relationship with her was also highly sexualized. She wore see-through nighties and talked to him as though he were a lover. He thought he recalled her inviting him into her bed when he was a child and had an image of himself sitting with his foot in her vagina while she masturbated. He had an exciting but furtive feeling about the scene. On the rare occasions when his father, who had retreated from his wife, was home, A. recalled himself as "perpetually short of breath," and trying to stop fights between his parents, who eventually divorced.

A. spent much of his childhood stifling his rage at both parents and often felt like a volcano about to burst. Intimate relationships seemed like forms of violence, in which others threatened to eat him alive, and yet by the time he had passed through childhood, it was for women who promised to do just that, and them alone, that he had acquired an erotic taste.

Human beings exhibit an extraordinary degree of sexual plasticity compared with other creatures. We vary in what we like to do with our partners in a sexual act. We vary where in our bodies we experience sexual excitement and satisfaction. But most of all we vary in whom or what we are attracted to. People often say they find a particular "type" attractive, or a "turn-on," and these types vary immensely from person to person.

For some, the types change as they go through different periods and have new experiences. One homosexual man had successive relations with men from one race or ethnic group, then with those from another, and in each period he could be attracted only to men in the group that was currently "hot." After one period was over, he could never be attracted to a man from the old group again. He acquired a taste for these "types" in quick succession and seemed more smitten by the person's category or type (i.e., "Asians" or "African-Americans") than by the individual. The plasticity of this man's sexual taste exaggerates a general truth: that the human libido is not a hardwired, invariable biological urge but can be curiously fickle, easily altered by our psychology and the history of our sexual encounters. And our libido can also be finicky. Much scientific writing implies otherwise and depicts the sexual instinct as a biological imperative, an ever-hungry brute, always demanding satisfaction—a glutton, not a gourmet. But human beings are more like gourmets and are drawn to types and have strong preferences; having a "type" causes us to defer satisfaction until we find what we are looking for, because attraction to a type is restrictive: the person who is "really turned on by blondes" may tacitly rule out brunettes and redheads.

Even sexual preference can occasionally change. Though some scientists increasingly emphasize the inborn basis of our sexual preferences, it is also true that some people have heterosexual attractions for part of their lives—with no history of bisexuality—and then "add on" a homosexual attraction and vice versa.

Sexual plasticity may seem to have reached its height in those who have had many different partners, learning to adapt to each new lover; but think of the plasticity required of the aging married couple with a good sex life. They looked very different in their twenties, when they met, than they do in their sixties, yet their libidos adjust, so they remain attracted.

But sexual plasticity goes further still. Fetishists desire inanimate objects. The male fetishist can be more excited by a high-heeled shoe with a fur trim, or by a woman's lingerie, than by a real woman. Since ancient times some human beings in rural areas have had intercourse with animals. Some people seem to be attracted not so much to people as to complex sexual scripts, where partners play roles, involving various perversions, combining sadism, masochism, voyeurism, and exhibitionism. When they place an ad in the personals, the description of what they are looking for in a lover often sounds more like a job description than like that of a person they would like to know.

Given that sexuality is an instinct, and instinct is traditionally defined as a hereditary behavior unique to a species, varying little from one member to the next, the variety of our sexual tastes is curious. Instincts generally resist change and are thought to have a clear, nonnegotiable, hardwired purpose, such as survival. Yet the human sexual "instinct" seems to have broken free of its core purpose, reproduction, and varies to a bewildering extent, as it does not in other animals, in which the sexual instinct seems to behave itself and act like an instinct.

No other instinct can so satisfy without accomplishing its biological purpose, and no other instinct is so disconnected from its purpose. Anthropologists have shown that for a long time humanity did not know that sexual intercourse was required for reproduction. This "fact of life" had to be learned by our ancestors, just as children must learn it today. This detachment from its primary purpose is perhaps the ultimate sign of sexual plasticity.

Love too is remarkably flexible, and its expression has changed through history. Though we speak of romantic love as the most *natural* of sentiments, in fact the concentration of our adult hopes for intimacy, tenderness, and lust in one person until death do us part is not common to all societies and has only recently become widespread in our own. For millennia most marriages were arranged by parents for practical reasons. Certainly, there are unforgettable stories of romantic love linked to marriage in the Bible, as in the Song of Songs, and linked to disaster in medieval troubadour poetry and, later, in Shakespeare. But romantic love began to gain social approval in the aristocracies and courts of Europe only in the twelfth century—originally between an unmarried man and a married woman, either adulterous or unconsummated, usually ending badly. Only with the spread of democratic ideals of individualism did the idea that lovers ought to be able to choose spouses for themselves take firmer hold and gradually begin to seem completely natural and inalienable.

It is reasonable to ask whether our sexual plasticity is related to neuroplasticity. Research has shown that neuroplasticity is neither ghettoized within certain departments in the brain nor confined to the sensory, motor, and cognitive processing areas we have already explored. The brain structure that regulates instinctive behaviors, including sex, called the hypothalamus, is plastic, as is the amygdala, the structure that processes emotion and anxiety. While some parts of the brain, such as the cortex, may have more plastic potential because there are more neurons and connections to be altered, even noncortical areas display plasticity. It is a property of all brain tissue. Plasticity exists in the hippocampus (the area that turns our memories from short-term to long-term ones) as well as in areas that control our breathing, process primitive sensation, and process pain. It exists in the spinal cord—as scientists have shown; actor Christopher Reeve, who suffered a severe spinal injury, demonstrated such plasticity, when he was able, through relentless exercise, to recover some feeling and mobility seven years after his accident.

Merzenich puts it this way: "You cannot have plasticity in isolation . . . it's an absolute impossibility." His experiments have shown that if one brain system changes, those systems connected to it change as well. The same "plastic rules"—use it or lose it, or neurons that fire together wire together—apply throughout. Different areas of the brain wouldn't be able to function together if that weren't the case.

Do the same plastic rules that apply to brain maps in the sensory, motor, and language cortices apply to more complex maps, such as those that represent our relationships, sexual or otherwise? Merzenich has also shown that complex brain maps are governed by the same plastic principles as simpler maps. Animals exposed to a simple tone will develop a single brain map region to process it. Animals exposed to a complex pattern, such as a melody of six tones, will not simply link together six different map regions but will develop a region that encodes the *entire* melody. These more complex melody maps obey the same plastic principles as maps for single tones.

"The sexual instincts," wrote Freud, "are noticeable to us for their plasticity, their capacity for altering their aims." Freud was not the first to argue that sexuality was plastic—Plato, in his dialogue on love, argued that human Eros took many forms—but Freud laid the foundations for a neuroscientific understanding of sexual and romantic plasticity.

One of his most important contributions was his discovery of critical periods for sexual plasticity. Freud argued that an adult's ability to love intimately and sexually unfolds in stages, beginning in the infant's first passionate attachments to its parents. He learned from his patients, and from observing children, that early childhood, not puberty, was the first critical period for sexuality and intimacy, and that children are capable of passionate, protosexual feelings—crushes, loving feelings, and in some cases even sexual excitement, as A. was. Freud discovered that the sexual abuse of children is harmful because it influences the critical period of sexuality in childhood, shaping our later attractions and thoughts about sex. Children are needy and typically develop passionate attachments to their parents. If the parent is warm, gentle, and reliable, the child will frequently develop a taste for that kind of relationship later on; if the parent is disengaged, cool, distant, self-involved, angry, ambivalent, or erratic, the child may seek out an adult mate who has similar tendencies. There are exceptions, but a significant body of research now confirms Freud's basic insight that early patterns of relating and attaching to others, if problematic, can get "wired" into our brains in childhood and repeated in adulthood. Many aspects of the sexual script that A. played out when he first came to see me were repetitions of his traumatic childhood situation, thinly disguised—such as his being attracted to an unstable woman who crossed normal sexual boundaries in furtive relationships, where hostility and sexual excitement were merged, while the woman's official partner was cuckolded and threatening to reenter the scene.

The idea of the critical period was formulated around the time Freud started writing about sex and love, by embryologists who observed that in the embryo the nervous system develops in stages, and that if these stages are disturbed, the animal or person will be harmed, often catastrophically, for life. Though Freud didn't use the term, what he said about the early stages of sexual development conforms to what we know about critical periods. They are brief windows of time when new brain systems and maps develop with the help of stimulation from the people in one's environment.

Traces of childhood sentiments in adult love and sexuality are detectable in everyday behaviors. When adults in our culture have tender foreplay, or express their most intimate adoration, they often call each other "baby" or "babe." They use terms of endearment that their mothers used with them as children, such as "honey" and "sweetie pie," terms that evoke the earliest months of life when the mother expressed her love by feeding, caressing, and talking sweetly to her baby—what Freud called the oral phase, the first critical period of sexuality, the essence of which is summed up in the words "nurturance" and "nourish"—tenderly caring for, loving, *and* feeding. The baby feels merged with the mother, and its trust of others develops as the baby is held and nurtured with a sugary food, milk. Being loved, cared for, and fed are mentally associated in the mind and wired together in the brain in our first formative experience after birth.

When adults talk baby talk, using words such as "sweetie pie" and "baby" to address each other, and give their conversation an oral flavor, they are, according to Freud, "regressing," moving from mature mental states of relating to earlier phases of life. In terms of plasticity, such regression, I believe, involves unmasking old neuronal pathways that then trigger all the associations of that earlier phase. Regression can be pleasant and harmless, as in adult foreplay, or it can be problematic, as when infantile aggressive pathways are unmasked and an adult has a temper tantrum.

Even "talking dirty" shows traces of infantile sexual stages. After all, why should sex be thought "dirty" at all? This attitude reflects a child's view of sex from a stage when it is conscious of toilet training, urination, and defecation and is surprised to learn that the genitals, which are involved in urination, and so close to the anus, are also involved in sex, and that Mommy permits Daddy to insert his "dirty" organ in a hole that is very close to her bottom. Adults are not generally bothered by this, because in adolescence they have gone through another critical period of sexual plasticity in which their brains reorganized again, so that the pleasure of sex becomes intense enough to override any disgust.

Freud showed that many sexual mysteries can be understood as critical-period fixations. After Freud, we are no longer surprised that the girl whose father left her as a child pursues unavailable men old enough to be her father, or that people raised by ice-queen mothers often seek such people out as partners, sometimes becoming "icy" themselves, because, never having experienced empathy in the critical period, a whole part of their brains failed to develop. And many perversions can be explained in terms of plasticity and the persistence of childhood conflicts. But the main point is that in our critical periods we can acquire sexual and romantic tastes and inclinations that get wired into our brains and can have a powerful impact for the rest of our lives. And the fact that we can acquire different sexual tastes contributes to the tremendous sexual variation between us.

The idea that a critical period helps shape sexual desire in adults contradicts the currently popular argument that what attracts us is less the product of our personal history than of our common biology. Certain people—models and movie stars, for instance—are widely regarded as beautiful or sexy. A certain strand of biology teaches us that these people are attractive because they exhibit biological signs of robustness, which promise fertility and strength: a clear complexion and symmetrical features mean a potential mate is free from disease; an hourglass figure is a sign a woman is fertile; a man's muscles predict he will be able to protect a woman and her offspring.

But this simplifies what biology really teaches. Not everyone falls in love with the body, as when a woman says, "I knew, when I first heard *that* voice, that he was for me," the music of the voice being perhaps a better indication of a man's soul than his body's surface. And sexual taste has changed over the centuries. Rubens's beauties were large by current standards, and over the decades the vital statistics of *Playboy* centerfolds and fashion models have varied from voluptuous to androgynous. Sexual taste is obviously influenced by culture and experience and is often acquired and then wired into the brain.

"Acquired tastes" are by definition learned, unlike "tastes," which are inborn. A baby needn't acquire a taste for milk, water, or sweets; these are immediately perceived as pleasant. Acquired tastes are initially experienced with indifference or dislike but later become pleasant—the odors of cheeses, Italian bitters, dry wines, coffees, patés, the hint of urine in a fried kidney. Many delicacies that people pay dearly for, that they must "develop a taste for," are the very foods that disgusted them as children.

In Elizabethan times lovers were so enamored of each other's body odors that it was common for a woman to keep a peeled apple in her armpit until it had absorbed her sweat and smell. She would give this "love apple" to her lover to sniff at in her absence. We, on the other hand, use synthetic aromas of fruits and flowers to mask our body odor from our lovers. Which of these two approaches is acquired and which is natural is not so easy to determine. A substance as "naturally" repugnant to us as the urine of cows is used by the

Masai tribe in East Africa as a lotion for their hair—a direct consequence of the cow's importance in their culture. Many tastes we think "natural" are acquired through learning and become "second nature" to us. We are unable to distinguish our "second nature" from our "original nature" because our neuroplastic brains, once rewired, develop a new nature, every bit as biological as our original.

The current porn epidemic gives a graphic demonstration that sexual tastes can be acquired. Pornography, delivered by high-speed Internet connections, satisfies every one of the prerequisites for neuroplastic change.

Pornography seems, at first glance, to be a purely instinctual matter: sexually explicit pictures trigger instinctual responses, which are the product of millions of years of evolution. But if that were true, pornography would be unchanging. The same triggers, bodily parts and their proportions, that appealed to our ancestors would excite us. This is what pornographers would have us believe, for they claim they are battling sexual repression, taboo, and fear and that their goal is to liberate the natural, pent-up sexual instincts.

But in fact the content of pornography is a *dynamic* phenomenon that perfectly illustrates the progress of an acquired taste. Thirty years ago "hardcore" pornography usually meant the *explicit* depiction of sexual intercourse between two aroused partners, displaying their genitals. "Softcore" meant pictures of women, mostly, on a bed, at their toilette, or in some semiromantic setting, in various states of undress, breasts revealed.

Now hardcore has evolved and is increasingly dominated by the sadomasochistic themes of forced sex, ejaculations on women's faces, and angry anal sex, all involving scripts fusing sex with hatred and humiliation. Hardcore pornography now explores the world of perversion, while softcore is now what hardcore was a few decades ago, explicit sexual intercourse between adults, now available on cable TV. The comparatively tame softcore pictures of yesteryear—women in various states of undress—now show up on mainstream media all day long, in the pornification of everything, including television, rock videos, soap operas, advertisements, and so on.

Pornography's growth has been extraordinary; it accounts for 25 percent of video rentals and is the fourth most common reason people give for going online. An MSNBC.com survey of viewers in 2001 found that 80 percent felt they were spending so much time on pornographic sites that they were putting their relationships or jobs at risk. Softcore pornography's influence is now most profound because, now that it is no longer hidden, it influences young people with little sexual experience and especially plastic minds, in the process of forming their sexual tastes and desires. Yet the plastic influence of pornography on adults can also be profound, and those who use it have no sense of the extent to which their brains are reshaped by it.

During the mid- to late 1990s, when the Internet was growing rapidly and pornography was exploding on it, I treated or assessed a number of men who all had essentially the same story. Each had acquired a taste for a kind of pornography that, to a greater or lesser degree, troubled or even disgusted him, had a disturbing effect on the pattern of his sexual excitement, and ultimately affected his relationships and sexual potency.

None of these men were fundamentally immature, socially awkward, or withdrawn from the world into a massive pornography collection that was a substitute for relationships with real women. These were pleasant, generally thoughtful men, in reasonably successful relationships or marriages.

Typically, while I was treating one of these men for some other problem, he would report, almost as an aside and with telling discomfort, that he found himself spending more and more time on the Internet, looking at pornography and masturbating. He might try to ease his discomfort by asserting that everybody did it. In some cases he would begin by looking at a *Playboy*-type site or at a nude picture or video clip that someone had sent him as a lark. In other cases he would visit a harmless site, with a suggestive ad that redirected him to risqué sites, and soon he would be hooked.

A number of these men also reported something else, often in passing, that caught my attention. They reported increasing difficulty in being turned on by their actual sexual partners, spouses or girlfriends, though they still considered them objectively attractive. When I asked if this phenomenon had any relationship to viewing pornography, they answered that it initially helped them get more excited during sex but over time had the opposite effect. Now, instead of using their senses to enjoy being in bed, in the present, with their partners, lovemaking increasingly required them to fantasize that they were part of a porn script. Some gently tried to persuade their lovers to act like porn stars, and they were increasingly interested in "fucking" as opposed to "making love." Their sexual fantasy lives were increasingly dominated by the scenarios that they had, so to speak, downloaded into their brains, and these new scripts were often more primitive and more violent than their previous sexual fantasies. I got the impression that any sexual creativity these men had was dying and that they were becoming addicted to Internet porn.

The changes I observed are not confined to a few people in therapy. A social shift is occurring. While it is usually difficult to get information about private sexual mores, this is not the case with pornography today, because its use is increasingly public. This shift coincides with the change from calling it "pornography" to the more casual term "porn." For his book on American campus life, *I Am Charlotte Simmons,* Tom Wolfe spent a number of years observing students on university campuses. In the book one boy, Ivy Peters, comes into the male residence and says, "Anybody got porn?"

Wolfe goes on, "This was not an unusual request. Many boys

spoke openly about how they masturbated at least once every day, as if this were some sort of prudent maintenance of the psychosexual system." One of the boys tells Ivy Peters, "Try the third floor. They got some one-hand magazines up there." But Peters responds, "I've built up a tolerance to magazines . . . I need videos." Another boy says, "Oh, f'r Chrissake, I.P., it's ten o'clock at night. In another hour the cum dumpsters will start coming over here to spend the night . . . And you're looking for porn videos and a knuckle fuck." Then Ivy "shrugged and turned his palms up as if to say, 'I want porn. What's the big deal?' "

The big deal is his tolerance. He recognizes that he is like a drug addict who can no longer get high on the images that once turned him on. And the danger is that this tolerance will carry over into relationships, as it did in patients whom I was seeing, leading to potency problems and new, at times unwelcome, tastes. When pornographers boast that they are pushing the envelope by introducing new, harder themes, what they don't say is that they must, because their customers are building up a tolerance to the content. The back pages of men's risqué magazines and Internet porn sites are filled with ads for Viagra-type drugs—medicine developed for older men with erectile problems related to aging and blocked blood vessels in the penis. Today young men who surf porn are tremendously fearful of impotence, or "erectile dysfunction" as it is euphemistically called. The misleading term implies that these men have a problem in their penises, but the problem is in their heads, in their sexual brain maps. The penis works fine when they use pornography. It rarely occurs to them that there may be a relationship between the pornography they are consuming and their impotence. (A few men, however, tellingly described their hours at computer porn sites as time spent "masturbating my brains out.")

One of the boys in Wolfe's scene describes the girls who are coming over to have sex with their boyfriends as "cum dumpsters." He too is influenced by porn images, for "cum dumpsters," like many women in porn films, are always eager, available receptacles and therefore devalued.

The addictiveness of Internet pornography is not a metaphor. Not all addictions are to drugs or alcohol. People can be seriously addicted to gambling, even to running. All addicts show a loss of control of the activity, compulsively seek it out despite negative consequences, develop tolerance so that they need higher and higher levels of stimulation for satisfaction, and experience withdrawal if they can't consummate the addictive act.

All addiction involves long-term, sometimes lifelong, neuroplastic change in the brain. For addicts, moderation is impossible, and they must avoid the substance or activity completely if they are to avoid addictive behaviors. Alcoholics Anonymous insists that there are no "former alcoholics" and makes people who haven't had a drink for decades introduce themselves at a meeting by saying, "My name is John, and I am an alcoholic." In terms of plasticity, they are often correct.

In order to determine how addictive a street drug is, researchers at the National Institutes of Health (NIH) in Maryland train a rat to press a bar until it gets a shot of the drug. The harder the animal is willing to work to press the bar, the more addictive the drug. Cocaine, almost all other illegal drugs, and even nondrug addictions such as running make the pleasure-giving neurotransmitter dopamine more active in the brain. Dopamine is called the reward transmitter, because when we accomplish something—run a race and win—our brain triggers its release. Though exhausted, we get a surge of energy, exciting pleasure, and confidence and even raise our hands and run a victory lap. The losers, on the other hand, who get no such dopamine surge, immediately run out of energy, collapse at the finish line, and feel awful about themselves. By hijacking our dopamine system, addictive substances give us pleasure without our having to work for it.

Dopamine, as we saw in Merzenich's work, is also involved in plastic change. The same surge of dopamine that thrills us also consolidates the neuronal connections responsible for the behaviors that led us to accomplish our goal. When Merzenich used an electrode to stimulate an animal's dopamine reward system while playing a sound, dopamine release stimulated plastic change, enlarging the representation for the sound in the animal's auditory map. An important link with porn is that dopamine is also released in sexual excitement, increasing the sex drive in both sexes, facilitating orgasm, and activating the brain's pleasure centers. Hence the addictive power of pornography.

Eric Nestler, at the University of Texas, has shown how addictions cause permanent changes in the brains of animals. A single dose of many addictive drugs will produce a protein, called ΔFosB (pronounced "delta Fos B"), that accumulates in the neurons. Each time the drug is used, more ΔFosB accumulates, until it throws a genetic switch, affecting which genes are turned on or off. Flipping this switch causes changes that persist long after the drug is stopped, leading to irreversible damage to the brain's dopamine system and rendering the animal far more prone to addiction. Nondrug addictions, such as running and sucrose drinking, also lead to the accumulation of ΔFosB and the same permanent changes in the dopamine system.

Pornographers promise healthy pleasure and relief from sexual tension, but what they often deliver is an addiction, tolerance, and an eventual decrease in pleasure. Paradoxically, the male patients I worked with often craved pornography but didn't like it.

The usual view is that an addict goes back for more of his fix because he likes the pleasure it gives and doesn't like the pain of withdrawal. But addicts take drugs when there is *no* prospect of pleasure, when they know they have an insufficient dose to make them high, and will crave more even before they begin to withdraw. Wanting and liking are two different things.

An addict experiences cravings because his plastic brain has become sensitized to the drug or the experience. Sensitization is different from tolerance. As tolerance develops, the addict needs more and more of a substance or porn to get a pleasant effect; as sensitization develops, he needs less and less of the substance to crave it intensely. So sensitization leads to increased wanting, though not necessarily liking. It is the accumulation of ΔFosB, caused by exposure to an addictive substance or activity, that leads to sensitization.

Pornography is more exciting than satisfying because we have two separate pleasure systems in our brains, one that has to do with exciting pleasure and one with satisfying pleasure. The exciting system relates to the "appetitive" pleasure that we get imagining something we desire, such as sex or a good meal. Its neurochemistry is largely dopamine-related, and it raises our tension level.

The second pleasure system has to do with the satisfaction, or consummatory pleasure, that attends actually having sex or having that meal, a calming, fulfilling pleasure. Its neurochemistry is based on the release of endorphins, which are related to opiates and give a peaceful, euphoric bliss.

Pornography, by offering an endless harem of sexual objects, hyperactivates the appetitive system. Porn viewers develop new maps in their brains, based on the photos and videos they see. Because it is a use-it-or-lose-it brain, when we develop a map area, we long to keep it activated. Just as our muscles become impatient for exercise if we've been sitting all day, so too do our senses hunger to be stimulated.

THE MEN AT THEIR COMPUTERS looking at porn were uncannily like the rats in the cages of the NIH, pressing the bar to get a shot of dopamine or its equivalent. Though they didn't know it, they had been seduced into pornographic training sessions that met all the conditions required for plastic change of brain maps. Since neurons that fire together wire together, these men got massive amounts of practice wiring these images into the pleasure centers of the brain, with the rapt attention necessary for plastic change. They imagined these images when away from their computers, or while having sex with their girlfriends, reinforcing them. Each time they felt sexual excitement and had an orgasm when they masturbated, a "spritz of dopamine," the reward neurotransmitter, consolidated the connections made in the brain during the sessions. Not only did the reward facilitate the behavior; it provoked none of the embarrassment they felt purchasing *Playboy* at a store. Here was a behavior with no "punishment," only reward.

The content of what they found exciting changed as the Web sites introduced themes and scripts that altered their brains without their awareness. Because plasticity is competitive, the brain maps for new, exciting images increased at the expense of what had previously attracted them—the reason, I believe, they began to find their girlfriends less of a turn-on.

The story of Sean Thomas, first published in England's *Spectator,* is a remarkable account of a man descending into a porn addiction, and it sheds light on how porn changes brain maps and alters sexual taste, as well as the role of critical-period plasticity in the process. Thomas wrote, "I never used to like pornography, not really. Yes, in my teens in the Seventies I used to have the odd copy of *Playboy* under my pillow. But on the whole I didn't really go for skin mags or blue movies. I found them tedious, repetitive, absurd, and very embarrassing to buy." He was repelled by the bleakness of the porn scene and the garishness of the mustachioed studs who inhabited it. But in 2001, shortly after he first went online, he got curious about the porn everyone said was taking over the Internet. Many of the sites were free—teasers, or "gateway sites," to get people into the harder stuff. There were galleries of naked girls, of common types of sexual fantasies and attractions, designed to press a button in the brain of the surfer, even one he didn't know he had. There were pictures of lesbians in a Jacuzzi, cartoon porn, women on the toilet smoking, coeds, group sex, and men ejaculating over submissive Asian women. Most of the pictures told a story.

Thomas found a few images and scripts that appealed to him, and they "dragged me back for more the next day. And the next. And the next." Soon he found that whenever he had a spare minute, he would "start hungrily checking out Net Porn."

Then one day he came across a site that featured spanking images. To his surprise, he got intensely excited. Thomas soon found all sorts of related sites, such as "Bernie's Spanking Pages" and the "Spanking College."

"This was the moment," he writes, "that the real addiction set in. My interest in spanking got me speculating: What other kinks was I harboring? What other secret and rewarding corners lurked in my sexuality that I would now be able to investigate in the privacy of my home? Plenty, as it turned out. I discovered a serious penchant for, inter alia, lesbian gynecology, interracial hardcore, and images of Japanese girls taking off their hotpants. I was also into netball players with no knickers, drunk Russian girls exposing themselves, and convoluted scenarios where submissive Danish actresses were intimately shaved by their dominant female partners in the shower. The Net had, in other words, revealed to me that I had an unquantifiable variety of sexual fantasies and quirks and that the process of satisfying these desires online only led to more interest."

Until he happened upon the spanking pictures, which presumably tapped into some childhood experience or fantasy about being punished, the images he saw interested him but didn't compel him. Other people's sexual fantasies bore us. Thomas's experience was similar to that of my patients: without being fully aware of what they were looking for, they scanned hundreds of images and scenarios until they hit upon an image or sexual script that touched some buried theme that really excited them.

Once Thomas found that image, he changed. That spanking image had his *focused attention,* the condition for plastic change. And unlike a real woman, these porn images were available all day, every day on the computer.

Now Thomas was hooked. He tried to control himself but was spending at least five hours a day on his laptop. He surfed secretly, sleeping only three hours a night. His girlfriend, aware of his exhaustion, wondered if he was seeing someone else. He became so sleep deprived that his health suffered, and he got a series of infections that landed him in a hospital emergency room and finally caused him to take stock. He began inquiring among his male friends and found that many of them were also hooked.

Clearly there was something about Thomas's sexuality, outside his awareness, that had suddenly surfaced. Does the net simply reveal quirks and kinks, or does it also help create them? I think it creates new fantasies out of aspects of sexuality that have been outside the surfer's conscious awareness, bringing these elements together to form new networks. It is not likely that thousands of men have witnessed, or even imagined, submissive Danish actresses intimately shaved by their dominant female partners in the shower. Freud discovered that such fantasies take hold of the mind because of the *individual* components in them. For instance, some heterosexual men are interested in porn scenarios where older, dominant women initiate younger women into lesbian sex. This may be because boys in early childhood often feel dominated by their mothers, who are the "boss," and dress, undress, and wash them. In early childhood some boys may pass through a period when they strongly identify with their mothers and feel "like a girl," and their later interest in lesbian sex can express their residual unconscious female identification. Hardcore porn unmasks some of the early neural networks that formed in the critical periods of sexual development and brings all these early, forgotten, or repressed elements together to form a new network, in which all the features are wired together. Porn sites generate catalogs of common kinks and mix them together in images. Sooner or later the surfer finds a killer combination that presses a number of his sexual buttons at once. Then he reinforces the network by viewing the images repeatedly, masturbating, releasing dopamine and strengthening these networks. He has created a kind of "neosexuality," a rebuilt libido that has strong roots in his buried sexual tendencies. Because he often develops tolerance, the pleasure of sexual discharge must be supplemented with the pleasure of an aggressive release, and sexual and aggressive images are increasingly mingled—hence the increase in sadomasochistic themes in hardcore porn.

Critical periods lay the groundwork for our types, but falling in love in adolescence or later provides an opportunity for a second round of massive plastic change. Stendhal, the nineteenth-century novelist and essayist, understood that love could lead to radical changes in attraction. Romantic love triggers such powerful emotion that we can reconfigure what we find attractive, even overcoming "objective" beauty. In *On Love* Stendhal describes a young man, Alberic, who meets a woman more beautiful than his mistress. Yet Alberic is far more drawn to his mistress than to this woman because his mistress promises him so much more happiness. Stendhal calls this "Beauty Dethroned by Love." Love has such power to change attraction that Alberic is turned on by a minor defect on his mistress's face, her pockmark. It excites him because "he has experienced so many emotions in the presence of that pockmark, emotions for the most part exquisite and of the most absorbing interest, that whatever his emotions may have been, they are renewed with incredible vividness at the sight of this sign, even observed on the face of another woman . . . in this case ugliness becomes beauty."

This transformation of taste can happen because we do not fall in love with looks alone. Under normal circumstances finding another person attractive can prompt a readiness to fall in love, but that person's character and a host of other attributes, including his ability to make us feel good about ourselves, crystallize the process of falling in love. Then being in love triggers an emotional state so pleasurable that it can make even pockmarks attractive, plastically rewiring our aesthetic sense. Here is how I believe it works.

In 1950 "pleasure centers" were discovered in the limbic system, a part of the brain heavily involved in processing emotion. In Dr. Robert Heath's experiments on humans—an electrode was implanted into the septal region of the limbic system and turned on—these patients experienced a euphoria so powerful that when the researchers tried to end the experiment, one patient pleaded with them not to. The septal region also fired when pleasant subjects were discussed with the patients and during orgasm. These pleasure centers were found to be part of the brain's reward system, the mesolimbic dopamine system. In 1954 James Olds and Peter Milner showed that when they inserted electrodes into an animal's pleasure center while teaching it a task, it learned more easily because learning felt so pleasurable and was rewarded.

When the pleasure centers are turned on, everything we experience gives us pleasure. A drug like cocaine acts on us by lowering the threshold at which our pleasure centers will fire, making it easier for them to turn on. It is not simply the cocaine that gives us pleasure. It is the fact that our pleasure centers now fire so easily that makes whatever we experience feel great. It is not just cocaine that can lower the threshold at which our pleasure centers fire. When people with bipolar disorder (formerly called manic depression) begin to move toward their manic highs, their pleasure centers begin firing more easily. And falling in love also lowers the threshold at which the pleasure centers will fire.

When a person gets high on cocaine, becomes manic, or falls in love, he enters an enthusiastic state and is optimistic about everything, because all three conditions lower the firing threshold for the *appetitive* pleasure system, the dopamine-based system associated with the pleasure of anticipating something we desire. The addict, the manic, and the lover are increasingly filled with hopeful anticipation and are sensitive to anything that might give pleasure—flowers and fresh air inspire them, and a slight but thoughtful gesture makes them delight in all mankind. I call this process "globalization."

Globalization is intense when falling in love and is, I believe, one of the main reasons that romantic love is such a powerful catalyst for plastic change. Because the pleasure centers are firing so freely, the enamored person falls in love not only with the beloved but with the world and romanticizes his view of it. Because our brains are experiencing a surge of dopamine, which consolidates plastic change, any pleasurable experiences and associations we have in the initial state of love are thus wired into our brains.

Globalization not only allows us to take more pleasure in the world, it also makes it harder for us to experience pain and displeasure or aversion. Heath showed that when our pleasure centers fire, it is more difficult for the nearby pain and aversion centers to fire too. Things that normally bother us don't. We love being in love not only because it makes it easy for us to be happy but also because it makes it harder for us to be unhappy.

Globalization also creates an opportunity for us to develop new tastes in what we find attractive, like the pockmark that gave Alberic such pleasure. Neurons that fire together wire together, and feeling pleasure in the presence of this normally unappealing pockmark causes it to get wired into the brain as a source of delight. A similar mechanism occurs when a "reformed" cocaine addict passes the seedy alleyway where he first took the drug and is overwhelmed with cravings so powerful that he goes back to it. The pleasure he felt during the high was so intense that it caused him to experience the ugly alleyway as enticing, by association.

There is thus a literal chemistry of love, and the stages of romance reflect the changes in our brain during not only the ecstasies but also love's throes. Freud, one of the first people to describe the psychic effects of cocaine and, as a young man, the first to discover its medical uses, got a glimpse of this chemistry. Writing to his fiancée, Martha, on February 2, 1886, he described taking cocaine while composing the letter. Because cocaine acts on the system so quickly, the letter, as it unfolds, gives us a marvelous window into its effects. He first describes how it makes him talkative and confessional. His initial self-deprecatory remarks vanish as the letter goes on, and soon he feels fearless, identifiying with his brave ancestors defending the Temple in Jerusalem. He likens cocaine's ability to cure his fatigue to the magical cure he gets from being with Martha romantically. In another letter he writes that cocaine reduces his shyness and depression, makes him euphoric, enhances his energy, self-esteem, and enthusiasm, and has an aphrodisiac effect. He is describing a state akin to "romantic intoxication," when people feel the initial high, talk all night, and have increased energy, libido, self-esteem, and enthusiasm, but because they think everything is good, they may also have impaired judgment—all of which occurs with a dopamine-promoting drug like cocaine. Recent fMRI (functional magnetic resonance imaging) scans of lovers looking at photos of their sweethearts show that a part of the brain with great concentrations of dopamine is activated; their brains looked like those of people on cocaine.

But the pains of love also have a chemistry. When separated for too long, lovers crash and experience withdrawal, crave their beloved, get anxious, doubt themselves, lose their energy, and feel run-down if not depressed. Like a little fix, a letter, an e-mail, or a telephone message from the beloved provides an instant shot of energy. Should they break up, they get depressed—the opposite of the manic high. These "addictive symptoms"—the highs, crashes, cravings, withdrawal, and fixes—are subjective signs of plastic changes occurring in the structure of our brains, as they adapt to the presence or absence of the beloved.

A tolerance, akin to tolerance for a drug, can develop in happy lovers as they get used to each other. Dopamine likes novelty. When monogamous mates develop a tolerance for each other and lose the romantic high they once had, the change may be a sign, not that either of them is inadequate or boring, but that their plastic brains have so well adapted to each other that it's harder for them to get the same buzz they once got from each other.

Fortunately, lovers can stimulate their dopamine, keeping the high alive, by injecting novelty into their relationship. When a couple go on a romantic vacation or try new activities together, or wear new kinds of clothing, or surprise each other, they are using novelty to turn on the pleasure centers, so that everything they experience, *including each other,* excites and pleases them. Once the pleasure centers are turned on and globalization begins, the new image of the beloved again becomes associated with unexpected pleasures and is plastically wired into the brain, which has evolved to respond to novelty. We must be learning if we are to feel fully alive, and when life, or love, becomes too predictable and it seems like there is little left to learn, we become restless—a protest, perhaps, of the plastic brain when it can no longer perform its essential task.

Love creates a generous state of mind. Because love allows us to experience as pleasurable situations or physical features that we otherwise might not, it also allows us to unlearn negative associations, another plastic phenomenon.

The science of unlearning is a very new one. Because plasticity is competitive, when a person develops a neural network, it becomes efficient and self-sustaining and, like a habit, hard to unlearn. Recall that Merzenich was looking for "an eraser" to help him speed up change and unlearn bad habits.

Different chemistries are involved in learning than in unlearning. When we learn something new, neurons fire together and wire together, and a chemical process occurs at the neuronal level called "long-term potentiation," or LTP, which strengthens the connections between the neurons. When the brain unlearns associations and disconnects neurons, another chemical process occurs, called "long-term depression," or LTD (which has nothing to do with a depressed mood state). Unlearning and weakening connections between neurons is just as plastic a process, and just as important, as learning and strengthening them. If we only strengthened connections, our neuronal networks would get saturated. Evidence suggests that unlearning existing memories is necessary to make room for new memories in our networks.

Unlearning is essential when we are moving from one developmental stage to the next. When at the end of adolescence a girl leaves home to go to college in another state, for example, both she and her parents undergo grief and massive plastic change, as they alter old emotional habits, routines, and self-images.

Falling in love for the first time also means entering a new developmental stage and demands a massive amount of unlearning. When people commit to each other, they must radically alter their existing and often selfish intentions and modify all other attachments, in order to integrate the new person in their lives. Life now involves ongoing cooperation that requires a plastic reorganization of the brain centers that deal with emotions, sexuality, and the self. Millions of neural networks have to be obliterated and replaced with new ones—one reason that falling in love feels, for so many people, like a loss of identity. Falling in love may also mean falling out of love with a past love; this too requires unlearning at a neural level.

A man's heart is broken by his first love when his engagement breaks off. He looks at many women, but each pales in comparison to the fiancée he came to believe was his one true love and whose image haunts him. He cannot unlearn the pattern of attraction to his first love. Or a woman married for twenty years becomes a young widow and refuses to date. She cannot imagine she will ever fall in love again, and the idea of "replacing" her husband offends her. Years pass, and her friends tell her it is time to move on, to no avail.

Often such people cannot move on because they cannot yet grieve; the thought of living without the one they love is too painful to bear. In neuroplastic terms, if the romantic or the widow is to begin a new relationship without baggage, each must first rewire billions of connections in their brains. The work of mourning is piecemeal, Freud noted; though reality tells us our loved one is gone, "its orders cannot be obeyed at once." We grieve by calling up one memory at a time, reliving it, and then letting it go. At a brain level we are turning on each of the neural networks that were wired together to form our perception of the person, experiencing the memory with exceptional vividness, then saying good-bye one network at a time. In grief, we *learn* to live without the one we love, but the reason this lesson is so hard is that we first must *unlearn* the idea that the person exists and can still be relied on.

Walter J. Freeman, a professor of neuroscience at Berkeley, was the first to argue that there is a connection between love and massive unlearning. He has assembled a number of compelling biological facts that point toward the conclusion that massive neuronal reorganization occurs at two life stages: when we fall in love and when we begin parenting. Freeman argues that massive plastic brain reorganization—far more massive than in normal learning or unlearning—becomes possible because of a brain neuromodulator.

Neuromodulators are different from neurotransmitters. While neurotransmitters are released in the synapses to excite or inhibit neurons, neuromodulators enhance or diminish the *overall* effectiveness of the synaptic connections and bring about enduring change. Freeman believes that when we commit in love, the brain neuromodulator oxytocin is released, allowing existing neuronal connections to melt away so that changes on a large scale can follow.

Oxytocin is sometimes called the commitment neuromodulator because it reinforces bonding in mammals. It is released when lovers connect and make love—in humans oxytocin is released in both sexes during orgasm—and when couples parent and nurture their children. In women oxytocin is released during labor and breast-feeding. An fMRI study shows that when mothers look at photos of their children, brain regions rich in oxytocin are activated. In male mammals a closely related neuromodulator called vasopressin is released when they become fathers. Many young people who doubt they will be able to handle the responsibilities of parenting are not aware of the extent to which oxytocin may change their brains, allowing them to rise to the occasion.

Studies of a monogamous animal called the prairie vole have shown that oxytocin, which is normally released in their brains during mating, makes them pair off for life. If a female vole has oxytocin injected into her brain, she will pair-bond for life with a nearby male. If a male vole is injected with vasopressin, it will cuddle with a nearby female. Oxytocin appears also to attach children to parents, and the neurons that control its secretion may have a critical period of their own. Children reared in orphanages without close loving contact often have bonding problems when older. Their oxytocin levels remain low for several years after they have been adopted by loving families.

Whereas dopamine induces excitement, puts us into high gear, and triggers sexual arousal, oxytocin induces a calm, warm mood that increases tender feelings and attachment and may lead us to lower our guard. A recent study shows that oxytocin also triggers trust. When people sniff oxytocin and then participate in a financial game, they are more prone to trust others with their money. Though there is still more work to be done on oxytocin in humans, evidence suggests that its effect is similar to that in prairie voles: it makes us commit to our partners and devotes us to our children.

But oxytocin, Freeman argues, works in a unique way, related to unlearning. In sheep, oxytocin is released in the olfactory bulb, a part of the brain involved in odor perception, with each new litter. Sheep and many other animals bond with, or "imprint" on, their offspring by scent. They mother their own lambs and reject the unfamiliar. But if oxytocin is injected into a mother ewe when exposed to an unfamiliar lamb, she will mother the strange lamb too.

Oxytocin is not, however, released with the first litter—only with those litters that follow—suggesting to Freeman that the oxytocin plays the role of *wiping out* the neural circuits that bonded the mother with her first litter, so she can bond with her second. (Freeman suspects that the mother bonds with her first litter using other neurochemicals.) Oxytocin's "ability" to wipe out learned behavior has led some scientists to call it an amnestic hormone. Freeman proposes that oxytocin melts down existing neuronal connections that underlie existing attachments, so new attachments can be formed. Oxytocin, in this theory, does not teach parents to parent. Nor does it make lovers cooperative and kind; rather, it makes it possible for them to learn new patterns.

There is some dispute over the idea that oxytocin is solely responsible for this new burst of learning, for changes in our existing attachments, or how it might facilitate these changes. Neuroscientist Jaak Pankseep argues that oxytocin, in combination with other brain chemicals, is so overwhelmingly good at reducing our feelings of separation-distress that the pain of losing previous attachments makes less of an impression than it would otherwise. This relative lack of distress may also free us to learn new things and form new bonds, while partially reconfiguring our existing relationships.

Freeman's theory helps to explain how love and plasticity affect each other. Plasticity allows us to develop brains so unique—in response to our individual life experiences—that it is often hard to see the world as others do, to want what they want, or to cooperate. But the successful reproduction of our species requires cooperation. What nature provides, in a neuromodulator like oxytocin, is the ability for two brains in love to go through a period of heightened plasticity, allowing them to mold to each other and shape each other's intentions and perceptions. The brain for Freeman is fundamentally an organ of socialization, and so there must be a mechanism that, from time to time, undoes our tendency to become overly individualized, overly self-involved, and too self-centered.

As Freeman says, "The deepest meaning of sexual experience lies not in pleasure, or even in reproduction, but in the opportunity it affords to surmount the solipsistic gulf, opening the door, so to speak, whether or not one undertakes the work to go through. It is the afterplay, not the foreplay, that counts in building trust."

Freeman's concept reminds us of many variations on love: the insecure man who leaves a woman quickly after making love during the night, because he fears being overly influenced by her should he stay through the morning; the woman who tends to fall in love with whomever she has sex with. Or the sudden transformation of the man who barely noticed children into a devoted father; we say "he's matured" and "the kids come first," but he may have had some help from oxytocin, which allowed him to go beyond his deep-seated patterns of selfish concern. Contrast him with the inveterate bachelor who never falls in love and becomes more eccentric and rigid with each passing year, plastically reinforcing his routines through repetition.

Unlearning in love allows us to change our image of ourselves—for the better, if we have an adoring partner. But it also helps account for our vulnerability when we fall in love and explains why so many self-possessed young men and women, who fall in love with a manipulative, undermining, or devaluing person, often lose all sense of self and become plagued with self-doubt, from which it may take years to recover.

Understanding unlearning, and some of the fine points of brain plasticity, turned out to be crucial in the treatment of my patient A. By the time A. went to college, he found himself replaying his critical-period experience and being attracted to emotionally disturbed, already attached women very much like his mother, feeling it was his job to love and rescue them.

A. was caught in two plastic traps.

The first was that a relationship with a thoughtful, stable woman who might have helped him unlearn his love for problem women, and teach him a new way to love, simply didn't turn him on, though he wished it would. So he was stuck with a destructive attraction, formed in his critical period.

His second, related trap can also be understood plastically. One of his most tormenting symptoms was the almost perfect fusion in his mind of sex with aggression. He felt that to love someone was to consume her, to eat her alive, and that to be loved was to be eaten alive. And his feeling that sexual intercourse was a violent act upset him greatly, yet excited him. Thoughts of sexual intercourse immediately led to thoughts of violence, and thoughts of violence, to sex. When he was effective sexually, he *felt* he was dangerous. It was as though he lacked separate brain maps for sexual and violent feelings.

Merzenich has described a number of "brain traps" that occur when two brain maps, meant to be separate, merge. As we have seen, he found that if a monkey's fingers were sewn together and so forced to move at the same time, the maps for them would fuse, because their neurons fired together and hence wired together. But he also discovered that maps fuse in everyday life. When a musician uses two fingers together frequently enough while playing an instrument, the maps for the two fingers sometimes fuse, and when the musician tries to move only one finger, the other moves too. The maps for the two different fingers are now "dedifferentiated." The more intensely the musician tries to produce a single movement, the more he will move both fingers, strengthening the merged map. The harder the person tries to get out of the brain trap, the deeper he gets into it, developing a condition called "focal dystonia." A similar brain trap occurs in Japanese people who, when speaking English, can't hear the difference between *r* and *l* because the two sounds are not differentiated in their brain maps. Each time they try to say the sounds properly, they say them incorrectly, reinforcing the problem.

This is what I believe A. experienced. Each time he thought of sex, he thought of violence. Each time he thought of violence, he thought of sex, reinforcing the connection in the merged map.

Merzenich's colleague Nancy Byl, who works in physical medicine, teaches people who can't control their fingers to redifferentiate their finger maps. The trick is not to try to move the fingers separately, but to relearn how to use their hands the way they did as babies. When treating guitarists with focal dystonias who have lost control of their fingers, for example, she first instructs them to stop playing guitar for a while, to weaken the merged map. Then they just hold an unstrung guitar for a few days. Then a single string with a different feel from a normal guitar string is put on the guitar, and they feel it carefully, but with only one finger. Finally they use a second finger, on a separate string. Eventually the fused brain maps for their fingers separate into two distinct maps, and they can play again.

A. came into psychoanalysis. Early on we sorted out why love and aggression had fused, tracing the roots of his brain trap to his experience with his drunken mother who often gave free rein to sexual and violent feelings simultaneously. But when he still couldn't change what attracted him, I did something similar to what Merzenich and Byl do to redifferentiate maps. For a long period in the therapy, whenever A. expressed any kind of physical tenderness outside the sexual arena untainted by aggression, I pointed it out and asked him to observe it closely, reminding him that he was capable of a positive feeling and capable of intimacy.

When violent thoughts came up, I got him to search his experience to find even a single instance in which aggression or violence was untainted with sex or was even praiseworthy, as in justified self-defense. Whenever these areas came up—a pure physical tenderness, or aggression that wasn't destructive—I drew his attention to them. As time passed, he was able to form two different brain maps, one for physical tenderness, which had nothing to do with the seductiveness he experienced with his mother, and another for aggression—including healthy assertiveness—which was quite different from the senseless violence he'd experienced when his mother was drunk.

Separating sex and violence in his brain maps allowed him to feel better about relationships and sex, and improvement followed in stages. While he wasn't immediately able to fall in love with or become excited by a healthy woman, he did fall in love with a woman who was a bit healthier than his previous girlfriend, and he benefited from the learning and unlearning that that love provided. This experience allowed him to enter progressively healthier relationships, unlearning more each time. By the end of therapy he was in a healthy, satisfying, happy marriage; his character, and his sexual type, had been radically transformed.

The rewiring of our pleasure systems, and the extent to which our sexual tastes can be acquired, is seen most dramatically in such perversions as sexual masochism, which turns physical pain into sexual pleasure. To do this the brain must make pleasant that which is inherently unpleasant, and the impulses that normally trigger our pain system are plastically rewired into our pleasure system.

People with perversions often organize their lives around activities that mix aggression and sexuality, and they often celebrate and idealize humiliation, hostility, defiance, the forbidden, the furtive, the lusciously sinful, and the breaking of taboos; they feel special for not being merely "normal." These "transgressive" or defiant attitudes are essential to the enjoyment of perversion. The idealization of the perverse, and the devaluation of "normalcy," is brilliantly captured in Vladimir Nabokov's novel *Lolita,* in which a middle-aged man idolizes and has sex with a prepubescent, twelve-year-old girl, while showing contempt for all older females.

Sexual sadism illustrates plasticity in that it fuses two familiar tendencies, the sexual and the aggressive, each of which can give pleasure separately, and brings them together so when they are discharged, the pleasure is doubled. But masochism goes much further because it takes something inherently unpleasant, pain, and turns it into a pleasure, altering the sexual drive more fundamentally and more vividly demonstrating the plasticity of our pleasure and pain systems.

For years the police, through raids on S&M establishments, knew more about serious perversions than most clinicians. While patients with milder perversions often come for treatment of such problems as anxiety or depression, those with serious perversions seldom seek therapy because, generally, they enjoy them.

Robert Stoller, M.D., a California psychoanalyst, did make important discoveries through visits to S&M and B&D (bondage and discipline) establishments in Los Angeles. He interviewed people who practiced hardcore sadomasochism, which inflicts real pain on the flesh, and discovered that masochistic participants had all had serious physical illnesses as children and had undergone regular, terrifying, painful medical treatment. "As a result," writes Stoller, "they had to be confined severely and for long periods [in hospitals] without the chance to unload their frustration, despair and rage openly and appropriately. Hence the perversions." As children, they consciously took their pain, their inexpressible rage, and reworked it in daydreams, in altered mental states, or in masturbation fantasies, so they could replay the story of the trauma with a happy ending and say to themselves, *This time, I win.* And the way they won was by erotizing their agony.

The idea that an "inherently" painful feeling can become pleasurable may at first strike us as hard to believe, because we tend to assume that each of our sensations and emotions is inherently either pleasurable (joy, triumph, and sexual pleasure) or painful (sadness, fear, and grief). But in fact this assumption does not hold up. We can cry tears of happiness and have bittersweet triumphs; and in neuroses people may feel guilty about sexual pleasure, or no pleasure at all, where others would feel delight. An emotion that we think inherently unpleasurable, such as sadness, can, if beautifully and subtly articulated in music, literature, or art, feel not only poignant but sublime. Fear can be exciting in frightening movies or on roller coasters. The human brain seems able to attach many of our feelings and sensations either to the pleasure system or to the pain system, and each of these links or mental associations requires a novel plastic connection in the brain.

The hardcore masochists whom Stoller interviewed must have formed a pathway that linked the painful sensations they had endured to their sexual pleasure systems, resulting in a new composite experience, voluptuous pain. That they all suffered in early childhood strongly suggests that this rewiring occurred during the critical periods of sexual plasticity.

In 1997 a documentary appeared that sheds light on plasticity and masochism: *Sick: The Life and Death of Bob Flanagan, Supermasochist.* Bob Flanagan performed his masochistic acts in public as a performance artist and exhibitionist and was articulate, poetic, and at times very funny.

In Flanagan's opening scenes we see him naked, humiliated, pies being thrown in his face, fed with a funnel. But images flash of his being physically hurt and choked, hinting at far more disturbing forms of pain.

Bob was born in 1952 with cystic fibrosis, a genetic disorder of the lungs and pancreas in which the body produces an excessive amount of abnormally thick mucus that clogs the air passages, making it impossible to breathe normally, and leads to chronic digestive problems. He had to fight for every breath and often turned blue from lack of oxygen. Most patients born with this disease die as children or in their early twenties.

Bob's parents noticed he was in pain from the moment he came home from the hospital. When he was eighteen months old, doctors discovered pus between his lungs and began treating him by inserting needles deep into his chest. He began to dread these procedures and screamed desperately. Throughout childhood he was hospitalized regularly and confined nearly naked inside a bubblelike tent so doctors could monitor his sweat—one of the ways cystic fibrosis is diagnosed—while he felt mortified that his body was visible to strangers. To help him breathe and fight infections, doctors inserted all sorts of tubes into him. He was also aware of the severity of his problem: two of his younger sisters had also had cystic fibrosis; one died at six months, the other at twenty-one years.

Despite the fact that he had become a poster boy for the Orange County Cystic Fibrosis Society, he began to live a secret life. As a young child, when his stomach hurt relentlessly, he would stimulate his penis to distract himself. By the time he was in high school, he would lie naked at night and secretly cover himself with thick glue, for he knew not what reason. He hung himself from a door with belts in painful positions. Then he began to insert needles into the belts to pierce his flesh.

When he was thirty-one, he fell in love with Sheree Rose, who came from a very troubled family. In the film we see Sheree's mother openly belittle her husband, Sheree's father, who, Sheree claims, was passive and never showed her affection. Sheree describes herself as being bossy since childhood. She is Bob's sadist.

In the film Sheree uses Bob, with his consent, as her slave. She humiliates him, cuts into the skin near his nipples with an X-Acto knife, puts clamps on his nipples, force-feeds him, chokes him with a cord till he turns blue, forces a large steel ball—as big as a billiard ball—into his anus, and puts needles in his erogenous zones. His mouth and lips are sutured shut with stitches. He writes of drinking Sheree's urine from a baby bottle. We see him with feces on his penis. His every orifice is invaded or defiled. These activities give Bob erections and lead to great orgasms in the sex that often follows.

Bob survives both his twenties and his thirties and in his early forties has become the oldest living survivor of cystic fibrosis. He takes his masochism on the road, to S&M clubs and art museums, where he enacts his masochistic rituals in public, always wearing his oxygen mask to breathe.

In one of the final scenes a naked Bob Flanagan takes a hammer and nails his penis, right through its center, to a board. He then matter-of-factly removes the nail so that blood spurts all over the camera lens, like a fountain, from the deep hole through his penis.

It is important to describe precisely what Flanagan's nervous system could endure, in order to understand the extent to which completely novel brain circuits can develop, linking the pain system to the pleasure system.

Flanagan's idea that his pain must be made pleasurable colored his fantasies from early childhood. His remarkable history confirms that his perversion developed out of his unique life experience and is linked to his traumatic memories. As an infant, he was tied into the crib in the hospital so he couldn't escape and hurt himself. By age seven his confinement had turned into a love of constriction. As an adult, he loved bondage and being handcuffed or tied up and hung for long periods in positions that torturers might use to break their victims. As a child, he was required to endure the powerful nurses and doctors who hurt him; as an adult, he voluntarily gave this power to Sheree, becoming her slave, whom she could abuse while practicing pseudomedical procedures on him. Even subtle aspects of his childhood relationship to his doctors were repeated in adulthood. The fact that Bob gave Sheree his consent repeated an aspect of the trauma because, after a certain age, when the doctors took blood, pierced his skin, and hurt him, he gave them permission, knowing his life depended on it.

This mirroring of childhood traumas through the repetition of such subtle details is typical of perversions. Fetishists—who are attracted to objects—have the same trait. A fetish, Robert Stoller said, is an object that tells a story, that captures scenes from childhood trauma and eroticizes them. (One man who developed a fetish for rubber underwear and raincoats was a childhood bedwetter, forced to sleep on rubber sheets, which he found humiliating and uncomfortable. Flanagan had a number of fetishes, for medical paraphernalia and the blunt metals from hardware stores—screws, nails, clamps, and hammers—all of which he used, at various times, for erotic-masochistic stimulation, to penetrate, pinch, or pound his flesh.)

Flanagan's pleasure centers were no doubt rewired in two ways. First, emotions such as anxiety that are normally unpleasant became pleasant. He explains that he is constantly flirting with death because he was promised an early death and is trying to master his fear. In his 1985 poem, "Why," he makes clear that his supermasochism allows him to feel triumphant, courageous, and invulnerable after a life of vulnerability. But he goes beyond simply mastering fear. Humiliated by doctors who stripped him and put him in a plastic tent to measure his sweat, he now proudly strips in museums. To master his feelings of being exposed and humiliated as a child, he becomes a triumphant exhibitionist. Shame is made into a pleasure, converted into shamelessness.

The second aspect of his rewiring is that physical pain becomes pleasure. Metal in flesh now feels good, gives him erections, and makes him have orgasms. Some people under great physical stress release endorphins, the opiumlike analgesics that our bodies make to dull our pain and that can make us euphoric. But Flanagan explains he is not dulled to pain—he is drawn to it. The more he hurts himself, the more sensitized to pain he becomes, and the more pain he feels. Because his pain and pleasure systems are connected, Flanagan feels real, intense pain, and it feels good.

Children are born helpless and will, in the critical period of sexual plasticity, do anything to avoid abandonment and to stay attached to adults, even if they must learn to love the pain and trauma that adults inflict. The adults in little Bob's world inflicted pain on him "for his own good." Now, by becoming a supermasochist, he ironically treats pain as though it is good for him. He is utterly aware that he is stuck in the past, reliving infancy, and says he hurts himself "because I am a big baby, and I want to stay that way." Perhaps the fantasy of *staying* the tortured baby is an imaginary way of keeping himself from the death that awaits him should he allow himself to grow up. If he can stay Peter Pan, endlessly "tormented" by Sheree, at least he will never grow up and die prematurely.

At the end of the film we see Flanagan dying. He stops making jokes and begins to look like a cornered animal, overwhelmed with fear. The viewer sees how terrified he must have been as a little boy, before he discovered the masochistic solution to tame his pain and terror. At this point, we learn from Bob that Sheree has been talking of splitting up—evoking every suffering child's worst fear, abandonment. Sheree says the problem is that Bob is no longer submitting to her. He looks utterly brokenhearted—and in the end, she stays, and nurses him tenderly.

In his final moments, almost in shock, he asks plaintively, "Am I dying? I don't understand it . . . What is going on? . . . I'd never believe this." So powerful were his masochistic fantasies, games, and rituals, in which he embraced painful death, that it seems he thought he had actually beaten it.

As for the patients who became involved in porn, most were able to go cold turkey once they understood the problem and how they were plastically reinforcing it. They found eventually that they were attracted once again to their mates. None of these men had addictive personalities or serious childhood traumas, and when they understood what was happening to them, they stopped using their computers for a period to weaken their problematic neuronal networks, and their appetite for porn withered away. Their treatment for sexual tastes acquired later in life was far simpler than that for patients who, in their critical periods, acquired a preference for problematic sexual types. Yet even some of these men were able, like A., to change their sexual type, because the same laws of neuroplasticity that allow us to acquire problematic tastes also allow us, in intensive treatment, to acquire newer, healthier ones and in some cases even to lose our older, troubling ones. It's a use-it-or-lose-it brain, even where sexual desire and love are concerned.

5

Midnight Resurrections

Stroke Victims Learn to Move and Speak Again

Michael Bernstein, M.D., an eye surgeon and tennis buff who played six times a week, was in the prime of life at fifty-four and married with four children when he had an incapacitating stroke. He completed a new neuroplastic therapy, recovered, and was back at work when I met him in his office in Birmingham, Alabama. Because of the many rooms in his office suite, I thought he must have a number of physicians working with him. No, he explained, he has a lot of rooms because he has a lot of elderly patients, and instead of making them move, he goes to them.

"These older patients, some of them, they don't move so well. They've had strokes." He laughed.

The morning of his stroke Dr. Bernstein had operated on seven patients, doing his usual cataract, glaucoma, and refractive surgeries —delicate procedures within the eye.

Afterward, when Dr. Bernstein rewarded himself by playing tennis, his opponent told him his balance was off and that he wasn't playing his usual game. After tennis he drove to do an errand at the bank, and when he tried to raise his leg to get out of his low-slung sports car, he couldn't. When he got back to his office, his secretary told him he didn't look right. His family physician, Dr. Lewis, who worked in the building, knew that Dr. Bernstein was mildly diabetic, that he had a cholesterol problem, and that his mother had had several strokes, making him a possible candidate for an early stroke. Dr. Lewis gave Dr. Bernstein a shot of heparin to keep his blood from clotting, and Dr. Bernstein's wife drove him to the hospital.

During the next twelve to fourteen hours the stroke worsened, and the entire left side of his body became completely paralyzed, a sign that a significant part of his motor cortex had been damaged.

An MRI brain scan confirmed the diagnosis—doctors saw a defect in the right part of the brain that governs movement on the left side. He spent a week in intensive care, where he showed some recovery. After a week of physical therapy, occupational therapy, and speech therapy in the hospital, he was transferred to a rehabilitation facility for two weeks, then sent home. He got three more weeks of rehabilitation as an outpatient and was told his treatment was finished. He had received typical poststroke care.

But his recovery was incomplete. He still needed a cane. His left hand barely functioned. He couldn't put his thumb and first finger together in a pincer movement. Though he was born right-handed, he had been ambidextrous and before his stroke could do a cataract operation with his left hand. Now he couldn't use it at all. He couldn't hold a fork, bring a spoon to his mouth, or button his shirt. At one point during rehab he was wheeled onto a tennis court and given a tennis racket to see if he could hold it. He couldn't and began to believe he'd never play tennis again. Though he was told he'd never drive his Porsche again, he waited until no one was home, "got into a

$50,000 car, and backed it out of the garage. And I got down to the end of the driveway, and I looked both ways, and I was like a teenage kid stealing a car. And I went to the dead end of the street, and the car stalled. The key is on the left side of the steering column in a Porsche. I couldn't turn the key with my left hand. I had to reach across and turn the key with my right hand to get the car started, because there was no way I was going to leave the car there and have to call home and tell them to come get me. And of course my left leg was limited and pushing the clutch was not easy."

Dr. Bernstein was one of the first people to go to the Taub Therapy Clinic, for Edward Taub's constraint-induced (CI) movement therapy, when the program was still in its research phases. He figured he had nothing to lose.

His progress with CI therapy was very rapid. He described it: "It was unrelenting. They start at eight o'clock in the morning, and it is nonstop till you are finished at four-thirty. It even went on at lunch. There were just two of us, because it was the initial stages of the therapy. The other patient was a nurse, younger than I, probably forty-one or forty-two. She'd had a stroke after a baby. And she was competitive with me, for some reason"—he laughs—"but we got along great, and we sort of fed off each other. There were a lot of menial tasks they would have you do, like lifting cans from one shelf to the next. And she was short, so I'd put the cans up as high as I could."

They washed tabletops and cleaned lab windows to engage their arms in a circular movement. To strengthen the brain networks for their hands and develop control, they stretched thick rubber bands over their weak fingers, then opened them against the resistance of the bands. "Then I'd have to sit there and do my ABCs, writing with my left hand." In two weeks he'd learned to print and then to write with his afflicted left hand. Toward the end of his stay he was able to play Scrabble, picking up the small tiles with his left hand and placing them appropriately on the board. His fine motor skills were coming back. When he got home, he continued to do the exercises and continued to improve. And he got another treatment, electrical stimulation on his arm, to fire up his neurons.

He is now back at work running his busy office. He is also playing tennis three days a week. He still has some trouble running and is working out to strengthen a weakness in his left leg that wasn't fully treated at the Taub clinic—which has since begun a special program for people with paralyzed legs.

He has a few residual problems. He finds that his left arm doesn't quite feel normal, as is typical after CI therapy. Function returns, but not quite to its former level. Yet when I had him write his ABCs with his left hand, they looked well shaped, and I never would have guessed he'd had a stroke or that he was right-handed.

Even though he'd gotten better by rewiring his brain and felt ready to return to performing surgery, he decided not to, but only because if someone were to sue him for malpractice, the first thing the lawyers would say is that he had had a stroke and shouldn't have been operating. Who would believe that Dr. Bernstein could make as complete a recovery as he had?

Stroke is a sudden, calamitous blow. The brain is punched out from within. A blood clot or bleed in the brain's arteries cuts off oxygen to the brain's tissues, killing them. The most stricken of its victims end up mere shadows of who they once were, often warehoused in impersonal institutions, trapped in their bodies, fed like babies, unable to care for themselves, move, or speak. Stroke is one of the leading causes of disability in adults. Though it most often affects the elderly, it can occur in people in their forties or earlier. Doctors in an emergency room may be able to prevent a stroke from getting worse by unblocking the clot or stopping the bleeding, but once the damage is done, modern medicine is of little help—or was until Edward Taub invented his plasticity-based treatment. Until CI therapy, studies of chronic stroke patients with paralyzed arms concluded that no existing treatment was effective. There were rare anecdotal reports of stroke recoveries, like that of Paul Bach-y-Rita's father. Some people made spontaneous recoveries on their own, but once they stopped improving, traditional therapies weren't much help. Taub's treatment changed all this by helping stroke patients rewire their brains. Patients who had been paralyzed for years and were told they would never get better began to move again. Some regained their ability to speak. Children with cerebral palsy gained control of their movements. The same treatment shows promise for spinal cord injuries, Parkinson's, multiple sclerosis, and even arthritis.

Yet few have heard of Taub's breakthroughs, even though he first conceived of and laid the foundation for them over a quarter century ago, in 1981. He was delayed from sharing them because he became one of the most maligned scientists of our time. The monkeys he worked with became among the most famous lab animals in history, not because of what his experiments with them demonstrated but because of the allegations that they had been mistreated—allegations that kept him from working for years. These charges seemed plausible because Taub was so far ahead of his peers that his claim that chronic stroke patients could be helped by a plasticity-based treatment seemed incredible.

Edward Taub is a neat, conscientious man who pays close attention to details. He is over seventy, though he looks much younger, is smartly dressed, and his every hair is in place. In conversation Taub is learned and speaks in a soft voice, correcting himself as he goes along to make sure he has said things accurately. He lives in Birmingham, Alabama, where, at the university, he is finally free to develop his treatment for stroke patients. His wife, Mildred, was a soprano, recorded with Stravinsky, and sang with the Metropolitan Opera. She is still a belle, with a magnificent mane of hair and southern feminine warmth.

Taub was born in Brooklyn in 1931, went to the public schools, and graduated from high school when he was only fifteen. At Columbia University he studied "behaviorism" with Fred Keller. Behaviorism was dominated by the Harvard psychologist B. F. Skinner, and Keller was Skinner's intellectual lieutenant. Behaviorists of the time believed that psychology should be an "objective" science and should examine only what can be seen and measured: observable behaviors. Behaviorism was a reaction against psychologies that focused on the mind because to behaviorists, thoughts, feelings, and desires were merely "subjective" experience that wasn't objectively measurable. They were equally uninterested in the physical brain, arguing that it, like the mind, is a "black box." Skinner's mentor, John B. Watson, wrote derisively, "Most of the psychologists talk quite volubly about the formation of new pathways in the brain, as though there were a group of tiny servants of Vulcan there who run through the nervous system with hammer and chisel digging new trenches and deepening old ones." For behaviorists, it didn't matter what went on inside either the mind or the brain. One could discover the laws of behavior simply by applying a stimulus to an animal or a person and observing the response.

At Columbia the behaviorists experimented mostly with rats. While still a graduate student Taub developed a way of observing rats and recording their activities by using a sophisticated "rat diary." But when he used this method to test a certain theory of his mentor, Fred Keller, he, to his horror, disproved it. Taub loved Keller and hesitated to discuss the experiment's results, but Keller found out and told Taub he must always "call the data the way they lay."

Behaviorism at the time, by insisting that all behavior is a response to a stimulus, portrayed human beings as passive and so was particularly weak in explaining how we may do things voluntarily. Taub realized that the mind and brain must be involved in initiating many behaviors, and that behaviorism's dismissal of the mind and brain was a fatal flaw. Though an unthinkable choice for a behaviorist in that era, he took a job as a research assistant in an experimental neurology lab, to better understand the nervous system. In the lab they were doing "deafferentation" experiments with monkeys.

Deafferentation is an old technique, used by the Nobel Prize winner Sir Charles Sherrington in 1895. An "afferent nerve," in this context, means a "sensory nerve," one that conveys sensory impulses to the spine and then the brain. Deafferentation is a surgical procedure in which the incoming sensory nerves are cut so none of their input can make this trip. A deafferented monkey cannot sense where its affected limbs are in space, or feel any sensation or pain in them when touched. Taub's next feat—while still a graduate student—was to overturn one of Sherrington's most important ideas and thus lay the foundation for his stroke treatment.

Sherrington supported the idea that *all* of our movement occurs in response to some stimulus and that we move, not because our brains command it, but because our spinal reflexes keep us moving. This idea was called the "reflexological theory of movement" and had come to dominate neuroscience.

A spinal reflex does not involve the brain. There are many spinal reflexes but the simplest example is the knee reflex. When the doctor taps your knee, a sensory receptor beneath the skin picks up the tap and conveys an impulse along the sensory neuron in your thigh and into the spine, which conveys it to a motor neuron *in the spine,* which sends an impulse back to your thigh muscle, making it contract and making your leg jerk forward involuntarily. In walking, movement in one leg triggers movement of the other, reflexly.

This theory was soon used to explain all movement. Sherrington based his belief that reflexes were the foundation of all movement on a deafferentation experiment that he did with F. W. Mott. They deafferented the sensory nerves in a monkey's arm, cutting them before they entered the spinal cord, so no sensory signals could pass to the monkey's brain, and found that the monkey stopped using the limb. This seemed strange, because they had cut *sensory* nerves (which transmit feeling), not the *motor* nerves from the brain to the muscles (which stimulate movement). Sherrington understood why the monkeys couldn't feel but not why they couldn't move. To solve this problem, he proposed that movement is based on, and initiated by, the sensory part of the spinal reflex, and that his monkeys couldn't move because he had destroyed the sensory part of their reflex by deafferentation.

Other thinkers soon generalized his idea, arguing that all movement, and indeed everything we do, even complex behavior, is built up from chains of reflexes. Even such voluntary movements as writing require the motor cortex to modify *preexisting* reflexes. Though behaviorists opposed study of the nervous system, they embraced the idea that all movements are based on reflex responses to previous stimuli, because it left the mind and the brain out of behavior. This in turn supported the idea that all behavior is predetermined by what has happened to us before and that free will is an illusion. The Sherrington experiment became standard teaching in medical schools and universities.

Taub, working with a neurosurgeon, A. J. Berman, wanted to see if he could replicate Sherrington's experiment on a number of monkeys, and he expected to get Sherrington's result. Going a step further than Sherrington, he decided not only to deafferent one of the monkey's arms but to put the monkey's good arm in a sling to restrain it. It had occurred to Taub that the monkeys might not be using their deafferented arms because they could use their good ones more easily. Putting the good one in a sling might force a monkey to use the deafferented arm to feed itself and move around.

It worked. The monkeys, unable to use their good arms, started using their deafferented arms. Taub said, "I remember it vividly. I realized that I had been seeing the monkeys using their limbs for several weeks, and I hadn't verbalized it because I wasn't expecting it."

Taub knew his finding had major implications. If the monkeys could move their deafferented arms without having feeling or sensation in them, then Sherrington's theory, and Taub's teachers, were wrong. There must be independent motor programs in the brain that could initiate voluntary movement; behaviorism and neuroscience had been going down a blind alley for seventy years. Taub also thought his finding might have implications for stroke recovery because the monkeys, like stroke patients, had seemed utterly unable to move their arms. Perhaps some stroke patients, like the monkeys, might also move their limbs if forced to.

Taub was soon to find that not all scientists were as gracious about having their theories disproved as Keller was. Devout followers of Sherrington began finding fault with the experiment, its methodology, and Taub's interpretation. Granting agencies argued about whether the young graduate student should be allowed further money. Taub's professor at Columbia, Nat Schoenfeld, had built a well-known behaviorist theory on the basis of Sherrington's deafferentation experiments. When it came time for Taub to defend his Ph.D., the hall, usually empty, was packed. Keller, Taub's mentor, was away, and Schoenfeld was present. Taub presented his data and his interpretation of it. Schoenfeld argued against him and walked out. Then came the final exam. Taub, by this time, had more grants than many of the teaching faculty and chose to work on two major grant applications during the week of the final, expecting to take it later. When he was denied a makeup and failed for his "insolence," he decided to complete his Ph.D. at New York University. Most scientists in his field refused to believe his findings. He was attacked at scientific meetings and received no scientific recognition or awards. Yet at NYU Taub was happy. "I was in heaven. I was doing research. There was nothing more that I wanted."

Taub was pioneering a new kind of neuroscience that merged the best of behaviorism, cleansed of some of its more doctrinaire ideas, and brain science. In fact, it was a fusion anticipated by Ivan Pavlov, the founder of behaviorism, who—though it is not widely known—had attempted in his later years to integrate his findings with brain science, and even argued that the brain is plastic. Ironically, behaviorism had in one way prepared Taub to make important plastic discoveries. Because behaviorists were so uninterested in the structure of the brain, they had not concluded, as had most neuroscientists, that the brain lacked plasticity. Many believed they could train an animal to do almost anything, and though they didn't speak of "neuroplasticity," they believed in behavioral plasticity.

Open to this idea of plasticity, Taub forged ahead with deafferentation. He reasoned that if both arms were deafferented, a monkey should soon be able to move them both, because it would have to to survive. So he deafferented both limbs and, in fact, the monkeys did move both.

This finding was paradoxical: if one arm was deafferented, the monkey couldn't use it. If both arms were deafferented, the monkey could use both!

Then Taub deafferented the whole spinal cord, so that there wasn't a single spinal reflex left in the body and the monkey could not receive sensory input from any of its limbs. Still it used its limbs. Sherrington's reflexological theory was dead.

Then Taub had another epiphany, the one that would transform the treatment of strokes. He proposed that the reason a monkey didn't use its arm after a single limb was deafferented was because it had *learned* not to use it in the period right after the operation when the spinal cord was still in "spinal shock" from the surgery.

Spinal shock can last from two to six months, a period when the neurons have difficulty firing. An animal in spinal shock will try to move its affected arm and fail many times during those months. Without positive reinforcement, the animal gives up and instead uses its good arm to feed itself, getting positive reinforcement each time it succeeds. And thus the motor map for the deafferented arm—which includes programs for common arm movements—begins to weaken and atrophy, according to the plasticity principle of use it or lose it. Taub called this phenomenon "learned nonuse." He reasoned that monkeys that had both arms deafferented were able to use them because they'd never had the opportunity to learn that they didn't work well; they had to use them to survive.

But Taub thought he still had only indirect evidence for his theory of learned nonuse, so in a series of ingenious experiments he tried to prevent monkeys from "learning" nonuse. In one, he deafferented a monkey's arm; then, instead of putting a sling on the good arm to restrain it, he put it on the deafferented arm. That way the monkey would not be able to "learn" that it was of no use in the period of spinal shock. And indeed, when he removed the restraint at three months, long after the shock had worn off, the monkey was soon able to use the deafferented limb. Taub next began investigating what success he could have teaching animals to overcome learned nonuse. He then tested whether he could correct learned nonuse several years after it had developed, by forcing a monkey to use the deafferented arm. It worked and led to improvements that lasted the rest of the monkey's life. Taub now had an animal model that both mimicked the effects of strokes when nerve signals are interrupted and limbs cannot be moved, and a possible way of overcoming the problem.

Taub believed these discoveries meant that people who had had strokes or other kinds of brain damage, even years earlier, might be suffering from learned nonuse. He knew the brains of some stroke patients with minimal damage went into an equivalent of spinal shock, "cortical shock," which can last for several months. During this period each attempt to move the hand is met with failure, possibly leading to learned nonuse.

Stroke patients with extensive brain damage in the motor area fail to improve for a long period and, when they do, only recover partially. Taub reasoned that any treatment for stroke would have to address both massive brain damage and learned nonuse. Because learned nonuse might be masking a patient's ability to recover, only by overcoming learned nonuse first could one truly gauge a patient's prospects. Taub believed that even after a stroke, there was a good chance that motor programs for movement were present in the nervous system. Thus the way to unmask motor capacity was to do to human beings what he did to monkeys: constrain the use of the good limb and force the affected one to begin moving.

In his early work with monkeys, Taub had learned an important lesson. If he simply offered them *a reward* for using their bad arms to reach for food—if he tried to do what behaviorists call "conditioning"—the monkeys made no progress. He turned to another technique called "shaping," which molds a behavior in very small steps. So a deafferented animal would get a reward not only for successfully reaching for the food but for making the first, most modest gesture toward it.

In May 1981 Taub was forty-nine, heading up his own lab, the Behavioral Biology Center in Silver Spring, Maryland, with grand plans to transform the work he was doing with monkeys into a treatment for stroke, when Alex Pacheco, a twenty-two-year-old political science student at George Washington University, in Washington, D.C., volunteered to work in his lab.

Pacheco told Taub he was considering becoming a medical researcher. Taub found him personable and eager to help. Pacheco did not tell him that he was the cofounder and president of People for the Ethical Treatment of Animals (PETA), the militant animal rights group. The other PETA cofounder was Ingrid Newkirk, thirty-one, once the pound master of the Washington dog pound. Newkirk and Pacheco were romantically involved and ran PETA out of their D.C.-area apartment.

PETA was and is against *all* medical research involving animals, even research to cure cancers, heart disease, and AIDS (once it was discovered). It fervently opposes all eating of animals (by human beings, not by other animals), the production of milk and honey (described as the "exploitation" of cows and bees), and the keeping of pets (described as "slavery"). When Pacheco volunteered to work with Taub, his goal was to free the seventeen "Silver Spring monkeys" and make them a rallying cry for an animal rights campaign.

While deafferentation isn't generally painful, it isn't pretty either. Because the deafferented monkeys couldn't feel pain in their arms, when they bumped against something, they could injure themselves. When their injured arms were bandaged, the monkeys sometimes reacted as though their arms were foreign objects and tried to bite them.

In 1981, while Taub was away for a three-week summer holiday, Pacheco broke into the lab and took photographs that seemed to show the monkeys suffering gratuitously, injured and neglected, and that suggested they were forced to eat from pans dirtied by their own feces.

Armed with the photos, Pacheco persuaded Maryland authorities and police to raid the lab and seize the monkeys, on Friday, September 11, 1981. Taub could be targeted because, unlike the laws in other states, the Maryland statute covering cruelty to animals could be interpreted as making no exception for medical research.

When Taub returned to the lab, he was stunned by the media circus that greeted him and by its repercussions. A few miles down the road the administrators at the National Institutes of Health (NIH), the nation's leading medical research institution, heard about the raid and became frightened. The NIH labs conduct more biomedical experimentation on animals than any other institution in the world and could clearly be PETA's next target. NIH had to decide whether to defend Taub and take on PETA or argue that he was a bad apple and distance themselves. They turned against Taub.

PETA posed as a great defender of the law, even though Pacheco is alleged to have said that arson, property destruction, burglary, and theft are acceptable "when they directly alleviate the pain and suffering of an animal." Taub's case became the cause célèbre of Washington society. The *Washington Post* covered the controversy, and its columnists pilloried Taub. Taub was demonized by animal rights activists in a campaign that depicted him as a torturer, like the Nazi Dr. Mengele. The publicity generated by the "Silver Spring monkeys" was enormous and made PETA the largest animal rights organization in the United States and Edward Taub a hated figure.

He was arrested and put on trial for cruelty to animals, charged with 119 counts. Before his trial two-thirds of Congress, its members besieged by angry constituents, voted for a Sense of Congress resolution to stop his funding. He suffered professional isolation; lost his salary, his grants, and his animals; was prevented from experimenting; and was driven from his home in Silver Spring. His wife was stalked, and he and she were hounded by death threats. At one point someone followed Mildred to New York City, phoned Taub, and gave him a detailed account of her activities. Shortly after that, Taub got another call from a man saying he was a Montgomery County police officer and that he had just been informed by the NYPD that Mildred had had "an unfortunate accident." It was a lie, but Taub couldn't know that.

Taub spent the next six years of his life working sixteen hours a day, seven days a week, to clear himself, often functioning as his own lawyer. Before his trials began, he had $100,000 in life savings. By the end he had $4,000. Because he was blackballed, he couldn't get a job at a university. But gradually, trial by trial, appeal by appeal, charge by charge, he refuted PETA.

Taub claimed that there was something fishy about the photos and that there were signs of complicity between PETA and the Montgomery County authorities. Taub has always contended that Pacheco's photos were staged, the captions fabricated, and that, for instance, in one picture a monkey that normally sat comfortably in a testing chair was positioned grimacing, straining, and stooped, in a way that could have occurred only if a number of nuts and bolts had been undone and the chair readjusted. Pacheco has denied they were staged.

One bizarre aspect of the raid is that the police turned the monkeys from Taub's lab over to Lori Lehner, a member of PETA, to keep in her basement, in effect giving away official evidence. Then suddenly the entire colony of monkeys disappeared. Taub and his supporters have never doubted that PETA and Pacheco were behind the removal of the monkeys, but Pacheco has been coy when discussing the matter. *New Yorker* author Caroline Fraser asked Pacheco if they had been taken, as was alleged, to Gainesville, Florida, and he said, "That's a pretty good guess."

When it became clear that Taub couldn't be prosecuted without the monkeys and that the theft of court evidence was a felony, the monkeys suddenly returned as mysteriously as they had disappeared and were briefly given back to Taub. No one was charged, but Taub has steadfastly maintained that blood tests showed the animals were extremely stressed by their two-thousand-mile round trip and had a condition called transport fever, and soon after, one, Charlie, was attacked and bitten by another very agitated monkey. Charlie was then given an overdose of medication by a court-appointed vet and died.

By the end of Taub's first trial before a judge, in November 1981, 113 of the 119 charges against him had been dismissed. There was a second trial, in which he made further progress, followed by an appeal in which the Maryland Court of Appeals found that the state anticruelty law was never intended by the Maryland legislature to apply to researchers. Taub was exonerated in a unanimous decision.

The tide seemed to turn. Sixty-seven American professional societies made representations on Taub's behalf to the NIH, which reversed its decision not to support him, now arguing there was no good evidence for the original charges.

But Taub still didn't have his monkeys or a job, and his friends told him that no one would want him. When he was finally hired by the University of Alabama in 1986, there were demonstrations against him and protesters threatened to stop all animal research at the university. But Carl McFarland, the head of the psychology department, and others who knew his work, stood by him.

Given his first break in years, Taub got a grant to study strokes and opened a clinic.

Mitts and slings are the first things you see at the Taub clinic: grown-ups, indoors, wearing mitts on their good hands, slings on their good arms, 90 percent of their waking hours.

The clinic has many small rooms and one big one, where Taub-inspired exercises take place. Taub developed these exercises working with a physiotherapist, Jean Crago. Some appear to be more intensive versions of the everyday tasks that conventional rehabilitation centers use. The Taub clinic always uses the behavioral technique of "shaping," taking an incremental approach to all tasks. Adults play what look like children's games: some patients push large pegs into pegboards, or grasp large balls; others pick the pennies out of a pile of pennies and beans and put them in a piggy bank. The gamelike quality is no accident—these people are relearning how to move, going through the small steps we all went through as babies, in order to retrieve the motor programs that Taub believes are still in the nervous system, even after many strokes, illnesses, or accidents.

Conventional rehab usually lasts for an hour, and sessions are three times a week. Taub patients drill six hours a day, for ten to fifteen days straight. They get exhausted and often have to nap. Patients do ten to twelve tasks a day, repeating each task ten times apiece. Improvement begins rapidly, then lessens progressively. Taub's original studies showed that treatment works for virtually all stroke survivors who are left with some ability to move their fingers—about half of patients who have had chronic strokes. The Taub clinic has since learned how to train people to use completely paralyzed hands. Taub began by treating people who had had milder strokes, but he has now shown, using control studies, that 80 percent of stroke patients who have lost arm function can improve substantially. Many of these people have had severe chronic strokes and showed very large improvements. Even patients who had had their strokes, on average, more than four years before beginning CI therapy benefited significantly.

One such patient, Jeremiah Andrews (not his real name), a fifty-three-year-old lawyer, had his stroke forty-five years before he went to the Taub clinic and was still helped, a half century after his childhood catastrophe. He had his stroke when he was only seven years old, in first grade, while playing baseball. "I was standing on the sideline," he told me, "and all of a sudden I dropped to the ground and said, 'I have no arm, I have no leg.' My dad carried me home." He'd lost feeling on his right side, couldn't lift his right foot, or use his arm, and developed a tremor. He had to learn to write with his left hand because his right was weak and incapable of fine motor movements. He got conventional rehab after the stroke but continued to have major difficulties. Though he walked with a cane, he fell constantly. By the time he was in his forties, he was falling about 150 times a year, breaking, at different times, his hand, his foot, and, when he was forty-nine, his hip. After he broke his hip, conventional rehab helped him reduce his falls to about thirty-six a year. Subsequently he went to Taub's clinic and had two weeks of training for his right hand, then three weeks for his leg, and improved his balance significantly. In this short period his hand had so improved that "they had me writing my name with my right hand with a pencil so that I could recognize it—which is amazing." He continues to do his exercises and continues to improve; three years after leaving the clinic he has fallen only seven times. "I have continued to improve three years after," he says, "and because of the exercises I'm in better shape than when I left Taub, by a huge, huge margin."

Jeremiah's improvement at Taub's clinic demonstrates that because the brain is plastic and capable of reorganization, we should be slow to predict how far a motivated patient with a stroke in a sensory or motor area may progress, regardless of how long the patient has lived with the disability. Because it is a use-it-or-lose-it brain, we might assume that the key areas of Jeremiah's brain for balance, walking, and hand use would have completely faded away, so that further treatment would be pointless. Though they did fade, his brain, given the appropriate input, was able to reorganize itself and find a new way to perform the lost functions—as we can now confirm with brain scans.

Taub, Joachim Liepert, and colleagues from the University of Jena, Germany, have demonstrated that after a stroke the brain map for an affected arm shrinks by about half, so a stroke patient has only half the original number of neurons to work with. Taub believes that this is why stroke patients report that using the affected arm requires more effort. It is not only muscle atrophy that makes movement harder but also brain atrophy. When CI therapy restores the motor area of the brain to its normal size, using the arm becomes less tiring.

Two studies confirm that CI therapy restores the reduced brain map. One measured the brain maps of six stroke patients who had had arm and hand paralysis for an average of six years—long after any spontaneous recovery could be expected. After CI therapy the size of the brain map that governed hand movement doubled. The second study showed that changes could be seen in both hemispheres of the brain, demonstrating how extensive the neuroplastic changes were. These are the first studies to demonstrate that brain structure can be changed in stroke patients in response to CI treatment, and they give us a clue as to how Jeremiah recovered.

Currently Taub is studying what length of training is best. He has begun to get reports from clinicians that three hours a day may produce good results and that increasing the number of movements per hour is better than undergoing the exhausting six hours of treatment.

What rewires patients' brains is not mitts and slings, of course. Though they force the patients to practice using their damaged arms, the essence of the cure is the *incremental* training or shaping, increasing in difficulty over time. "Massed practice"—concentrating an extraordinary amount of exercise in only two weeks—helps rewire their brains by triggering plastic changes. Rewiring is not perfect after there has been massive brain death. New neurons have to take over the lost functions, and they may not be quite as effective as the ones they replace. But improvements can be as significant as those seen in Dr. Bernstein—and in Nicole von Ruden, a woman who was afflicted not with a stroke but with another kind of brain damage.

Nicole von Ruden, I was told, is the kind of person who lights up the room the moment she walks in. Born in 1967, she has worked as an elementary school teacher and as a producer for CNN and for the television show *Entertainment Tonight*. She did volunteer work at a school for the blind, with children who had cancer and with children who had AIDS because they had been raped or born infected. She was hardy and active. She loved whitewater rafting and mountain biking, had run a marathon, and had gone to Peru to hike the Inca trail.

One day when she was thirty-three, engaged to be married and living in Shell Beach, California, she went to an eye doctor for double vision that had been bothering her for a couple of months. Alarmed, he sent her for an MRI scan the same day. When the scan was done, she was admitted to the hospital. The next morning, January 19, 2000, she was told she had a rare inoperable brain tumor, called a glioma, in the brain stem, a narrow area that controls breathing, and that she had between three and nine months to live.

Nicole's parents immediately took her to the hospital at the University of California at San Francisco. That evening the head of neurosurgery told her that her only hope of staying alive was massive doses of radiation. A surgeon's knife in that small area would kill her. On the morning of January 21 she got her first dose of radiation and then, over the next six weeks, received the maximum amount a human being can tolerate, so much that she can never have radiation again. She also was given high doses of steroids to reduce swelling in her brain stem, which can also be fatal.

The radiation saved her life but was the beginning of new woes. "About two or three weeks into the radiation," Nicole says, "I started having tingling in my right foot. With time it climbed up the right side of my body, up to my knee, hips, torso, and arms, and then my face." She was soon paralyzed and without sensation on her whole right side. She is right-handed, so the loss of that hand was critical. "It got so bad," she says, "I couldn't sit up or even turn in bed. It was like when your leg falls asleep, and you can't stand up on it, and it collapses." The doctors soon determined that it was not a stroke but a rare and severe side effect of the radiation that had damaged her brain. "One of life's little ironies," she says.

From the hospital she was taken to her parents' home. "I had to be pushed in a wheelchair, pulled out of bed and carried, and helped into or out of a chair." She was able to eat with her left hand but only after her parents tied her into a chair with a sheet, to prevent her from falling—especially dangerous because she couldn't reach out to break a fall with her arms. With continued immobility and doses of steroids, she went from 125 pounds to 190 and developed what she calls a "pumpkin face." The radiation also made patches of her hair fall out.

She was psychologically devastated and especially upset by the grief her illness was causing others. For six months Nicole became so depressed that she stopped speaking or even sitting up in bed. "I remember this period, but I don't understand it. I remember watching the clock, waiting for time to go by or getting up for my meals, as my parents were adamant that I got up for three meals a day."

Her parents had been in the Peace Corps and had a can-do attitude. Her father, a general practitioner, quit his medical practice and stayed home to nurse her, despite her protests. They took her to movies or out along the ocean in her wheelchair to keep her connected to life. "They told me I'd get through it," she said, "to ride the ride, and this would pass." Meanwhile, friends and family sought information about possible treatments. One of them told Nicole about the Taub clinic, and she decided to undergo CI therapy.

There she was given a mitt to wear, so she wouldn't be able to use her left hand. She found the staff unyielding on this point. She laughs and says, "They did a funny thing the first night." When the phone rang at the hotel where she was staying with her mother, Nicole threw off her mitt and picked up after one ring. "I instantly got scolded by my therapist. She was checking on me and knew that if I picked up on one ring, I was obviously not using my affected arm. I was instantly busted."

Not only did she have a mitt. "Because I talk with my hands, and I'm a storyteller, they had to strap my mitt to my leg with a Velcro strip, which I found very funny. You definitely lower your pride on that one.

"We were each assigned one therapist. I was assigned Christine. That was an instant connection." Mitt on her good hand, Nicole soon was trying to write on a white board or type on a keyboard with her paralyzed hand. One exercise began by putting poker chips into a large oatmeal can. By the end of the week she was putting the chips into a small slit in a tennis ball can. Again and again she stacked rainbow-colored baby rings on a rod, clipped clothespins to a yardstick, or tried to stick a fork into Play-Doh and bring it to her mouth. At first the staff helped her. Then she did the exercises while Christine timed her with a stopwatch. Each time Nicole completed a task and said, "That was the best I could do," Christine would say, "No, it's not."

Nicole says, "It's really incredible, the amount of improvement that occurred in just five minutes! And then over two weeks—it's earth-shattering. They do not allow you to say the word 'can't,' which Christine called 'the four-letter word.' Buttoning was insanely frustrating for me. Just one button seemed like an impossible task. I had rationalized that I could get through life without ever doing that again. And what you learn at the end of the two weeks, as you are buttoning and unbuttoning a lab coat rapidly, is that your whole mind-set can shift about what you are able to do."

One night in the middle of the two-week course of therapy, all the patients went out for dinner in a restaurant. "We definitely made a mess at the table. The waiters had seen Taub clinic patients before, and they knew what to expect. Food was flying, with us all trying to eat with our affected arms. There were sixteen of us. It was pretty funny. By the end of the second week, I was actually making the pot of coffee with my affected arm. If I wanted coffee, they said, 'Guess what? You get to make it.' I had to scoop it out and put it in the machine and fill it with water, the whole thing with my affected arm. I don't know how drinkable it was."

I asked her how she felt when she was leaving.

"Completely rejuvenated, even more mentally than physically. It gave me the will to improve, and have normalcy in my life." She hadn't hugged anyone with her affected arm for three years, but now she could do so again. "I am now known for having a wimpy handshake, but I do it. I'm not throwing a javelin with the arm, but I can open up the refrigerator door, turn off a light or a faucet, and put shampoo on my head." These "little" improvements allow her to live alone and drive to work on the freeway with two hands on the wheel. She's started swimming, and the week before she and I spoke, she'd gone parallel skiing without poles in Utah.

Throughout her ordeal her bosses and coworkers at both CNN and *Entertainment Tonight* followed her progress and helped financially. When a freelance job in entertainment at CNN New York came up, she took it. By September she was working full-time again. On September 11, 2001, she was at her desk looking out the window and saw the second plane hit the World Trade Center. In the crisis she was assigned to the newsroom and to stories that, under other circumstances, might have been simplified out of sensitivity to her "special needs." But they weren't. The attitude was "You've got a good mind, use it." This, she says, "was probably the best thing for me."

When that job came to an end, Nicole returned to California and to teaching elementary school. The children embraced her immediately. They even had a "Miss Nicole von Ruden Day," when the children got out of their schoolbuses wearing cooking mitts, like those at the Taub clinic, and kept them on all day. They joked about her writing and her weak right hand, so she had them write with their weaker or less dominant hands. "And," says Nicole, "they weren't allowed to use the word 'can't.' I actually had little therapists. My first graders had me raise my hand over my head while they counted. Every day I had to hold it up longer . . . They were tough."

Nicole is now working full-time as a producer for *Entertainment Tonight*. Her job includes script-writing, fact-checking, and coordinating shoots. (She was in charge of the Michael Jackson trial coverage.) The woman who couldn't roll herself over in bed now gets to work at five A.M. and works a fifty-hour-plus week. She's back to her old weight of 126 pounds. She still has some residual tingling and weakness on her right side, but she can carry things in her right hand, raise it, get dressed, and take care of herself in general. And she has returned to helping kids who have AIDS.

The principles of constraint-induced therapy have been applied by a team headed by Dr. Friedemann Pulvermüller in Germany, which worked with Taub to help stroke patients who have damage to Broca's area and have lost the ability to speak. About 40 percent of patients who have a left hemisphere stroke have this speech aphasia. Some, like Broca's famous aphasia patient, "Tan," can use only one word; others have more words but are still severely limited. Some do get better spontaneously or get some words back, but it has generally been thought that those who didn't improve within a year couldn't.

What is the equivalent of putting a mitt on the mouth or a sling on speech? Patients with aphasia, like those with arm paralysis, tend to fall back on the equivalent of their "good" arm. They use gestures or draw pictures. If they can speak at all, they tend to say what is easiest over and over.

The "constraint" imposed on aphasiacs is not physical, but it's just as real: a series of language rules. Since behavior must be shaped, these rules are introduced slowly. Patients play a therapeutic card game. Four people play with thirty-two cards, made up of sixteen different pictures, two of each picture. A patient with a card with a rock on it must ask the others for the same picture. At first, the only requirement is that they not point to the card, so as not to reinforce learned nonuse. They are allowed to use any kind of circumlocution, as long as it is verbal. If they want a card with a picture of the sun and can't find the word, they are permitted to say "The thing that makes you hot in the day" to get the card they want. Once they get two of a kind, they can discard them. The winner is the player who gets rid of his cards first.

The next stage is to name the object correctly. Now they must ask a precise question, such as "Can I have the dog card?" Next they must add the person's name and a polite remark: "Mr. Schmidt, may I please have a copy of the sun card?" Later in the training more complex cards are used. Colors and numbers are introduced—a card with three blue socks and two rocks, for instance. At the beginning patients are praised for accomplishing simple tasks; as they progress, only for more difficult ones.

The German team took on a very challenging population—patients who had had their strokes on average 8.3 years before, the very ones whom most had given up on. They studied seventeen patients. Seven in a control group got conventional treatment, simply repeating words; the other ten got CI therapy for language and had to obey the rules of the language game, three hours a day for ten days. Both groups spent the same number of hours, then were given standard language tests. In the ten days of treatment, after only thirty-two hours, the CI therapy group had a 30 percent increase in communication. The conventional treatment group had none.

Based on his work with plasticity, Taub has discovered a number of training principles: training is more effective if the skill closely relates to everyday life; training should be done in increments; and work should be concentrated into a short time, a training technique Taub calls "massed practice," which he has found far more effective than long-term but less frequent training.

Many of these same principles are used in "immersion" learning of a foreign language. How many of us have taken language courses over years and not learned as much as when we went to the country and "immersed" ourselves in the language for a far shorter period? Our time spent with people who don't speak our native tongue, forcing us to speak theirs, is the "constraint." Daily immersion allows us to get "massed practice." Our accent suggests to others that they may have to use simpler language with us; hence we are incrementally challenged, or shaped. Learned nonuse is thwarted, because our survival depends on communication.

Taub has applied CI principles to a number of other disorders. He has begun working with children with cerebral palsy—a complex, tragic disability that can be brought on by damage in the developing brain caused by stroke, infection, lack of oxygen during birth, and other problems. These children often cannot walk and are confined to wheelchairs for life, cannot speak clearly or control their movements, and have impaired or paralyzed arms. Before CI therapy, treatment of paralyzed arms in these children was generally considered ineffective. Taub did a study in which half the children got conventional cerebral palsy rehab and half got CI therapy, with their better-functioning arm placed in a light fiberglass cast. The CI therapy included popping soap bubbles with their affected fingers, pounding balls into a hole, and picking up puzzle pieces. Each time the children succeeded, they were heaped with praise and then, in the next game, encouraged to improve accuracy, speed, and fluidity of motion, even if they were very tired. The children showed extraordinary gains in a three-week training period. Some began to crawl for the first time. An eighteen-month-old was able to crawl up steps and use his hand to put food in his mouth for the first time. One four-and-a-half-year-old boy, who had *never* used his arm or hand, began to play ball. And then there was Frederick Lincoln.

Frederick had a massive stroke when he was in his mother's womb. When he was four and a half months old, it became clear to his mother that something was not right. "I noticed he wasn't doing what other boys in day care were doing. They could sit up and hold their bottle, and my child could not. I knew something was wrong but didn't know where to turn." The entire left side of his body was affected: his arm and leg didn't function well. His eye drooped, and he couldn't form sounds or words because his tongue was partially paralyzed. Frederick couldn't crawl or walk when other children did. He couldn't talk until he was three.

When Frederick was seven months old, he had a seizure, and his left arm was drawn up to his chest and couldn't be pulled away. He was given an MRI brain scan that, the doctor told his mother, showed that "one-quarter of his brain was dead," and that "he would probably never crawl, walk, or talk." The doctor believed the stroke had occurred about twelve weeks after Frederick was conceived.

He was diagnosed with cerebral palsy, with paralysis on the left side of his body. His mother, who worked in the Federal District Court, quit her job to devote all her time to Frederick, causing a major financial strain on the family. Frederick's disability also affected his eight-and-a-half-year-old sister.

"I had to explain to his sister," his mother says, "that her new brother would not be able to take care of himself, and that Mama would have to do it, and that we didn't know how long that would last. We didn't even know if Frederick would ever be able to do things by himself." When Frederick was eighteen months old, his mother heard about the Taub clinic for adults and asked if Frederick could be treated. But it would be several years before the clinic developed a program for children.

By the time he went to Taub's clinic, Frederick was four. He had made some progress using conventional approaches. He could walk with a leg brace and could talk with difficulty, but his progress had plateaued. He could use his left arm but not his left hand. Because he had no pincer grasp and couldn't touch his thumb to any of his fingers, he couldn't pick up a ball and hold it in his palm. He had to use the palm of his right hand and the back of his left.

At first Frederick didn't want to participate in the Taub treatment and rebelled, eating his mashed potatoes with the hand that had a cast on it instead of trying to use his affected one.

To make sure that Frederick got twenty-one uninterrupted days of treatment, the CI therapy was not done at the Taub clinic. "At our convenience," says his mother, "it was done at day care, home, church, Grandma's, anywhere we were. The therapist rode to church with us, and while she did, she worked on his hand in the car. Then she'd go to Sunday school class with him. She worked around our plans. The majority of Monday through Friday was spent in Frederick's day care, though. He knew we were trying to make 'lefty' better, because that is what we call it."

A mere nineteen days into the therapy, "lefty" developed a pincer grasp. "Now," says his mother, "he can do anything with that left hand, but it is weaker than the right. He can open a Ziploc bag, and he can hold a baseball bat. He continues to improve every day. His motor skills are dramatically improved. That improvement started during the project with Taub and has continued ever since. I can't think of anything I do for him other than being a typical parent, as far as assisting him goes." Because Frederick became more independent, his mother was able to go back to work.

Frederick is now eight, and he doesn't think of himself as disabled. He can run. He plays a number of sports, including volleyball, but he has always loved baseball best. So that he can keep his glove on, his mother sewed Velcro inside it, which fastens to the Velcro on a small brace he wears on his arm.

Frederick's progress has been phenomenal. He tried out for the regular baseball team—not one for handicapped children—and made the cut. "He played so well on the team," says his mother, "that he was chosen by the coaches for the all-star team. I cried for two hours when they told me that." Frederick is right-handed and holds the bat normally. Occasionally he loses his left-hand grip, but his right hand is now so strong that he can swing one-handed.

"In 2002," she says, "he played in the five-to-six-year-old division baseball, and he played in five all-star games. He had the winning play in three of the five games—he won the championship with his winning RBI. It was awesome. I've got it on video."

The tale of the Silver Spring monkeys and neuroplasticity was not yet finished. Years had passed since the monkeys were removed from Taub's lab. But in the meantime neuroscientists had begun to appreciate what Taub, so often ahead of his time, had been discovering. This new interest in Taub's work, and in the monkeys themselves, would lead to one of the single most important plasticity experiments ever performed.

Merzenich, in his experiments, showed that when sensory input from a finger was cut off, brain map changes typically occurred in 1 to 2 millimeters of the cortex. Scientists thought that the probable explanation for this amount of plastic change was the growth of individual neuronal branches. Brain neurons, when damaged, might send out small sprouts, or branches, to connect to other neurons. If one neuron died or lost input, the branches of an adjacent neuron had the ability to grow 1 to 2 millimeters to compensate. But if this was the mechanism by which plastic change occurred, then change was limited to the few neurons close to the damage. There could be plastic change between nearby sectors of the brain but not between sectors that lay farther apart.

Merzenich's colleague at Vanderbilt, Jon Kaas, worked with a student named Tim Pons, who was troubled by the 1-to-2-millimeter limit. Was that really the upper limit of plastic change? Or did Merzenich observe that amount of change because of his technique, which in some key experiments involved cutting only a single nerve?

Pons wondered what would happen in the brain if all the nerves in the hand were cut. Would more than 2 millimeters be affected? And would changes be seen between sectors?

The animals that could answer that question were the Silver Spring monkeys, because they alone had spent twelve years without sensory input to their brain maps. Ironically, PETA's interference for so many years had made them increasingly valuable to the scientific community. If any creature had massive cortical reorganization that could be mapped, it would be one of them.

But it wasn't clear who owned the animals, though they were in NIH custody. The agency at times insisted it didn't own them—they were hot potatoes—and didn't dare experiment with them because they were the focus of PETA's campaign to have them released. By now, however, the serious scientific community, including NIH, was growing fed up with witch hunts. In 1987 PETA brought a custody case to the Supreme Court, but the Court declined to hear it.

As the monkeys aged, their health deteriorated, and one of them, Paul, lost a lot of weight. PETA began lobbying NIH to have him euthanized—a mercy killing—and sought a court order to bring it about. By December 1989 another monkey, Billy, was also suffering and dying.

Mortimer Mishkin, head of the Society for Neuroscience and chief of the Laboratory of Neuropsychology at the NIH's Institute of Mental Health, had many years before inspected Taub's first deafferentation experiment that had overturned Sherrington's reflexological theory. Mishkin had stood up for Taub during the Silver Spring monkey affair and was one of the very few who had opposed ending Taub's NIH grant. Mishkin met with Pons and agreed that when the monkeys were to be euthanized, a final experiment could be done. It was a brave decision, since Congress had gone on record as favoring PETA. The scientists were well aware that PETA might go berserk, so they left the government out of it and arranged to have the experiment funded privately.

In the experiment the monkey Billy was to be anesthetized and a microelectrode analysis of the brain map for his arm was to be done, just before he was euthanized. Because there was so much pressure on the scientists and surgeons, they did in four hours what would normally have taken more than a day. They removed part of the monkey's skull, inserted electrodes into 124 different spots in the sensory cortex area for the arm, and stroked the deafferented arm. As expected, the arm sent no electrical impulses to the electrodes. Then Pons stroked the monkey's face—knowing that the brain map for the face is adjacent to the map for the arm.

To his amazement, as he touched the face, the neurons in the monkey's deafferented arm map also began to fire—confirming that the facial map had taken over the arm map. As Merzenich had seen in his own experiments, when a brain map is not used, the brain can reorganize itself so that another mental function takes over that processing space. Most surprising was the scope of the reorganization. Fourteen millimeters, or over half an inch of the "arm" map, had rewired itself to process facial sensory input—the largest amount of rewiring that had ever been mapped.

Billy was given a lethal injection. Six months later the experiment was repeated on three other monkeys, with the same results.

The experiment gave a tremendous boost to Taub, a coauthor of the paper that followed, and to other neuroplasticians who were hoping to rewire the brains of people who had large amounts of brain damage. Not only could the brain respond to damage by having single neurons grow new branches *within* their own small sectors, but, the experiment showed, reorganization could occur *across* very large sectors.

Like many neuroplasticians, Taub has his hand in numerous collaborative experiments. He has a computer version of CI therapy for people who cannot come to the clinic, called Auto-CITE (Automated CI Therapy), that is showing promising results. CI therapy is now being assessed in national trials throughout the United States. Taub is also on a team developing a machine to help people who are totally paralyzed with amyotrophic lateral sclerosis—the illness Stephen Hawking has. The machine would transmit their thoughts through brain waves that direct a computer cursor to select letters and spell words to form short sentences. He is involved in a cure for tinnitus, or ringing in the ears, that can be caused by plastic changes in the auditory cortex. Taub also wants to find out whether stroke patients can develop completely normal movement with CI therapy. Patients now receive treatment for only two weeks; he wants to know what would happen with a year of the therapy.

But perhaps his greatest contribution is that his approach to brain damage and problems in the nervous system applies to so many conditions. Even a nonneurological disease like arthritis may lead to learned nonuse because after an attack patients often stop using the limb or joint. CI therapy might help them get their movement back.

In all of medicine, few conditions are as terrifying as a stroke, when a part of our brain dies. But Taub has shown that even in this state, as long as there is adjacent living tissue, because that tissue is plastic, there may be hope that it might take over. Few scientists have gathered so much immediately practical knowledge from their experimental animals. Ironically, the only episode of pointless physical distress to animals in the entire Silver Spring affair occurred when, while in PETA's hands, they suspiciously disappeared. For that was when they appear to have been taken on a two-thousand-mile round trip to Florida and back, which left them so physically disturbed and agitated.

Edward Taub's work daily transforms people, most of whom were struck down in the midnight of their lives. Each time they learn to move their paralyzed bodies and speak, they resurrect not only themselves but the brilliant career of Edward Taub.

6

Brain Lock Unlocked

Using Plasticity to Stop Worries, Obsessions, Compulsions, and Bad Habits

All of us have worries. We worry because we are intelligent beings. Intelligence predicts, that is its essence; the same intelligence that allows us to plan, hope, imagine, and hypothesize also allows us to worry and anticipate negative outcomes. But there are people who are "great worriers," whose worrying is in a class of its own. Their suffering, though "all in the head," goes far beyond what most people experience precisely *because* it is all in the head and is thus inescapable. Such people are so constantly traumatized by their own brains that they often consider suicide. In one case a desperate college student felt so trapped by his obsessive worries and compulsions that he put a gun in his mouth and pulled the trigger. The bullet passed into his frontal lobe, causing a frontal lobotomy, which was at the time a treatment for obsessive-compulsive disorder. He was found still alive, his disorder cured, and he returned to college.

There are many kinds of worriers and many types of anxiety—phobias, post-traumatic stress disorders, and panic attacks. But among the people who suffer most are those with obsessive-compulsive disorder, or OCD, who are terrified that some harm will come, or has come, to them or to those they love. Though they may have been fairly anxious as children, at some later point, often as young adults, they have an "attack" that takes their worrying to a new level. Once self-possessed adults, they now feel like anguished, terrified children. Ashamed that they've lost control, they often hide their worry from others, sometimes for years, before they seek help. In the worst cases they cannot awaken from these nightmares for months at a time or even years. Medications may quell their anxieties but often don't eliminate the problem.

OCD often worsens over time, gradually altering the structure of the brain. A patient with OCD may try to get relief by focusing on his worry—making sure he's covered all the bases and left nothing to chance—but the more he thinks about his fear, however, the more he worries about it, because with OCD, worry begets worry.

There is often an emotional trigger for the first major attack. A person might remember that it is the anniversary of his mother's death, hear about a rival's car accident, feel an ache or lump in his body, read about a chemical in the food supply, or see an image of burned hands in a film. Then he begins to worry that he is approaching the age that his mother was when she died and, though not generally superstitious, now feels he is doomed to die that day; or that his rival's early death awaits him too; or that he has discovered the first symptoms of an untreatable disease; or that he has already been poisoned because he was not vigilant enough about what he ate.

We all experience such thoughts fleetingly. But people with OCD lock onto the worry and can't let it go. Their brains and minds march them through various dread scenarios, and though they try to resist thinking about them, they cannot. The threats feel so real, they think they must attend to them. Typical obsessions are fears of contracting a terminal illness, being contaminated by germs, being poisoned by chemicals, being threatened by electromagnetic radiation, or even being betrayed by one's own genes. Sometimes obsessionals get preoccupied with symmetry: they are bothered when pictures are not perfectly level or their teeth are not perfectly straight, or when objects are not kept in perfect order, and they can spend hours lining them up properly. Or they become superstitious about certain numbers and can set an alarm clock or volume control only on an even number. Sexual or aggressive thoughts—a fear they have hurt loved ones—might intrude into their minds, but where these thoughts come from they do not know. A typical obsessional thought might be "The thud that I heard while driving means I may have run somebody over." If they are religious, blasphemous thoughts might arise, causing guilt and worry. Many people with OCD have obsessive doubts and are always second-guessing themselves: have they turned off the stove, locked the door, or hurt someone's feelings inadvertently?

The worries can be bizarre—and make no conceivable sense even to the worrier—but that doesn't make them any less tormenting. A loving mother and wife worries, "I am going to harm my baby," or, "I will get up in my sleep and stab my husband with a butcher knife in the chest while he's sleeping." A husband has the obsessive thought that there are razor blades attached to his fingernails, so he cannot touch his children, make love to his wife, or pat his dog. His eyes see no blades, but his mind insists they are there, and he keeps asking his wife for reassurance that he hasn't hurt her.

Often obsessives fear the future because of some mistake they may have made in the past. But it is not only the mistakes that have happened that haunt them. Mistakes that they *imagine* they could make, should they let their guard down for a moment—which they, being human, eventually will—also generate a sense of dread that cannot be turned off. The agony of the obsessive worrier is that whenever something bad is remotely possible, it *feels* inevitable.

I have had several patients whose worries about their health were so intense that they felt as though they were on death row, each day awaiting their execution. But their drama does not end there. Even if they are told their health is fine, they may feel only the briefest flash of relief before they harshly diagnose themselves as "crazy" for all they have put themselves through—though, often, this "insight" is obsessional second-guessing in a new guise.

Soon after obsessive worries begin, OCD patients typically do something to diminish the worry, a compulsive act. If they feel they have been contaminated by germs, they wash themselves; when that doesn't make the worry go away, they wash all their clothing, the floors, and then the walls. If a woman fears she will kill her baby, she wraps the butcher knife in cloth, packs it in a box, locks it in the basement, then locks the door to the basement. The UCLA psychiatrist Jeffrey M. Schwartz describes a man who feared being contaminated by the battery acid spilled in car accidents. Each night he lay in bed listening for sirens that would signal an accident nearby. When he heard them, he would get up, no matter what the hour, put on special running shoes, and drive until he found the site. After the police left, he would scrub the asphalt with a brush for hours, then skulk home and throw out the shoes he had worn.

Obsessive doubters often develop "checking compulsions." If they doubt they've turned off the stove or locked the door, they go back to check and recheck often a hundred or more times. Because the doubt never goes away, it might take them hours to leave the house.

People who fear that a thud they heard while driving might mean they ran someone over will drive around the block just to make sure there is no corpse in the road. If their obsessional fear is of a dread disease, they will scan and rescan their body for symptoms or make dozens of visits to the doctor. After a while these checking compulsions are ritualized. If they feel they have been dirtied, they must clean themselves in a precise order, putting on gloves to turn on the tap and scrubbing their bodies in a particular sequence; if they have blasphemous or sexual thoughts, they may invent a ritual way of praying a certain number of times. These rituals are probably related to the magical and superstitious beliefs most obsessionals have. If they have managed to avoid disaster, it is only because they checked themselves in a certain way, and their only hope is to keep checking in the same way each time.

Obsessive-compulsives, so often filled with doubt, may become terrified of making a mistake and start compulsively correcting themselves and others. One woman took hundreds of hours to write brief letters because she felt so unable to find words that didn't feel "mistaken." Many a Ph.D. dissertation stalls—not because the author is a perfectionist, but because the doubting writer with OCD can't find words that don't "feel" totally wrong.

When a person tries to resist a compulsion, his tension mounts to a fever pitch. If he acts on it, he gets temporary relief, but this makes it more likely that the obsessive thought and compulsive urge will only be worse when it strikes again.

OCD has been very difficult to treat. Medication and behavior therapy are only partially helpful for many people. Jeffrey M. Schwartz has developed an effective, plasticity-based treatment that helps not only those with obsessive-compulsive disorder but also those of us with more everyday worries, when we start stewing about something and can't stop even though we know it's pointless. It can help us when we get mentally "sticky" and hold on to worries or when we become compulsive and driven by such "nasty habits" as compulsive nail biting, hair pulling, shopping, gambling, and eating. Even some forms of obsessive jealousy, substance abuse, compulsive sexual behaviors, and excessive concern about what others think about us, self-image, the body, and self-esteem can be helped.

Schwartz developed new insights into OCD by comparing brain

scans of people with OCD and those without it, then used these insights to develop his new form of therapy—the first time, to my knowledge, that such brain scans as the PET helped doctors both to understand a disorder and to develop a psychotherapy for it. He then tested this new treatment by doing brain scans on his patients before and after their psychotherapy and showed that their brains normalized with treatment. This was another first—a demonstration that a talking therapy could change the brain.

Normally, when we make a mistake, three things happen. First, we get a "mistake feeling," that nagging sense that something is wrong. Second, we become anxious, and that anxiety drives us to correct the mistake. Third, when we have corrected the mistake, an automatic gearshift in our brain allows us to move on to the next thought or activity. Then both the "mistake feeling" and the anxiety disappear.

But the brain of the obsessive-compulsive does not move on or "turn the page." Even though he has corrected his spelling mistake, washed the germs off his hands, or apologized for forgetting his friend's birthday, he continues to obsess. His automatic gearshift does not work, and the mistake feeling and its pursuant anxiety build in intensity.

We now know, from brain scans, that three parts of the brain are involved in obsessions.

We detect mistakes with our *orbital frontal cortex*, part of the frontal lobe, on the underside of the brain, just behind our eyes. Scans show that the more obsessive a person is, the more activated the orbital frontal cortex is.

Once the orbital frontal cortex has fired the "mistake feeling," it sends a signal to the *cingulate gyrus*, located in the deepest part of the cortex. The cingulate triggers the dreadful anxiety that something bad is going to happen unless we correct the mistake and sends signals to both the gut and the heart, causing the physical sensations we associate with dread.

The "automatic gearshift," the *caudate nucleus*, sits deep in the center of the brain and allows our thoughts to flow from one to the next unless, as happens in OCD, the caudate becomes extremely "sticky."

Brain scans of OCD patients show that all three brain areas are hyperactive. The orbital frontal cortex and the cingulate turn on and stay on as though locked in the "on position" together—one reason that Schwartz calls OCD "brain lock." Because the caudate doesn't "shift the gear" automatically, the orbital frontal cortex and the cingulate continue to fire off their signals, increasing the mistake feeling and the anxiety. Because the person has already corrected the mistake, these are, of course, false alarms. The malfunctioning caudate is probably overactive because it is stuck and is still being inundated with signals from the orbital frontal cortex.

The causes of severe OCD brain lock vary. In many cases it runs in families and may be genetic, but it can also be caused by infections that swell the caudate. And, as we shall see, learning also plays a role in its development.

Schwartz set out to develop a treatment that would change the OCD circuit by unlocking the link between the orbital cortex and the cingulate and normalizing the functioning of the caudate. Schwartz wondered whether patients could shift the caudate "manually" by paying constant, effortful attention and actively focusing on something besides the worry, such as a new, pleasurable activity. This approach makes plastic sense because it "grows" a new brain circuit that gives pleasure and triggers dopamine release which, as we have seen, rewards the new activity and consolidates and grows new neuronal connections. This new circuit can eventually compete with the older one, and according to use it or lose it, the pathological networks will weaken. With this treatment we don't so much "break" bad habits as replace bad behaviors with better ones.

Schwartz divides the therapy into a number of steps, of which two are key.

The first step is for a person having an OCD attack to *relabel* what is happening to him, so that he realizes that what he is experiencing is not an attack of germs, AIDS, or battery acid but an episode of OCD. He should remember that brain lock occurs in the three parts of the brain. As a therapist, I encourage OCD patients to make the following summary for themselves: "Yes, I *do* have a real problem right now. But it is not germs, it is my OCD." This relabeling allows them to get some distance from the content of the obsession and view it in somewhat the same way Buddhists view suffering in meditation: they *observe* its effects on them and so slightly separate themselves from it.

The OCD patient should also remind himself that the reason the attack doesn't go away immediately is the faulty circuit. Some patients may find it helpful, in the midst of an attack, to look at the pictures of the abnormal OCD brain scan in Schwartz's book *Brain Lock,* and compare it with the more normal brain scans that Schwartz's patients developed with treatment, to remind themselves it is possible to change circuits.

Schwartz is teaching patients to distinguish between the universal *form* of OCD (worrisome thoughts and urges that intrude into consciousness) and the *content* of an obsession (i.e., the dangerous germs). The more patients focus on content, the worse their condition becomes.

For a long time therapists have focused on the content as well. The most common treatment for OCD is called "exposure and response prevention," a form of behavior therapy that helps about half of OCD patients make some improvement, though most don't get completely better. If a person fears germs, he is *incrementally exposed* to more of them, in an attempt to desensitize him. In practice this could mean making patients spend time in toilets. (The first time I heard of this treatment, the psychiatrist was asking a man to wear dirty underwear over his face.) Understandably, 30 percent of patients refused such treatments. Exposure to germs doesn't aim to "shift" the gear on to the next thought; it leads the patient to dwell more intensely on them—for a while, at least. The second part of the standard behavioral treatment is "response prevention," preventing the patient from acting on his compulsion. Another form of therapy, Cognitive Therapy, is based on the premise that problematic mood and anxiety states are caused by cognitive distortions—inaccurate or exaggerated thoughts. Cognitive therapists have their OCD patients write down their fears and then list reasons they don't make sense. But this procedure also immerses the patient in the content of his OCD. As Schwartz says, "To teach a patient to say, 'My hands are not dirty,' is just to repeat something she already knows . . . cognitive distortion is just not an intrinsic part of the disease; a patient basically knows that failing to count the cans in the pantry today won't really cause her mother to die a horrible death tonight. The problem is, she doesn't feel that way." Psychoanalysts too have focused on the content of the symptoms, many of which deal with troubling sexual and aggressive ideas. They have found that an obsessive thought, such as "I will hurt my child," might express a suppressed anger at the child, and that this insight might, in mild cases, be enough to make an obsession go away. But this often does not work with moderate or severe OCD. And while Schwartz believes that the origins of many obsessions relate to the kind of conflicts about sex, aggression, and guilt that Freud emphasized, these conflicts explain only the content, not the form of the disorder.

After a patient has acknowledged that the worry is a symptom of OCD, the next crucial step is to *refocus* on a positive, wholesome, ideally pleasure-giving activity the moment he becomes aware he is having an OCD attack. The activity could be gardening, helping someone, working on a hobby, playing a musical instrument, listening to music, working out, or shooting baskets. An activity that involves another person helps keep the patient focused. If OCD strikes while the patient is driving a car, he should be ready with an activity like a book on tape or a CD. It is essential to *do* something, to "shift" the gear manually.

This may seem like an obvious course of action, and may sound simple, but it is not for people with OCD. Schwartz assures his patients that though their "manual transmission" is sticky, with hard work it can be shifted using their cerebral cortex, one effortful thought or action at a time.

Of course, the gearshift is a machine metaphor, and the brain is not a machine; it is plastic and living. Each time patients try to shift gears, they begin fixing their "transmission" by growing new circuits and altering the caudate. By refocusing, the patient is learning not to get sucked in by the content of an obsession but to work around it. I suggest to my patients that they think of the use-it-or-lose-it principle. Each moment they spend thinking of the symptom—believing that germs are threatening them—they deepen the obsessive circuit. By bypassing it, they are on the road to losing it. With obsessions and compulsions, *the more you do it, the more you want to do it; the less you do it, the less you want to do it.*

Schwartz has found it essential to understand that *it is not what you feel while applying the technique that counts, it is what you do.* "The struggle is not to make the feeling go away; the struggle is *not to give in to the feeling*"—by acting out a compulsion, or thinking about the obsession. This technique won't give immediate relief because lasting neuroplastic change takes time, but it does lay the groundwork for change by exercising the brain in a new way. So at first one will still feel both the urge to enact the compulsion, and the tension and anxiety that come from resisting it. The goal is to "change the channel" to some new activity for fifteen to thirty minutes when one has an OCD symptom. (If one can't resist that long, any time spent resisting is beneficial, even if it is only for a minute. That resistance, that effort, is what appears to lay down new circuits.)

One can see that Schwartz's technique with OCD has parallels with Taub's CI approach to strokes. By forcing the patients to "change the channel" and refocus on a new activity, Schwartz is imposing a constraint like Taub's mitt. By getting his patients to concentrate on the new behavior intensively, in thirty-minute segments, he is giving them massed practice.

In chapter 3, "Redesigning the Brain," we learned two key laws of plasticity that also underlie this treatment. The first is that *Neurons that fire together wire together.* By doing something pleasurable in place of the compulsion, patients form a new circuit that is gradually reinforced instead of the compulsion. The second law is that *Neurons that fire apart wire apart.* By not acting on their compulsions, patients weaken the link between the compulsion and the idea it will ease their anxiety. This delinking is crucial because, as we've seen, while acting on a compulsion eases anxiety in the short term, it worsens OCD in the long term.

Schwartz has had good results with severe cases. Eighty percent of his patients get better when they use his method in combination with medication—typically an antidepressant such as Anafranil or a Prozac-type drug. The medication functions like training wheels on a bike, to ease anxiety or to lower it enough for patients to benefit from the therapy. In time many patients get off the medication, and some don't need it to start with.

I have seen the brain lock approach work well with such typical OCD problems as fear of germs, hand washing, checking compulsions, compulsive second-guessing, and incapacitating hypochondriacal fears. As patients apply themselves, the "manual gear shift" gets more and more automatic. The episodes become shorter and less frequent, and though patients can relapse during stressful times, they can quickly regain control using their newfound technique.

When Schwartz and his team scanned the brains of their improved patients, they found that the three parts of the brain that had been "locked" and, firing together in a hyperactive way, had begun to fire separately in a normal way. The brain lock was being relieved.

I was at a dinner party with a friend, whom I shall call Emma; her writer husband, Theodore; and several other writers.

Emma is now in her forties. When she was twenty-three, a spontaneous genetic mutation led to an illness called retinitis pigmentosa that caused her retinal cells to die. Five years ago she became totally blind and began using a seeing-eye dog, Matty, a Labrador.

Emma's blindness has reorganized her brain and her life. A number of us who were at the dinner are interested in literature, but since she has gone blind, Emma has done more reading than any of us. A computer program from Kurzweil Educational Systems reads books aloud to her in a monotone that pauses for commas, stops for periods, and rises in pitch for questions. This computer voice is so rapid, I cannot make out a single word. But Emma has gradually learned to listen at a faster and faster pace, so she is now reading at about 340 words a minute and is marching through all the great classics. "I get into an author, and I read everything he has ever written, and then I move on to another." She has read Dostoyevsky (her favorite), Gogol, Tolstoy, Turgenev, Dickens, Chesterton, Balzac, Hugo, Zola, Flaubert, Proust, Stendhal, and many others. Recently she read three Trollope novels in one day. She asked me how it might be possible for her to read so much more quickly than before she went blind. I theorized that her massive visual cortex, no longer processing sight, had been taken over for auditory processing.

That particular evening Emma asked me if I knew anything about needing to check things a lot. She told me that she often has a lot of trouble getting out of the house, because she keeps checking the stoves and the locks. Back when she was still going to her office, she might leave for work, get halfway there, and then have to go back to make sure she had locked the door properly. By the time she got back, she would feel obliged to check that the stove, electrical appliances, and water were turned off. She'd leave, then have to repeat the whole cycle several more times, all the while trying to fight the urge. She told me that her authoritarian father had made her anxious when she was growing up. When she left home, she'd lost that anxiety but noticed that it now seemed to have been replaced by this checking, which kept getting worse.

I explained the brain lock theory to her. I told her that often we check and recheck appliances without really concentrating. So I suggested she check once, and once only, with utmost care.

The next time I saw her, she was delighted. "I'm better," she said. "I check once, now, and I move on. I still feel the urge, but I resist it, and then it passes. And as I get more practice, it is passing more quickly."

She gave her husband a mock scowl. He had joked that it was not polite to bother the psychiatrist with her neuroses while we were at a party.

"Theodore," she said, "it's not that I'm *crazy*. It's just that my brain wasn't turning the page."

7

Pain

The Dark Side of Plasticity

When we wish to perfect our senses, neuroplasticity is a blessing; when it works in the service of pain, plasticity can be a curse.

Our guide to pain is one of the most inspiring of the neuroplasticians, V. S. Ramachandran. Vilayanur Subramanian Ramachandran was born in Madras, India. He is a neurologist, of Hindu background, and a proud relic of nineteenth-century science who tackles twenty-first-century dilemmas.

Ramachandran is an M.D., a specialist in neurology, with a Ph.D. in psychology from Trinity College, Cambridge. We met in San Diego, where he directs the Center for Brain and Cognition at the University of California. "Rama" has black, wavy hair and wears a black leather jacket. His voice booms. His accent is British, but when he is excited, his *r*'s are like a long drumroll.

Whereas many neuroplasticians work to help people develop or recover skills—to read, move, or overcome learning disabilities—Ramachandran uses plasticity to reconfigure the content of our minds. He shows that we can rewire our brains through comparatively brief, painless treatments that use imagination and perception.

His office is filled not with high-tech devices but rather with simple nineteenth-century machines, the little inventions that draw children to science. There is a stereoscope, an optical instrument that makes two pictures of the same scene look three-dimensional. There is a magnetic device that was once used to treat hysteria, some funhouse-type mirrors, magnifying glasses of early vintage, fossils, and the preserved brain of an adolescent. There is also a bust of Freud, a picture of Darwin, and some voluptuous Indian art.

This could only be the office of one man, the Sherlock Holmes of modern neurology, V. S. Ramachandran. He is a sleuth, solving mysteries one case at a time, as though utterly unaware that modern science is now occupied with large statistical studies. He believes that individual cases have everything to contribute to science. As he puts it, "Imagine I were to present a pig to a skeptical scientist, insisting it could speak English, then waved my hand, and the pig spoke English. Would it really make sense for the skeptic to argue, 'But that is just one pig, Ramachandran. Show me another, and I might believe you!' "

He has repeatedly shown that by explaining neurological "oddities," he can shed light on the functioning of normal brains. "I hate crowds in science," he tells me. He doesn't fancy large scientific meetings either. "I tell my students, when you go to these meetings, see what direction everyone is headed, so you can go in the opposite direction. Don't polish the brass on the bandwagon."

Beginning at age eight, Ramachandran tells me, he avoided sports and parties and progressed from one passion to another: paleontology (he collected rare fossils in the field), conchology (the study of sea shells), entomology (he had a special fondness for beetles), and botany (he cultivated orchids). His biography is scattered throughout his office, in the form of beautiful natural objects—fossils, shells, insects, and flowers. Were he not a neurologist, he tells me, he would be an archeologist studying ancient Sumer, Mesopotamia, or the Indus Valley.

These essentially Victorian pursuits reveal his fondness for the science of that period, the golden age of taxonomy, when the learned ranged around the world, using the naked eye and Darwinian detective work to catalog nature's variations and eccentricities and weave them into broad theories that explain the great themes of the living world.

Ramachandran approaches neurology the same way. In his early research he investigated patients who experienced mental illusions. He studied people who, after brain injuries, began to believe they were prophets, or others suffering from Capgras syndrome, who came to believe their parents and spouses were impostors, exact replicas of their real loved ones. He studied optical illusions and the eye's blind spots. As he figured out what was happening in each of these diseases—generally without the use of modern technology—he shed new light on how the normal brain works.

"I have a disdain," he says, "for complicated fancy equipment because it takes a lot of time to learn how to use, and I'm suspicious when the distance between the raw data and the final conclusion is too long. It gives you plenty of opportunity to massage that data, and human beings are notoriously susceptible to self-deception, whether scientists or not."

Ramachandran pulls out a large square box with a mirror standing inside it that looks like a child's magic trick. Using this box and his insights into plasticity, he solved the centuries-old mystery of phantom limbs and the chronic pain they engender.

There are a whole host of haunting pains that torment us for reasons we do not understand and that arrive from we know not where—pains without return address. Lord Nelson, the British admiral, lost his right arm in an attack on Santa Cruz de Tenerife in 1797. Soon afterward, Ramachandran points out, he vividly began to experience the presence of his arm, a phantom limb that he could feel but not see. Nelson concluded that its presence was "direct evidence for the existence of the soul," reasoning that if an arm can exist after being removed, so then might the whole person exist after the annihilation of the body.

Phantom limbs are troubling because they give rise to a chronic "phantom pain" in 95 percent of amputees that often persists for a lifetime. But how do you remove a pain in an organ that isn't there?

Phantom pains torment soldiers with amputations and people who lose limbs in accidents, but they are also part of a larger class of uncanny pains that have confused doctors for millennia, because they had no known source in the body. Even after routine surgery, some people are left with equally mysterious postoperative pains that last a lifetime. The scientific literature on pain includes stories of women who suffer menstrual cramps and labor pains *even after* their uteruses have been removed, of men who still feel ulcer pain *after* the ulcer and its nerve have been cut out, and of people who are left with chronic rectal and hemorrhoidal pain after their rectums have been removed. There are stories of people whose bladders were removed who still have an urgent, painful chronic need to urinate. These episodes are comprehensible if we remember that they too are phantom pains, the result of internal organs being "amputated."

Normal pain, "acute pain," alerts us to injury or disease by sending a signal to the brain, saying, "This is where you are hurt—attend to it." But sometimes an injury can damage both our bodily tissues *and* the nerves in our pain systems, resulting in "neuropathic pain," for which there is no external cause. Our pain maps get damaged and fire incessant false alarms, making us believe the problem is in our body when it is in our brain. Long after the body has healed, the pain system is still firing and the acute pain has developed an afterlife.

The phantom limb was first proposed by Silas Weir Mitchell, an American physician who tended the wounded at Gettysburg and became intrigued by an epidemic of phantoms. Civil War soldiers' wounded arms and legs often turned gangrenous, and in an age before antibiotics, the only way to save the soldier's life was to amputate the limb before the gangrene spread. Soon amputees began to report that their limbs had returned to haunt them. Mitchell first called these experiences "sensory ghosts," then switched to calling them "phantom limbs."

They are often very lively entities. Patients who have lost arms can sometimes feel them gesticulating when they talk, waving hello to friends, or reaching spontaneously for a ringing phone.

A few doctors thought the phantom was the product of wishful thinking—a denial of the painful loss of a limb. But most assumed that the nerve endings on the stump end of the lost limb were being stimulated or irritated by movement. Some doctors tried to deal with phantoms by serial amputations, cutting back the limbs—and nerves—farther and farther, hoping the phantom might disappear. But after each surgery it reemerged.

Ramachandran had been curious about phantoms since medical school. Then in 1991 he read the paper by Tim Pons and Edward Taub about the final operations on the Silver Spring monkeys. As you'll recall, Pons mapped the brains of the monkeys who had had all the sensory input from their arms to their brains eliminated by deafferentation and found that the brain map for the arm, instead of wasting away, had become active and now processed input from the face—which might be expected because, as Wilder Penfield had shown, the hand and facial maps are side by side.

Ramachandran immediately thought that plasticity might explain phantom limbs because Taub's monkeys and patients with phantom arms were similar. The brain maps for both the monkeys and the patients had been deprived of stimuli from their limbs. Was it possible that the face maps of amputees had invaded the maps for their missing arms, so that when the amputee was touched on the face, he felt his phantom arm? And where, Ramachandran wondered, did Taub's monkeys feel it when their faces were stroked—on their faces, or in their "deafferented" arm?

Tom Sorenson—a pseudonym—was only seventeen years old when he lost his arm in an automobile accident. As he was hurled into the air, he looked back and saw his hand, severed from his body, still grabbing the seat cushion. What remained of his arm had to be amputated just above the elbow.

About four weeks later he became aware of a phantom limb that did many of the things his arm used to. It reached out reflexively to break a fall or to pat his younger brother. Tom had other symptoms, including one that really irked him. He had an itch in his phantom hand that he couldn't scratch.

Ramachandran heard of Tom's amputation from colleagues and asked to work with him. To test his theory that phantoms were caused by rewired brain maps, he blindfolded Tom. Then he stroked parts of Tom's upper body with a Q-tip, asking Tom what he felt. When he got to Tom's cheek, Tom told him he felt it there but also in his phantom. When Ramachandran stroked Tom's upper lip, he felt it there but also in the index finger of his phantom. Ramachandran found that when he touched other parts of Tom's face, Tom felt it in other parts of his phantom hand. When Ramachandran put a drop of warm water on Tom's cheek, he felt a warm trickle move down his cheek and also down his phantom limb. Then after some experimentation Tom found that he could finally scratch the unscratchable itch that had plagued him for so long by scratching his cheek.

After Ramachandran's success with the Q-tip, he went high-tech with a brain scan called an MEG, or magnetoencephalography. When he mapped Tom's arm and hand, the scan confirmed that his hand map was now being used to process facial sensations. His hand and face maps had blurred together.

Ramachandran's finding in the Tom Sorenson case, at first controversial among clinical neurologists who doubted brain maps were plastic, is now widely accepted. Brain scan studies by the German team that Taub works with have also confirmed a correlation between the amount of plastic change and the degree of phantom pain people experience.

Ramachandran strongly suspects that one reason map invasion occurs is that the brain "sprouts" new connections. When a part of the body is lost, he believes, its surviving brain map "hungers" for incoming stimulation and releases nerve growth factors that invite neurons from nearby maps to send little sprouts into them.

Normally these little sprouts link up to similar nerves; nerves for touch link with other nerves for touch. But our skin, of course, conveys far more than touch; it has distinct receptors that detect temperature, vibration, and pain as well, each with its own nerve fibers that travel up to the brain, where they have their own maps, some of which are very near each other. Sometimes after an injury, because the nerves for touch, temperature, and pain are so close together, there can be cross-wiring errors. So, Ramachandran wondered, might a person who is touched, in cases of cross-wiring, feel pain or warmth? Could a person who was touched gently on the face feel pain in a phantom arm?

Another reason phantoms are so unpredictable and cause so much trouble is that brain maps are dynamic and changing: even under normal circumstances, as Merzenich showed, face maps tend to move around a bit in the brain. Phantom maps move because their input has been so radically changed. Ramachandran and others—Taub and his colleagues among them—have shown with repeated scans of brain maps that the contours of phantoms and their maps are constantly changing. He thinks one reason people get phantom pain is that when a limb is cut off, its map not only shrinks but gets disorganized and stops working properly.

Not all phantoms are painful. After Ramachandran published

his discoveries, amputees began to seek him out. Several leg amputees reported, with much shame, that when they had sex, they often experienced their orgasms in their phantom legs and feet. One man confessed that because his leg and foot were so much larger than his genitals, the orgasm was "much bigger" than it used to be. Though such patients might once have been dismissed as having overly rich imaginations, Ramachandran argued that the claim made perfect neuroscientific sense. The Penfield brain map shows the genitals next to the feet, and since the feet no longer receive input, the genital maps likely invade the foot maps, so when the genitals experience pleasure, so do the phantom feet. Ramachandran began to wonder whether some people's erotic preoccupation with feet, or foot fetishes, might be due in part to the proximity of feet and genitals on the brain map.

Other erotic enigmas fell into place. An Italian physician, Dr. Salvatore Aglioti, reported that some women who have had mastectomies experience sexual excitement when their ears, clavicles, and sternums are stimulated. All three are close to nipples on the brain map. Some men with carcinoma of the penis who have had their penises amputated experience not only phantom penises but phantom erections.

As Ramachandran examined more amputees, he learned that about half of them have the unpleasant feeling that their phantom limbs are frozen, hanging in a fixed paralyzed position, or encased in cement. Others feel they are lugging around a dead weight. And not only do images of paralyzed limbs get frozen in time, but in some horrific cases the original agony of losing a limb is locked in. When grenades blow up in soldiers' hands, they can develop a phantom pain that endlessly repeats the excruciating moment of the explosion. Ramachandran encountered a woman whose frostbitten thumb was amputated and whose phantom "froze" the agonizing frostbite pains in place. People are tortured by phantom memories of gangrene, ingrown toenails, blisters, and cuts felt in the limb before it was amputated, especially if that pain existed at the time of the amputation. These patients experience such agonies not as faint "memories" of pain but as happening in the present. Sometimes a patient can be pain free for decades, and then an event, perhaps a needle inserted in a trigger point, reactivates the pain months or years later.

When Ramachandran reviewed the histories of people with painful frozen arms, he discovered that they had all had their arms in slings or casts for several months before amputation. Their brain maps now seemed to record, for all time, the fixed position of the arm just prior to amputation. He began to suspect that it was the very fact that the limb did not exist that allowed the sensation of paralysis to persist. Normally, when the motor command center in the brain sends out an order to move the arm, the brain gets feedback from various senses, confirming that the order has been executed. But the brain of a person without a limb never gets confirmation that the arm has moved, since there are neither arm nor motion sensors in the arm to provide that feedback. Thus, the brain is left with the impression that the arm is frozen. Because the arm had been stuck in a cast or sling for months, the brain map developed a representation of the arm as unmoving. When the arm was removed, there was no new input to alter the brain map, so the mental representation of the limb as fixed became frozen in time—a situation similar to the learned paralysis that Taub discovered in stroke patients.

Ramachandran came to believe that the absence of feedback causes not only frozen phantoms but phantom pain. The brain's motor center might send commands for the hand muscles to contract but, getting no feedback confirming the hand has moved, escalates its command, as if to say: "Clench! You're not clenching enough! You haven't touched the palm yet! Clench as hard as you can!" These patients feel their fingernails are digging into their palms. While actual clenching caused pain when the arm was present, this imaginary clenching evokes pain because maximum contraction and pain are associated in memory.

Ramachandran next asked a most daring question: whether phantom paralysis and pain could be "unlearned." This was the sort of question psychiatrists, psychologists, and psychoanalysts might ask: how does one change a situation that has a psychic but not a material reality? Ramachandran's work began to blur the boundary between neurology and psychiatry, reality and illusion.

Ramachandran then hit on the wizardlike idea of fighting one illusion with another. What if he could send false signals to the brain to make the patient think that the nonexistent limb was moving?

That question led him to invent a mirror box designed to fool the patient's brain. It would show him the mirror image of his good hand in order to make him believe it was his amputated hand "resurrected."

The mirror box is the size of a large cake box, without a top, and is divided into two compartments, one on the left and one on the right. There are two holes in the front of the box. If the patient's left was amputated, he puts his good right hand through the hole and into the right compartment. Then he is told to imagine putting his phantom hand into the left compartment.

The divider that separates the two compartments is a vertical mirror facing the good hand. Because there is no top on the box, the patient can, by leaning a bit to the right, see a *mirror image* reflection of his good right hand, which will seem to be his left hand as it was before the amputation. As he moves his right hand back and forth, his "resurrected" left hand will also appear to move back and forth, superimposed on his phantom. Ramachandran hoped the patient's brain might get the impression that the phantom arm was moving.

To find subjects to test his mirror box, Ramachandran ran enigmatic ads in local newspapers saying, "Amputees needed." "Philip Martinez" responded.

About a decade before, Philip was hurled from his motorcycle while going forty-five miles per hour. All the nerves leading from his left hand and arm to his spine were torn out by the accident. His arm was still attached to his body, but no functioning nerves sent signals from his spine to his arm, and no nerves entered his spine to convey sensation to his brain. Philip's arm was worse than useless, an immovable burden he had to keep in a sling, and he eventually chose to have the arm amputated. But he was left with terrible phantom pain in his phantom elbow. The phantom arm also felt paralyzed, and he had the sense that if he could only somehow move it, he might relieve the pain. This dilemma so depressed him that he contemplated suicide.

When Philip put his good arm into the mirror box, he not only began to "see" his "phantom" move, but he felt it moving for the first time. Amazed and overwhelmed with joy, Philip said he felt his phantom arm "was plugged in again."

Yet the moment he stopped looking at the mirror image or closed his eyes, the phantom froze. Ramachandran gave Philip the mirror box to take home, to practice with, hoping that Philip might unlearn his paralysis by stimulating a plastic change that would rewire his brain map. Philip used the box for ten minutes a day, but it still seemed only to work when his eyes were open, looking at the mirror image of his good hand.

Then after four weeks Ramachandran got an excited call from Philip. Not only was his phantom arm permanently unfrozen, it was gone—even when he wasn't using the box. Gone too was his phantom elbow and its excruciating pain. Only painless phantom fingers were left, dangling from his shoulder.

V. S. Ramachandran, the neurological illusionist, had become the first physician to perform a seemingly impossible operation: the successful amputation of a phantom limb.

Ramachandran has used his box with a number of patients, about half of whom have lost their phantom pain, unfrozen their phantoms, and started to feel control over them. Other scientists have also found that patients who train with the mirror box get better. fMRI brain scans show that as these patients improve, the motor maps for their phantoms increase, the map shrinkage that accompanies amputation is reversed, and sensory and motor maps normalize.

The mirror box appears to cure pain by altering the patients' perception of their body image. This is a remarkable discovery because it sheds light both on how our minds work and on how we experience pain.

Pain and body image are closely related. We always experience pain as *projected* into the body. When you throw your back out, you say, "My back is killing me!" and not, "My pain system is killing me." But as phantoms show, we don't need a body part or even pain receptors to feel pain. We need only *a body image,* produced by our brain maps. People with actual limbs don't usually realize this, because the body images of our limbs are *perfectly projected* onto our actual limbs, making it impossible to distinguish our body image from our body. "Your own body is a phantom," says Ramachandran, "one that your brain has constructed purely for convenience."

Distorted body images are common and demonstrate that there is a difference between the body image and the body itself. Anorexics experience their bodies as fat when they are on the edge of starvation; people with distorted body images, a condition called "body dysmorphic disorder," can experience a part of the body that is perfectly within the norm as defective. They think their ears, nose, lips, breasts, penis, vagina, or thighs are too large or too small, or just "wrong," and they feel tremendous shame. Marilyn Monroe experienced herself as having many bodily defects. Such people often seek plastic surgery but still feel misshapen after their operations. What they need instead is "neuroplastic surgery" to change their body image.

Ramachandran's success with rewiring phantoms suggested to him that there may be ways to rewire distorted body images. To better understand what he meant by a body image, I asked him if he might demonstrate the difference between it, a mental construct, and the material body.

Taking out the type of fake rubber hand sold in novelty shops, he sat me at a table and placed the fake hand on it, its fingers parallel to the table edge in front of me, about an inch from the edge. He told me to put my hand on the table, parallel to the fake hand, but about eight inches from the table's edge. My hand and the fake were perfectly aligned, pointing in the same direction. Then he put a cardboard screen between the fake hand and my own, so I could see only the fake.

Then with his hand he stroked the fake hand, as I watched. With his other hand he simultaneously stroked my hand, hidden behind the screen. When he stroked the fake's thumb, he stroked my thumb. When he tapped the fake pinkie three times, he tapped my pinkie three times, in the same rhythm. When he stroked the fake middle finger, he stroked my middle finger.

Within moments my feeling that my own hand was being stroked disappeared, and I began to experience the feeling I was being stroked as if coming from the fake hand. The dummy hand had become part of my body image! This illusion works by the same principle that fools us into thinking that ventriloquist's dummies, or cartoons, or movie actors in films are actually talking because the lips move in sync with the sound.

Then Ramachandran performed an even simpler trick. He told me to put my right hand under the table, so my hand was hidden. Then he tapped the tabletop with one hand, while with his other he tapped mine under the table, where I couldn't see it, in an identical rhythm. When he moved the spot where he hit the tabletop, a bit to the left or right, he moved his hand under the table exactly the same way. After a few minutes I stopped experiencing him as tapping my hand under the table and instead—fantastic as it sounds—started to feel that the body image of my hand had merged with the tabletop, so that the sensation of being tapped seemed to come from the tabletop. He had created an illusion in which my sensory body image had now been expanded to include a piece of furniture!

Ramachandran has wired subjects to a galvanic skin response meter that measures stress responses during this table experiment. After stroking the tabletop and a patient's hand under the table until his body image included the table, he would pull out a hammer and bash the tabletop. The subject's stress response went through the roof, just as if Ramachandran had smashed the subject's actual hand.

According to Ramachandran, pain, like the body image, is created by the brain and projected onto the body. This assertion is contrary to common sense and the traditional neurological view of pain that says that when we are hurt, our pain receptors send a *one-way* signal to the brain's pain center and that the intensity of pain perceived is proportional to the seriousness of the injury. We assume that pain always files an accurate damage report. This traditional view dates back to the philosopher Descartes, who saw the brain as a passive recipient of pain. But that view was overturned in 1965, when neuroscientists Ronald Melzack (a Canadian who studied phantom limbs and pain) and Patrick Wall (an Englishman who studied pain and plasticity) wrote the most important article in the history of pain. Wall and Melzack's theory asserted that the pain system is spread throughout the brain and spinal cord, and far from being a passive recipient of pain, the brain always controls the pain signals we feel.

Their "gate control theory of pain" proposed a series of controls, or "gates," between the site of injury and the brain. When pain messages are sent from damaged tissue through the nervous system, they pass through several "gates," starting in the spinal cord, before they get to the brain. But these messages travel only if the brain gives them "permission," after determining they are important enough to be let through. If permission is granted, a gate will open and increase the feeling of pain by allowing certain neurons to turn on and transmit their signals. The brain can also close a gate and block the pain signal by releasing endorphins, the narcotics made by the body to quell pain.

The gate theory made sense of all sorts of pain experiences. For instance, when the U.S. troops landed in Italy in World War II, 70 percent of the men who were seriously wounded reported that they were not in pain and did not want pain-killers. Men wounded on the battlefield often don't feel pain and keep fighting; it's as if the brain closes the "gate," to keep the embattled soldier's attention riveted on how to get out of harm's way. Only when he is safe are the pain signals allowed to pass to the brain.

Physicians have long known that a patient who expects to get pain relief from a pill often does, even though it is a placebo containing no medication. fMRI brain scans show that during the placebo effect the brain turns down its own pain-responsive regions. When a mother soothes her hurt child, by stroking and talking sweetly to her, she is helping the child's brain turn down the volume on its pain. How much pain we feel is determined in significant part by our brains and minds—our current mood, our past experiences of pain, our psychology, and how serious we think our injury is.

Wall and Melzack showed that the neurons in our pain system are far more plastic than we ever imagined, that important pain maps in the spinal cord can change following injury, and that a chronic injury can make the cells in the pain system fire more easily—a plastic alteration—making a person hypersensitive to pain. Maps can also enlarge their receptive field, coming to represent more of the body's surface, increasing pain sensitivity. As the maps change, pain signals in one map can "spill" into adjacent pain maps, and we may develop "referred pain," when we are hurt in one body part but feel the pain in another. Sometimes a single pain signal reverberates throughout the brain, so that pain persists even after its original stimulus has stopped.

The gate theory led to new treatments for blocking pain. Wall coinvented "transcutaneous electrical nerve stimulation," or TENS, which uses electric current to stimulate neurons that *inhibit* pain, helping in effect to close the gate. The gate theory also made Western scientists less skeptical of acupuncture, which reduces pain by stimulating points of the body often far from the site where the pain is felt. It seemed possible that acupuncture turns on neurons that *inhibit* pain, closing gates and blocking pain perception.

Melzack and Wall had another revolutionary insight: that the pain system includes motor components. When we cut a finger, we reflexively squeeze it, a motor act. We instinctively guard an injured ankle by finding a safe position. Guarding commands, "Don't move a muscle until that ankle's better."

Extending the gate theory, Ramachandran developed his next idea: that pain is a complex system under the plastic brain's control. He summed this up as follows: "Pain is an opinion on the organism's state of health rather than a mere reflexive response to injury." The brain gathers evidence from many sources before triggering pain. He has also said that "pain is an illusion" and that "our mind is a virtual reality machine," which experiences the world indirectly and processes it at one remove, constructing a model in our head. So pain, like the body image, is a construct of our brain. Since Ramachandran could use his mirror box to modify a body image and eliminate a phantom and its pain, could he also use the mirror box to make chronic pain in a real limb disappear?

Ramachandran thought he might be able to remedy "type 1 chronic pain," experienced in a disorder called "reflex sympathetic dystrophy." This occurs when a minor injury, a bruise, or an insect bite on the fingertip makes an entire limb so excruciatingly painful that "guarding" prevents the patient from moving it. The condition can last long after the original injury and often becomes chronic, accompanied by burning discomfort and agonizing pain in response to a light brushing or stroking of the skin. Ramachandran theorized that the brain's plastic ability to rewire itself was leading to a pathological form of guarding.

When we guard, we prevent our muscles from moving and aggravating our injury. If we had to remind ourselves consciously not to move, we'd become exhausted and slip up, hurt ourselves, and feel pain. Now suppose, thought Ramachandran, the brain preempts the mistaken movement by triggering pain the moment *before* the movement takes place, between the time when the motor center issues the command to move and the time when the move is performed. What better way for the brain to prevent movement than to make sure the motor command itself triggers pain? Ramachandran came to believe that in these chronic pain patients the motor command got wired into the pain system, so that even though the limb had healed, when the brain sent out a motor command to move the arm, it still triggered pain.

Ramachandran called this "learned pain" and wondered whether the mirror box could help relieve it. All the traditional remedies had been tried on these patients—interrupting the nerve connection to the painful area, physiotherapy, pain-killers, acupuncture, and osteopathy—to no avail. In a study conducted by a team that included Patrick Wall, the patient was instructed to put both hands into the mirror box, sitting so he could see only his good arm and its reflection in the mirror. The patient then moved his good arm in whatever way he chose (and his affected one if possible) in the box for ten minutes, several times a day, for several weeks. Perhaps the moving reflection, which occurred without a motor command initiating it, was fooling the patient's brain into thinking his hurt arm could now move freely without pain, or perhaps this exercise was enabling the brain to learn that guarding was no longer necessary, so it would disconnect the neuronal link between the motor command to move the arm and the pain system.

Patients who had had the pain syndrome for only two months got better. The first day the pain lessened, and relief lasted even after a mirror session was over. After a month they no longer had any pain. Patients who had had the syndrome for between five months and a year didn't do quite as well, but they lost stiffness in their limbs and were able to go back to work. Those who had had the pain for longer than two years failed to get better.

Why? One thought was that these long-term patients had not moved their guarded limbs for so long that the motor maps for the affected limb had begun to waste away—once again use it or lose it. All that remained were the few links that were most active when the limb was last used, and unfortunately these were links to the pain system, just as patients who wore casts before amputations developed phantoms "stuck" where their arms were just before the amputation.

An Australian scientist, G. L. Moseley, thought he might be able to help the patients who hadn't improved by using the mirror box, often because their pain was so great they couldn't move their limbs in mirror therapy. Moseley thought that building up the affected limb's motor map with mental exercises might trigger plastic change. He asked these patients to simply *imagine* moving their painful limbs, without executing the movements, in order to activate brain networks for movement. The patients also looked at pictures of hands, to determine whether they were the left or right, until they could identify them quickly and accurately—a task known to activate the motor cortex. They were shown hands in various positions and asked to imagine them for fifteen minutes, three times a day. After practicing the visualization exercises they did the mirror therapy, and with twelve weeks of therapy, pain had diminished in some and had disappeared in half.

Think how remarkable this is—for a most excruciating, chronic pain, a whole new treatment that uses imagination and illusion to restructure brain maps plastically without medication, needles, or electricity.

The discovery of pain maps has also led to new approaches to surgery and the use of pain medication. Postoperative phantom pain can be minimized if surgical patients get local nerve blocks or local anesthetics that act on peripheral nerves *before* the general anesthetic puts them to sleep. Pain-killers, administered before surgery, not just afterward, appear to prevent plastic change in the brain's pain map that may "lock in" pain.

Ramachandran and Eric Altschuler have shown that the mirror box is effective on other nonphantom problems, such as the paralyzed legs of stroke patients. Mirror therapy differs from Taub's in that it fools the patient's brain into thinking he is moving the affected limb, and so it begins to stimulate that limb's motor programs. Another study showed that mirror therapy was helpful in preparing a severely paralyzed stroke patient, who had no use of one side of the body, for a Taub-like treatment. The patient recovered some use of his arm, the first occasion in which two novel plasticity-based approaches—mirror therapy and CI-like therapy—were used in sequence.

In India, Ramachandran grew up in a world where many things that seem fantastic to Westerners were commonplace. He knew about yogis who relieved suffering with meditation and walked barefoot across hot coals or lay down on nails. He saw religious people in trances putting needles through their chins. The idea that living things change their forms was widely accepted; the power of the mind to influence the body was taken for granted, and illusion was seen as so fundamental a force that it was represented in the deity Maya, the goddess of illusion. He has transposed a sense of wonder from the streets of India to Western neurology, and his work inspires questions that mingle the two. What is a trance but a closing down of the gates of pain within us? Why should we think phantom pain any less real than ordinary pain? And he has reminded us that great science can still be done with elegant simplicity.

8

Imagination

How Thinking Makes It So

I am in Boston in the laboratory for magnetic brain stimulation, at Beth Israel Deaconess Medical Center, part of Harvard Medical School. Alvaro Pascual-Leone is chief of the center, and his experiments have shown that we can change our brain anatomy simply by using our imaginations. He has just put a paddle-shaped machine on the left side of my head. The device emits transcranial magnetic stimulation, or TMS, and can influence my behavior. Inside the machine's plastic casing is a coil of copper wire, through which a current passes to generate a changing magnetic field that surges into my brain, into the cablelike axons of my neurons, and from there into the motor map of my hand in the outer layer of my cerebral cortex. A changing magnetic field induces an electric current around it, and Pascual-Leone has pioneered the use of TMS to make neurons fire. Each time he turns on the magnetic field, the fourth finger on my right hand moves because he is stimulating an area of about 0.5 cubic centimeter in my brain, consisting of millions of cells—the brain map for that finger.

TMS is an ingenious bridge into my brain. Its magnetic field passes painlessly and harmlessly through my body, inducing an electric current only when the field reaches my neurons. Wilder Penfield had to open the skull surgically and insert his electric probe into the brain to stimulate the motor or sensory cortex. When Pascual-Leone turns on the machine and makes my finger move, I experience *exactly* what Penfield's patients did when he cut open their skulls and prodded them with large electrodes.

Alvaro Pascual-Leone is young for all he has accomplished. He was born in 1961 in Valencia, Spain, and has conducted research both there and in the United States. Pascual-Leone's parents, both physicians, sent him to a German school in Spain, where, like many neuroplasticians, he studied the classical Greek and German philosophers before turning to medicine. He took his combined M.D. and Ph.D. in physiology in Freiburg, then went to the United States for further training.

Pascual-Leone has olive skin, dark hair, and an expressive voice, and he radiates a serious playfulness. His small office is dominated by the massive Apple computer screen he uses to display what he sees through his TMS window onto the brain. E-mails from collaborators pour in from far-flung parts of the world. There are books on electromagnetism shelved behind him, and papers everywhere.

He was the first to use TMS to map the brain. TMS can be used either to turn on a brain area or to block it from functioning, depending on the intensity and frequency used. To determine the function of a specific brain area, he fires bursts of TMS to temporarily block the area from working, then observes which mental function is lost.

He is also one of the great pioneers in the use of high-frequency "repetitive TMS," or rTMS. High-frequency repetitive TMS can activate neurons so much that they excite each other and keep firing even after the original burst of rTMS has stopped. This turns a brain area on for a while and can be used therapeutically. For instance, in some depressions the prefrontal cortex is partially turned off and underfunctioning. Pascual-Leone's group was the first to show that rTMS is effective in treating such severely depressed patients. Seventy percent of those who had failed all the traditional treatments improved with rTMS and had fewer side effects than with medication.

In the early 1990s, when Pascual-Leone was still a young medical fellow at the National Institute of Neurological Disorders and Stroke, he did experiments—celebrated among neuroplasticians for their elegance—that perfected a way to map the brain, made his imagination experiments possible, and taught us how we learn skills.

He studied how people learn new skills by using TMS to map the brains of blind subjects learning to read Braille. The subjects studied Braille for a year, five days a week, for two hours a day in class, followed by an hour of homework. Braille readers "scan" by moving their index fingers across a series of small raised dots, a motor activity. Then they feel the arrangement of the dots, a sensory activity. These findings were among the first to confirm that when human beings learn a new skill, plastic change occurs.

When Pascual-Leone used TMS to map the *motor* cortex, he found that the maps for people's "Braille reading fingers" were larger than the maps for their other index fingers and also those for the index fingers of non-Braille readers. Pascual-Leone also found that the motor maps increased in size as the subjects increased the number of words per minute they could read. But his most surprising discovery, one with major implications for learning any skill, was the way the plastic change occurred over the course of each week.

The subjects were mapped with TMS on Fridays (at the end of the week's training), and on Mondays (after they had rested for the weekend). Pascual-Leone found that the changes were different on Friday and Monday. From the beginning of the study, Friday maps showed very rapid and dramatic expansion, but by Monday these maps had returned to their baseline size. The Friday maps continued to grow for six months—stubbornly returning to baseline each Monday. After about six months the Friday maps were still increasing but not as much as in the first six months.

Monday maps showed an opposite pattern. They didn't begin to change until six months into the training; then they increased slowly and plateaued at ten months. The speed at which the subjects could read Braille correlated much better with the Monday maps, and though the changes on Mondays were never as dramatic as on Fridays, they were more stable. At the end of ten months the Braille students took two months off. When they returned, they were remapped, and their maps were unchanged from the last Monday mapping two months before. Thus daily training led to dramatic short-term changes during the week. But over the weekends, and months, more permanent changes were seen on Mondays.

Pascual-Leone believes that the differing results on Monday and Friday suggest differing plastic mechanisms. The fast Friday changes strengthen *existing* neuronal connections and unmask buried pathways. The slower, more permanent Monday changes suggest the formation of *brand-new* structures, probably the sprouting of new neuronal connections and synapses.

Understanding this tortoise-and-hare effect can help us understand what we must do to truly master new skills. After a brief period of practice, as when we cram for a test, it is relatively easy to improve because we are likely strengthening existing synaptic connections. But we quickly forget what we've crammed—because these are easy-come, easy-go neuronal connections and are rapidly reversed. Maintaining improvement and making a skill permanent require the slow steady work that probably forms new connections. If a learner thinks he is making no cumulative progress, or feels his mind is "like a sieve," he needs to keep at the skill until he gets "the Monday effect," which in Braille readers took six months. The Friday-Monday difference is probably why some people, the "tortoises," who seem slow to pick up a skill, may nevertheless learn it better than their "hare" friends—the "quick studies" who won't necessarily hold on to what they have learned without the sustained practice that solidifies the learning.

Pascual-Leone expanded his study to examine how Braille readers get so much information through their fingertips. It is well known that the blind can develop superior nonvisual senses and that Braille readers gain extraordinary sensitivity in their Braille-reading fingers. Pascual-Leone wanted to see if that increased skill was facilitated by an enlargement of the sensory map for touch or by plastic change in other parts of the brain, such as the visual cortex, which might be underused, since it wasn't getting input from the eyes.

He reasoned that if the visual cortex helped the subjects to read Braille, blocking it would interfere with Braille reading. And it did: when the team applied blocking TMS to the *visual* cortex of Braille readers to create a virtual lesion, the subjects could not read Braille or feel with the Braille-reading finger. The visual cortex had been recruited to process information derived from touch. Blocking TMS applied to the visual cortex of sighted people had *no effect* on their ability to feel, indicating that something unique was happening to the blind Braille readers: a part of the brain devoted to one sense had become devoted to another—the kind of plastic reorganization suggested by Bach-y-Rita. Pascual-Leone also showed that the better a person could read Braille, the more the visual cortex was involved. His next venture would break ground in a new way altogether, by showing that our thoughts can change the material structure of our brains.

He would study the way thoughts change the brain by using TMS to observe changes in the finger maps of people learning to play the piano. One of Pascual-Leone's heroes, the great Spanish neuroanatomist and Nobel laureate Santiago Ramón y Cajal, who spent his later life looking in vain for brain plasticity, proposed in 1894 that the "organ of thought is, within certain limits, malleable, and perfectible by well-directed mental exercise." In 1904 he argued that thoughts, repeated in "mental practice," must strengthen the existing neuronal connections and create new ones. He also had the intuition that this process would be particularly pronounced in neurons that control the fingers in pianists, who do so much mental practice.

Ramón y Cajal, using his imagination, had painted a picture of a plastic brain but lacked the tools to prove it. Pascual-Leone now thought he had a tool in TMS to test whether mental practice and imagination in fact lead to physical changes.

The details of the imagining experiment were simple and picked up Cajal's idea to use the piano. Pascual-Leone taught two groups of people, who had never studied piano, a sequence of notes, showing them which fingers to move and letting them hear the notes as they were played. Then members of one group, the "mental practice" group, sat in front of an electric piano keyboard, two hours a day, for five days, and *imagined* both playing the sequence and hearing it played. A second "physical practice" group actually played the music two hours a day for five days. Both groups had their brains mapped before the experiment, each day during it, and afterward. Then both groups were asked to play the sequence, and a computer measured the accuracy of their performances.

Pascual-Leone found that both groups learned to play the sequence, and both showed similar brain map changes. Remarkably, mental practice alone produced the same physical changes in the motor system as actually playing the piece. By the end of the fifth day, the changes in motor signals to the muscles were the same in both groups, and the imagining players were as accurate as the actual players were on their third day.

The level of improvement at five days in the mental practice group, however substantial, was not as great as in those who did physical practice. But when the mental practice group finished its mental training and was given a single two-hour physical practice session, its overall performance improved to the level of the physical practice group's performance at five days. Clearly mental practice is an effective way to prepare for learning a physical skill with minimal physical practice.

We all do what scientists call mental practice or mental rehearsing when we memorize answers for a test, learn lines for a play, or rehearse any kind of performance or presentation. But because few of us do it systematically, we underestimate its effectiveness. Some athletes and musicians use it to prepare for performances, and toward the end of his career the concert pianist Glenn Gould relied largely on mental practice when preparing himself to record a piece of music.

One of the most advanced forms of mental practice is "mental chess," played without a board or pieces. The players imagine the board and the play, keeping track of the positions. Anatoly Sharansky, the Soviet human rights activist, used mental chess to survive in prison. Sharansky, a Jewish computer specialist falsely accused of spying for the United States in 1977, spent nine years in prison, four hundred days of that time in solitary confinement in freezing, darkened five-by-six-foot punishment cells. Political prisoners in isolation often fall apart mentally because the use-it-or-lose-it brain needs external stimulation to maintain its maps. During this extended period of sensory deprivation, Sharansky played mental chess for months on end, which probably helped him keep his brain from degrading. He played both white and black, holding the game in his head, from opposite perspectives—an extraordinary challenge to the brain. Sharansky once told me, half joking, that he kept at chess thinking he might as well use the opportunity to become the world champion. After he was released, with the help of Western pressure, he went to Israel and became a cabinet minister. When the world champion Garry Kasparov played against the prime minister and leaders of the cabinet, he beat all of them except Sharansky.

We know from brain scans of people who use massive amounts of mental practice what was probably happening in Sharansky's brain while he was in prison. Consider the case of Rüdiger Gamm, a young German man of normal intelligence who turned himself into a mathematical phenomenon, a human calculator. Though Gamm was not born with exceptional mathematical ability, he can now calculate the ninth power or the fifth root of numbers and solve such problems as "What is 68 times 76?" in five seconds. Beginning at age twenty, Gamm, who worked in a bank, began doing four hours of computational practice a day. By the time he was twenty-six, he had become a calculating genius, able to make his living by performing on television. Investigators who examined him with a positron emission tomography (PET) brain scan while he was calculating found he was able to recruit five more brain areas for calculating than "normal" people. The psychologist Anders Ericsson, an expert in the development of expertise, has shown that people like Gamm rely on long-term memory to help them solve mathematical problems when others rely on short-term memory. Experts don't store the answers, but they do store key facts and strategies that help them get answers, and they have immediate access to them, as though they were in short-term memory. This use of long-term memory for problem solving is typical of experts in most fields, and Ericsson found that becoming an expert in most fields usually takes about a decade of concentrated effort.

One reason we can change our brains simply by imagining is that, from a neuroscientific point of view, imagining an act and doing it are not as different as they sound. When people close their eyes and visualize a simple object, such as the letter *a*, the primary visual cortex lights up, just as it would if the subjects were actually looking at the letter *a*. Brain scans show that in action and imagination many of the same parts of the brain are activated. That is why visualizing can improve performance.

In an experiment that is as hard to believe as it is simple, Drs. Guang Yue and Kelly Cole showed that imagining one is using one's muscles actually strengthens them. The study looked at two groups, one that did physical exercise and one that imagined doing exercise. Both groups exercised a finger muscle, Monday through Friday, for four weeks. The physical group did trials of fifteen maximal contractions, with a twenty-second rest between each. The mental group merely imagined doing fifteen maximal contractions, with a twenty-second rest between each, while also imagining a voice shouting at them, "Harder! Harder! Harder!"

At the end of the study the subjects who had done physical exercise increased their muscular strength by 30 percent, as one might expect. Those who only *imagined* doing the exercise, for the same period, increased their muscle strength by 22 percent. The explanation lies in the motor neurons of the brain that "program" movements. During these imaginary contractions, the neurons responsible for stringing together sequences of instructions for movements are activated and strengthened, resulting in increased strength when the muscles are contracted.

This research has led to the development of the first machines that actually "read" people's thoughts. Thought translation machines tap into motor programs in a person or animal imagining an act, decode the distinctive electrical signature of the thought, and broadcast an electrical command to a device that puts the thought into action. These machines work because the brain is plastic and physically changes its state and structure as we think, in ways that can be tracked by electronic measurements.

These devices are currently being developed to permit people who are completely paralyzed to move objects with their thoughts.

As the machines become more sophisticated, they may be developed into thought readers, which recognize and translate the content of a thought, and have the potential to be far more probing than lie detectors, which can only detect stress levels when a person is lying.

These machines were developed in a few simple steps. In the mid-1990s, at Duke University, Miguel Nicolelis and John Chapin began a behavioral experiment, with the goal of learning to read an animal's thoughts. They trained a rat to press a bar, electronically attached to a water-releasing mechanism. Each time the rat pressed the bar, the mechanism released a drop of water for the rat to drink. The rat had a small part of its skull removed, and a small group of microelectrodes were attached to its motor cortex. These electrodes recorded the activity of forty-six neurons in the motor cortex involved in planning and programming movements, neurons that normally send instructions down the spinal cord to the muscles. Since the goal of the experiment was to register thoughts, which are complex, the forty-six neurons had to be measured simultaneously. Each time the rat moved the bar, Nicolelis and Chapin recorded the firing of its forty-six motor-programming neurons, and the signals were sent to a small computer. Soon the computer "recognized" the firing pattern for bar pressing.

After the rat became used to pressing the bar, Nicolelis and Chapin disconnected the bar from the water release. Now when the rat pressed the bar, no water came. Frustrated, it pressed the bar a number of times, but to no avail. Next the researchers connected the water release to the computer that was connected to the rat's neurons. In theory, now, each time the rat had the thought "press the bar," the computer would recognize the neuronal firing pattern and send a signal to the water release to dispense a drop.

After a few hours, the rat realized it didn't have to touch the bar to get water. All it had to do was to imagine its paw pressing the bar, and water would come! Nicolelis and Chapin trained four rats to perform this task.

Then they began to teach monkeys to do even more complex thought translations. Belle, an owl monkey, was trained to use a joystick to follow a light as it moved across a video screen. If she succeeded, she got a drop of fruit juice. Each time she moved the joystick, her neurons fired, and the pattern was mathematically analyzed by a computer. The pattern of neuronal firing always occurred 300 milliseconds before Belle actually moved the joystick, because it took that long for her brain to send the command down her spinal cord to her muscles. When she moved it to the right, a "move your arm right" pattern occurred in her brain, and the computer detected it; when she moved her arm to the left, the computer detected that pattern. Then the computer converted these mathematical patterns into commands that were sent to a robotic arm, out of Belle's view. The mathematical patterns were also transmitted from Duke University to a second robotic arm in a lab in Cambridge, Massachusetts. Again, as in the rat experiment, there was no connection between the joystick and the robotic arms; the robotic arms connected to the computer, which read the patterns in Belle's neurons. The hope was that the robotic arms at Duke and in Cambridge would move exactly when Belle's arm did, 300 milliseconds after her thought.

As the scientists randomly changed the light patterns on the computer screen and Belle's real arm moved the joystick, so did the robotic arms, six hundred miles apart, powered only by her thoughts transmitted by computer.

The team has since taught a number of monkeys to use only their thoughts to move a robotic arm in any direction in three-dimensional space, in order to perform complex movements—such as reaching and grasping for objects. The monkeys also play video games (and seem to enjoy them) using only their thoughts to move a cursor on a video screen and zap a moving target.

Nicolelis and Chapin hoped their work would help patients with various kinds of paralysis. That happened in July 2006, when a team led by neuroscientist John Donoghue, from Brown University, used a similar technique with a human being. The twenty-five-year-old man, Matthew Nagle, had been stabbed in the neck and paralyzed in all four limbs by the resulting spinal cord injury. A tiny, painless silicone chip with a hundred electrodes was implanted in his brain and attached to a computer. After four days of practice he was able to move a computer cursor on a screen, open e-mail, adjust the channel and volume control on a television, play a computer game, and control a robotic arm using his thoughts. Patients with muscular dystrophy, strokes, and motor-neuron disease are scheduled to try the thought-translation device next. The goal in these approaches is ultimately to implant a small microelectrode array, with batteries and a transmitter the size of a baby's fingernail, in the motor cortex. A small computer could be connected either to a robotic arm or wirelessly to a wheelchair control or to electrodes implanted in muscles to trigger movements. Some scientists hope to develop a technology less invasive than microelectrodes to detect neuronal firing—possibly a variant of TMS, or a device Taub and colleagues are developing to detect changes in brain waves.

What these "imaginary" experiments show is how truly integrated imagination and action are, despite the fact that we tend to think of imagination and action as completely different and subject to different rules. But consider this: in some cases, the faster you can imagine something, the faster you can do it. Jean Decety of Lyon, France, has done different versions of a simple experiment. When you time how long it takes to imagine writing your name with your "good hand," and then actually write it, the times will be similar. When you imagine writing your name with your nondominant hand, it will take longer both to imagine it and to write it. Most people who are right-handed find that their "mental left hand" is slower than their "mental right hand." In studies of patients with stroke or Parkinson's disease (which causes people's movements to slow),

Decety observed that patients took longer to imagine moving the affected limb than the unaffected one. Both mental imagery and actions are thought to be slowed because they both are products of the *same* motor program in the brain. The speed with which we imagine is probably constrained by the neuronal firing rate of our motor programs.

Pascual-Leone has profound observations about how neuroplasticity, which promotes change, can also lead to rigidity and repetition in the brain, and these insights help solve this paradox: if our brains are so plastic and changeable, why do we so often get stuck in rigid repetition? The answer lies in understanding, first, how remarkably plastic the brain is.

Plasticina, he tells me, is the musical Spanish word for "plasticity," and it captures something the English word does not. *Plasticina,* in Spanish, is also the word for "Play-Doh" or "plasticine" and describes a substance that is fundamentally impressionable. For him our brains are so plastic that even when we do the same behavior day after day, the neuronal connections responsible are slightly different each time because of what we have done in the intervening time.

"I imagine," Pascual-Leone says, "that the brain activity is like Play-Doh one is playing with all the time." Everything that we do shapes this chunk of Play-Doh. But, he adds, "if you start out with a package of Play-Doh that is a square, and you then make a ball of it, it is possible to get back to a square. But it won't be the *same* square as you had to begin with." Outcomes that appear similar are not identical. The molecules in the new square are arranged differently than in the old one. In other words, similar behaviors, performed at different times, use different circuits. For him, even when a patient with a neurological or psychological problem is "cured," that cure never returns the patient's brain to its preexisting state.

"The system is plastic, not elastic," Pascual-Leone says in a booming voice. An elastic band can be stretched, but it always reverts to its former shape, and the molecules are not rearranged in the process. The plastic brain is perpetually altered by every encounter, every interaction.

So the question becomes, if the brain is so easily altered, how are we protected from endless change? Indeed, if the brain is like Play-Doh, how is it that we remain ourselves? Our genes help give us consistency, up to a point, and so does repetition.

Pascual-Leone explains this with a metaphor. The plastic brain is like a snowy hill in winter. Aspects of that hill—the slope, the rocks, the consistency of the snow—are, like our genes, a given. When we slide down on a sled, we can steer it and will end up at the bottom of the hill by following a path determined both by how we steer and the characteristics of the hill. Where exactly we will end up is hard to predict because there are so many factors in play.

"But," Pascual-Leone says, "what will definitely happen the *second time* you take the slope down is that you will more likely than not find yourself somewhere or another that is related to the path you took the first time. It won't be exactly that path, but it will be closer to that one than any other. And if you spend your entire afternoon sledding down, walking up, sledding down, at the end you will have some paths that have been used a lot, some that have been used very little . . . and there will be tracks that you have created, and it is very difficult now to get out of those tracks. And those tracks are not genetically determined anymore."

The mental "tracks" that get laid down can lead to habits, good or bad. If we develop poor posture, it becomes hard to correct. If we develop good habits, they too become solidified. Is it possible, once "tracks" or neural pathways have been laid down, to get out of those paths and onto different ones? Yes, according to Pascual-Leone, but it is difficult because, once we have created these tracks, they become "really speedy" and very efficient at guiding the sled down the hill. To take a different path becomes increasingly difficult. A roadblock of some kind is necessary to help us change direction.

In his next experiment Pascual-Leone developed the use of roadblocks and showed that alterations of established pathways and massive plastic reorganizations can occur at unexpected speed.

His work using roadblocks began when he heard about an unusual boarding school in Spain where teachers who instructed the blind went to study darkness. They were blindfolded for a week to experience blindness firsthand. A blindfold is a roadblock for the sense of sight, and within the week their tactile senses and their ability to judge space had become extremely sensitive. They were able to differentiate makes of motorcycles by the sounds of their engines and to distinguish objects in their paths by their echoes. When the teachers first removed their blindfolds, they were profoundly disoriented and couldn't judge space or see.

When Pascual-Leone heard about this school of darkness, he thought, "Let's take sighted people and make them *absolutely* blind."

He blindfolded people for five days, then mapped their brains with TMS. He found that when he blocked out all light—the road "block" had to be impermeable—the subjects' "visual" cortices began to process the sense of touch coming from their hands, like blind patients learning Braille. What was most astounding, however, was that the brain reorganized itself in just a few days. With brain scans Pascual-Leone showed that it could take as few as two days for the "visual" cortex to begin processing tactile and auditory signals. (As well, many of the blindfolded subjects reported that as they moved, or were touched, or heard sounds, they began having *visual* hallucinations of beautiful, complex scenes of cities, skies, sunsets, Lilliputian figures, cartoon figures.) Absolute darkness was essential to the change because vision is so powerful a sense that if any light got in, the visual cortex preferred to process it over sound and touch. Pascual-Leone discovered, as Taub had, that to develop a new pathway, you have to block or constrain its competitor, which is often the most commonly used pathway. After the blindfolds came off, the subjects' visual cortices stopped responding to tactile or auditory stimulation within twelve to twenty-four hours.

The *speed* with which the visual cortex switched to processing sound and touch posed a major question for Pascual-Leone. He believed there wasn't enough time in two days for the brain to re-wire itself so radically. When nerves are placed in a growth culture, they grow at most a millimeter a day. The "visual" cortex could have begun processing the other senses so quickly only if connections to those sources already existed. Pascual-Leone, working with Roy Hamilton, took the idea that preexisting paths were unmasked and pushed it one step further to propose a theory that the kind of radical brain reorganization seen in the school of darkness is not the exception but the rule. The human brain can reorganize so quickly because individual parts of the brain are not necessarily committed to processing particular senses. We can, and routinely do, use parts of our brains for many different tasks.

As we've seen, almost all current theories of the brain are localizationist and assume that the sensory cortex processes each sense—sight, sound, touch—in locations devoted to processing them alone. The phrase "visual cortex" assumes that the sole *purpose* of that area of the brain is to process vision, just as the phrases "auditory cortex" and "somatosensory cortex" assume a single purpose in other areas.

But, says Pascual-Leone, "our brains are not truly organized in terms of systems that process a given sensory modality. Rather, our brain is organized in a series of specific operators."

An operator is a processor in the brain that, instead of processing input from a single sense, such as vision, touch, or hearing, processes more abstract information. One operator processes information about *spatial relationships,* another *movement,* and another *shapes.* Spatial relationships, movement, and shapes are information that is processed by several of our senses. We can both feel and see spatial differences—how wide a person's hand is—as we can both feel and see movement and shapes. A few operators may be good for only a single sense (e.g., the color operator), but spatial, movement, and shape operators process signals from more than one.

An operator is selected by competition. The operator theory appears to draw on the theory of neuronal group selection developed in 1987 by Nobel Prize winner Gerald Edelman, who proposed that for any brain activity, the ablest group of neurons is selected to do the task. There is an almost Darwinian competition—a neural Darwinism, to use Gerald Edelman's phrase—going on all the time between operators to see which ones can most effectively process signals from a particular sense and in a particular circumstance.

This theory provides an elegant bridge between the localizationist emphasis on things tending to happen in certain typical locations, and the neuroplasticians' emphasis on the brain's ability to restructure itself.

What it implies is that people learning a new skill can recruit operators devoted to other activities, vastly increasing their processing power, provided they can create a roadblock between the operator they need and its usual function.

Someone presented with an overwhelming auditory task, such as memorizing Homer's *Iliad,* might blindfold himself to recruit operators usually devoted to sight, since the vast operators in the visual cortex can process sound. In Homer's time, long poems were composed and passed from generation to generation in oral form. (Homer, according to tradition, was himself blind.) Memorization was essential in preliterate cultures; indeed, illiteracy may have prompted people's brains to assign more operators to auditory tasks. Yet such feats of oral memory are possible in literate cultures if there is sufficient motivation. For centuries Yemenite Jews taught their children to memorize the entire Torah, and today children in Iran memorize the entire Koran.

We have seen that imagining an act engages the same motor and sensory programs that are involved in doing it. We have long viewed our imaginative life with a kind of sacred awe: as noble, pure, immaterial, and ethereal, cut off from our material brain. Now we cannot be so sure about where to draw the line between them.

Everything your "immaterial" mind imagines leaves material traces. Each thought alters the physical state of your brain synapses at a microscopic level. Each time you imagine moving your fingers across the keys to play the piano, you alter the tendrils in your living brain.

These experiments are not only delightful and intriguing, they also overturn the centuries of confusion that have grown out of the work of the French philosopher René Descartes, who argued that mind and brain are made of different substances and are governed by different laws. The brain, he claimed, was a physical, material thing, existing in space and obeying the laws of physics. The mind (or the soul, as Descartes called it) was immaterial, a thinking thing that did not take up space or obey physical laws. Thoughts, he argued, were governed by the rules of reasoning, judgment, and desires, not by the physical laws of cause and effect. Human beings consisted of this duality, this marriage of immaterial mind and material brain.

But Descartes—whose mind/body division has dominated science for four hundred years—could never credibly explain how the immaterial mind could influence the material brain. As a result, people began to doubt that an immaterial thought, or mere imagining, might change the structure of the material brain. Descartes's view seemed to open an unbridgeable gap between mind and brain.

His noble attempt to rescue the brain from the mysticism that surrounded it in his time, by making it mechanical, failed. Instead the brain came to be seen as an inert, inanimate machine that could be moved to action only by the immaterial, ghostlike soul Descartes placed within it, which came to be called "the ghost in the machine."

By depicting a mechanistic brain, Descartes drained the life out of it and slowed the acceptance of brain plasticity more than any other thinker. Any plasticity—any ability to change that we had—existed in the mind, with its changing thoughts, not in the brain.

But now we can see that our "immaterial" thoughts too have a physical signature, and we cannot be so sure that thought won't someday be explained in physical terms. While we have yet to understand exactly *how* thoughts actually change brain structure, it is now clear that they do, and the firm line that Descartes drew between mind and brain is increasingly a dotted line.

9

Turning Our Ghosts into Ancestors

Psychoanalysis as a Neuroplastic Therapy

Mr. L. had been suffering from recurring depressions for over forty years and had had difficulties in his relationships with women. He was in his late fifties and had recently retired when he sought help from me.

Few psychiatrists at the time, in the early 1990s, had any sense that the brain was plastic, and it was often thought that people approaching sixty were "too fixed in their ways" to benefit from a treatment that aimed not only to rid them of symptoms but to alter long-standing aspects of their character.

Mr. L. was always formal and polite. He was intelligent, subtle, and spoke in a clipped, spare way, without much music in his voice. He became increasingly remote when he spoke about his feelings.

In addition to his deep depressions, which had responded only partially to antidepressants, he suffered from a second strange mood state. Often he was struck—seemingly out of the blue—by a mysterious sense of paralysis, feeling numb and purposeless, as though time had stopped. He also reported that he drank too much.

He was particularly upset about his relations with women. As soon as he got romantically involved, he would start backing off, feeling that "there is a better woman elsewhere whom I'm being denied." He had been unfaithful to his wife on a number of occasions and as a result lost his marriage, an outcome he greatly regretted. Worse, he wasn't sure why he was unfaithful, because he had a lot of respect for his wife. He tried many times to get back with her, but she refused.

He was uncertain what love was, had never felt jealous or possessive of others, and always felt women wanted to "own" him. He avoided both commitment to and conflict with women. He was devoted to his children but felt attached through a sense of duty rather than joyful affection. This feeling pained him, because they were doting and affectionate to him.

When Mr. L. was twenty-six months old, his mother died giving birth to his younger sister. He didn't believe that her death had affected him significantly. He had seven siblings, and now their sole provider was their father, a farmer, who ran the isolated farm on which they lived without electricity or running water, in a destitute county during the Great Depression. A year later Mr. L. became chronically ill with a gastrointestinal problem that needed continual attention. When he was four, his father, unable to care for both him and his siblings, sent him to live with a married but childless aunt and her husband a thousand miles away. In two years everything in Mr. L.'s short life had changed. He had lost his mother, his father, his siblings, his health, his home, his village, and all his familiar physical surroundings—everything he cared about and had been attached to.

And because he grew up among people used to enduring hard times and to keeping a stiff upper lip, neither his father nor his adoptive family talked much about his losses with him.

Mr. L. said he had no memories from age four or earlier and very few from his teenage years. He felt no sadness about what had happened to him and never cried, even as an adult—about anything. Indeed, he spoke as though nothing that had happened to him had registered. Why should it? he asked. Aren't children's minds too poorly formed to record such early events?

Yet there were clues that his losses did register. As he told his story, he looked, all those years later, as though he were still in shock. He was also haunted by dreams in which he was always searching for something. As Freud discovered, recurring dreams, with a relatively unchanging structure, often contain memory fragments of early traumas.

Mr. L. described a typical dream as follows:

> I am searching for something, I know not what, an unidentified object, maybe a toy, which is beyond familiar territory . . . I'd like it back again.

His only comment was that the dream represented "a terrible loss." But remarkably, he did not link it to the loss of his mother or family.

Through understanding this dream Mr. L. would learn to love, change important aspects of his character, and rid himself of forty years of symptoms, in an analysis that lasted from age fifty-eight to sixty-two. This change was possible because psychoanalysis is in fact a neuroplastic therapy.

For years now it has been fashionable in some quarters to argue that psychoanalysis, the original "talking cure," and other psychotherapies are not serious ways to treat psychiatric symptoms and character problems. "Serious" treatments require drugs, not just "talking about thoughts and feelings," which could not possibly affect the brain or alter character, which was increasingly thought to be a product of our genes.

It was the work of the psychiatrist and researcher Eric Kandel that first interested me in neuroplasticity when I was a resident at the Columbia University Department of Psychiatry, where he taught and was a major influence on everyone present. Kandel was the first to show that as we learn, our individual neurons alter their structure and strengthen the synaptic connections between them. He was also first to demonstrate that when we form long-term memories, neurons change their anatomical shape and increase the number of synaptic connections they have to other neurons—work for which he won the Nobel Prize in 2000.

Kandel became both a physician and a psychiatrist, hoping to practice psychoanalysis. But several psychoanalyst friends urged him to study the brain, learning, and memory, something about which very little was known, in order to deepen the understanding of why psychotherapy is effective and how it might be improved. After some early discoveries Kandel decided to become a full-time laboratory scientist, but he never lost interest in how the mind and brain change in psychoanalysis.

He began to study a giant marine snail, called *Aplysia*, whose unusually large neurons—its cells are a millimeter wide and visible to the naked eye—might provide a window into how human nervous tissue functions. Evolution is conservative, and elementary forms of learning function the same way both in animals with simple nervous systems and in humans.

Kandel's hope was to "trap" a learned response in the smallest possible group of neurons he could find, and study it. He found a simple circuit in the snail, which he could partially remove from the animal by dissection and keep alive and intact in sea water. In this way he could study it, while it was alive, and while it learned.

A sea snail's simple nervous system has sensory cells that detect danger and send signals to its motor neurons, which act reflexively to protect it. Sea snails breathe by exposing their gills, which are covered by a fleshy tissue called a siphon. If the sensory neurons in the siphon detect an unfamiliar stimulus or danger, they send a message to six motor neurons that fire, causing the muscles around the gill to pull both siphon and gill safely back into the snail, where they are protected. This is the circuit that Kandel studied by inserting microelectrodes into the neurons.

He was able to show that as the snail learned to avoid shocks and withdraw its gill, its nervous system changed, enhancing the synaptic connections between its sensory and motor neurons and giving off more powerful signals detected by the microelectrodes. This was the first proof that learning led to neuroplastic strengthening of the connections between neurons.

If he repeated the shocks in a short period, the snails became "sensitized," so that they developed "learned fear" and a tendency to overreact even to more benign stimuli, as do humans who develop anxiety disorders. When the snails developed learned fear, the presynaptic neurons released more of the chemical messenger into the synapse, giving off a more powerful signal. Then he showed that the snails could be taught to recognize a stimulus as harmless. When the snail's siphon was touched gently over and over and not followed with a shock, the synapses leading to the withdrawal reflex weakened, and the snail eventually ignored the touch. Finally Kandel was able to show that snails can also learn to associate two different events and that their nervous systems change in the process. When he gave the snail a benign stimulus, followed immediately by a shock to the tail, the snail's sensory neuron soon responded to the benign stimulus as though it were dangerous, giving off very strong signals—even if not followed by the shock.

Kandel, working with Tom Carew, a physiological psychologist, next showed that the snails could develop both short- and long-term memories. In one experiment the team trained a snail to withdraw its gill after they'd touched it ten times. The changes in the neurons remained for several minutes—the equivalent of a short-term memory. When they touched the gill ten times, in four different training sessions, separated from one another by several hours to one day, the changes in the neurons lasted as long as three weeks. The animals developed primitive long-term memories.

Kandel next worked with his colleague the molecular biologist James Schwartz and geneticists to better understand the individual *molecules* that are involved in forming long-term memories in the snails. They showed that in the snails, for short-term memories to become long-term, a new protein had to be made in the cell. The team showed that a short-term memory becomes long-term when a chemical in the neuron, called protein kinase A, moves from the body of the neuron into its nucleus, where genes are stored. The protein turns on a gene to make a protein that alters the structure of the nerve ending, so that it grows new connections between the neurons. Then Kandel, Carew, and colleagues Mary Chen and Craig Bailey showed that when a single neuron develops a long-term memory for sensitization, it might go from having 1,300 to 2,700 synaptic connections, a staggering amount of neuroplastic change.

The same process occurs in humans. When we learn, we alter which genes in our neurons are "expressed," or turned on.

Our genes have two functions. The first, the "template function," allows our genes to replicate, making copies of themselves that are passed from generation to generation. The template function is beyond our control.

The second is the "transcription function." Each cell in our body contains all our genes, but not all those genes are turned on, or expressed. When a gene is turned on, it makes a new protein that alters the structure and function of the cell. This is called the transcription function because when the gene is turned on, information about how to make these proteins is "transcribed" or read from the individual gene. This transcription function is influenced by what we do and think.

Most people assume that our genes shape us—our behavior and our brain anatomy. Kandel's work shows that when we learn our minds also affect which genes in our neurons are transcribed. Thus we can shape our genes, which in turn shape our brain's microscopic anatomy.

Kandel argues that when psychotherapy changes people, "it presumably does so through learning, by producing changes in gene expression that alter the strength of synaptic connections, and structural changes that alter the anatomical pattern of interconnections between nerve cells of the brain." Psychotherapy works by going deep into the brain and its neurons and changing their structure by turning on the right genes. Psychiatrist Dr. Susan Vaughan has argued that the talking cure works by "talking to neurons," and that an effective psychotherapist or psychoanalyst is a "microsurgeon of the mind" who helps patients make needed alterations in neuronal networks.

These discoveries about learning and memory at the molecular level have their roots in Kandel's own history.

Kandel was born in 1929 in Vienna, a city of great cultural and intellectual richness. But Kandel was a Jew, and Austria at the time was a virulently anti-Semitic country. In March 1938, when Hitler rolled into Vienna, annexing Austria to the German Reich, he was welcomed by adoring crowds, and the Catholic archbishop of Vienna ordered all the churches to fly the Nazi flag. The next day all Kandel's classmates—except for one girl, the only other Jew in the class—stopped talking to him and began to bully him. By April all Jewish children had been expelled from the school.

On November 9, 1938—Kristallnacht, the "night of broken glass," when the Nazis destroyed all the synagogues in the German Reich, including Austria—they arrested Kandel's father. Austrian Jews were evicted from their homes, and thirty thousand Jewish men were sent to concentration camps the following day.

Kandel wrote, "I remember Kristallnacht even today, more than sixty years later, almost as if it were yesterday. It fell two days after my ninth birthday, on which I was showered with toys from my father's shop. When we returned to our apartment a week or so after having been evicted, everything of value was gone, including my toys . . . It is probably futile, even for someone trained in psychoanalytic thinking as I am, to attempt to trace the complex interests and actions of my later life to a few selected experiences of my youth. Nevertheless I cannot help but think that the experiences of my last year in Vienna helped to determine my later interests in the mind, in how people behave, the unpredictability of motivation, and the persistence of memory . . . I am struck, as others have been, at how deeply these traumatic events of my childhood became burned into memory." He was drawn to psychoanalysis, because he believed it "outlined by far the most coherent, interesting and nuanced view of the human mind" and, of all the psychologies, had the most comprehensive understanding of the contradictions of human behavior, of how civilized societies could suddenly release "such great viciousness in so many people," and of how a country as seemingly civilized as Austria could become "so radically dissociated."

Psychoanalysis (or "analysis") is a treatment that helps people who are deeply troubled not only by symptoms but by aspects of their own character. These problems occur when we have powerful internal conflicts in which, as Kandel says, parts of ourselves become radically "dissociated," or cut off from the rest of us.

Whereas Kandel's career took him from the clinic to the neuroscience laboratory, Sigmund Freud began his career as a laboratory neuroscientist, but because he was too impoverished to continue, he went in the opposite direction and became a neurologist in private practice, in order to have a sufficient income to support a family. One of his first endeavors was to fuse what he had learned about the brain as a neuroscientist with what he was learning about the mind while treating patients. As a neurologist, Freud quickly became disenchanted with the localizationism of the time, which was based on the work of Broca and others, and realized that the notion of the hardwired brain did not adequately explain how complex, culturally acquired mental activities such as reading and writing are possible. In 1891 he wrote a book titled *On Aphasia,* which showed the flaws in the existing evidence for "one function, one location," and proposed that complex mental phenomena such as reading and writing are not restricted to distinct cortical areas, and that it made no sense to argue, as localizationists had, that there is a brain "center" for literacy, since literacy is not innate. Rather, the brain in the course of our individual lives must dynamically reorganize itself, and its wiring, to perform such culturally acquired functions.

In 1895 Freud completed the "Project for a Scientific Psychology," one of the first comprehensive neuroscientific models to integrate brain and mind, still admired for its sophistication. Here Freud proposed the "synapse," several years before Sir Charles Sherrington, who bears the credit. In the "Project" Freud even gave a description of how synapses, which he called "contact barriers," might be changed by what we learn, anticipating Kandel's work. He also began proposing neuroplastic ideas.

The first plastic concept Freud developed is the law that neurons that fire together wire together, usually called Hebb's law, though Freud proposed it in 1888, sixty years before Hebb. Freud stated that when two neurons fire *simultaneously,* this firing facilitates their ongoing *association.* Freud emphasized that what linked neurons was their firing together *in time,* and he called this phenomenon the law of association by simultaneity. The law of association explains the importance of Freud's idea of "free association," in which psychoanalytic patients lie on the couch and "free-associate," or say everything that comes into their minds, regardless of how uncomfortable or trivial it seems. The analyst sits behind the patient, out of the patient's sight, and usually says little. Freud found that if he didn't interfere, many warded-off feelings and interesting connections emerged in the patient's associations—thoughts and feelings the patient normally pushed away. Free association is based on the understanding that all our mental associations, even seemingly "random" ones that appear to make no sense, are expressions of links formed in our memory networks. His law of association by simultaneity implicitly links changes in neuronal networks with changes in our memory networks, so that neurons that fired together years before wired together, and these original connections are often still in place and show up in a patient's free associations.

Freud's second plastic idea was that of the psychological critical period and the related idea of sexual plasticity. As we saw in chapter 4, "Acquiring Tastes and Loves," Freud was the first to argue that human sexuality and the ability to love have critical periods in early childhood that he called "phases of organization." What happens during these critical periods has an inordinate effect on our ability to love and relate later in life. If something goes awry, it is possible to make changes later in life, but plastic change is much harder to achieve after a critical period closes.

Freud's third idea was a plastic view of memory. The idea Freud inherited from his teachers was that events we experience can leave *permanent memory traces* in our minds. But when he started working with patients, he observed that memories are not written down once, or "engraved," to remain unchanged forever but can be altered by subsequent events and *retranscribed.* Freud observed that events could take on an altered meaning for patients years after they occurred, and that patients then altered their memories of those events. Children who were molested when very young and unable to understand what was being done to them were not always upset at the time, and their initial memories were not always negative. But once they matured sexually, they looked upon the incident anew and gave it new meaning, and their memory of the molestation changed. In 1896 Freud wrote that from time to time memory traces are subjected to "a *rearrangement* in accordance with fresh circumstances—to a *retranscription.* Thus what is essentially new about my theory is the thesis that memory is present not once but several times over." Memories are constantly remodeled, "analogous in every way to the process by which a nation constructs legends about its early history." To be changed, Freud argued, memories had to be conscious and become the focus of our conscious attention, as neuroscientists have since shown. Unfortunately, as was the case with Mr. L., certain traumatic memories of events that happened early in childhood are not easily accessible to consciousness, so they don't change.

Freud's fourth neuroplastic idea helped explain how it might be possible to make unconscious traumatic memories conscious and retranscribe them. He observed that in the mild sensory deprivation created by his sitting out of the patients' view, and commenting only when he had insights into their problems, patients began to regard him as they had important people in their past, usually their parents, especially in their critical psychological periods. It was as though the patients were reliving past memories without being aware of it. Freud called this unconscious phenomenon "transference" because patients were transferring scenes and ways of perceiving from the past onto the present. They were "reliving" them instead of "remembering" them. An analyst who is out of view and says little becomes a blank screen on which the patient begins to project his transference. Freud discovered that patients projected these "transferences" not only onto him but onto other people in their lives, without being aware of doing so, and that viewing others in a distorted way often got them into difficulty. Helping patients understand their transferences allowed them to improve their relationships. Most important, Freud discovered that transferences of early traumatic scenes could often be altered if he pointed out to the patient what was happening when the transference was activated and the patient was paying close attention. Thus, the underlying neuronal networks, and the associated memories, could be retranscribed and changed.

At twenty-six months, the age at which Mr. L. lost his mother, a child's plastic change is at its height: new brain systems are forming and strengthening neural connections, and maps are differentiating and completing their basic structure with the help of stimulation from and interaction with the world. The right hemisphere has just completed a growth spurt, and the left hemisphere is beginning a spurt of its own.

The right hemisphere generally processes nonverbal communication; it allows us to recognize faces and read facial expressions, and it connects us to other people. It thus processes the nonverbal visual cues exchanged between a mother and her baby. It also processes the musical component of speech, or tone, by which we convey emotion. During the right hemisphere's growth spurt, from birth until the second year, these functions undergo critical periods.

The left hemisphere generally processes the *verbal-linguistic* elements of speech, as opposed to the emotional-musical ones, and analyzes problems using *conscious* processing. Babies have a larger right hemisphere, up to the end of the second year, and because the left hemisphere is only beginning its growth spurt, our right hemisphere dominates the brain for the first three years of our lives. Twenty-six-month-olds are complex, "right-brained" emotional creatures but cannot talk about their experiences, a left-brain function. Brain scans show that during the first two years of life, the mother principally communicates nonverbally with her right hemisphere to reach her infant's right hemisphere.

A particularly important critical period lasts from approximately ten or twelve months to sixteen or eighteen months, during which a key area of the right frontal lobe is developing and shaping the brain circuits that will allow infants both to maintain human attachments and to regulate their emotions. This maturing area, the part of the brain behind our right eye, is called the *right orbitofrontal system.* (The orbitofrontal system has its central area in the orbitalfrontal cortex, which was discussed in chapter 6, "Brain Lock Unlocked," but the "system" includes links to the limbic system, which processes emotion.) This system allows us both to read people's facial expressions, and hence their emotions, and also to understand and control our own emotions. Little L. at twenty-six months would have finished orbitofrontal development but would not have had the opportunity to reinforce it.

A mother who is with her baby during the critical period for emotional development and attachment is constantly teaching her child what emotions are by using musical speech and nonverbal gestures. When she looks at her child who swallowed some air with her milk, she might say, "There, there, honey, you look so upset, don't be frightened, your tummy hurts because you ate too fast. Let Mommy burp you, and give you a hug, and you'll feel all right." She is telling the child the *name of the emotion* (fright), that it has a *trigger* (she ate too fast), that the emotion is *communicated by facial expression* ("you look so upset"), that it is associated with a *bodily sensation* (a tummy cramp), and that *turning to others for relief is often helpful* ("Let Mommy burp you and give you a hug"). That mother has given her child a crash course in the many aspects of emotion conveyed not only with words but with the loving music of her voice and the reassurance of her gestures and touch.

For children to know and regulate their emotions, and be socially connected, they need to experience this kind of interaction many hundreds of times in the critical period and then to have it reinforced later in life.

Mr. L. lost his mother only a few months after he completed development of his orbitofrontal system. So it fell on others, who were themselves grieving and probably less attuned to him than his mother would have been, to help him use and exercise his orbitofrontal system, lest it begin to weaken. The child who loses his mother at this young age is almost always struck two devastating blows: he loses his mother to death and the surviving parent to depression. If others cannot help him soothe himself and regulate his own emotions as his mother did, he learns to "autoregulate" by turning off his emotions. When Mr. L. sought treatment, he still had this tendency to turn emotions off and trouble maintaining attachments.

Long before brain scans of the orbitofrontal cortex were possible, psychoanalysts had observed the characteristics of children deprived of mothering in early critical periods. During World War II René Spitz studied infants reared by their own mothers in prison, comparing them with those reared in a foundling home, where one nurse was responsible for seven infants. The foundling infants stopped developing intellectually, were unable to control their emotions, and instead rocked endlessly back and forth, or made strange hand movements. They also entered "turned-off" states and were indifferent to the world, unresponsive to people who tried to hold and comfort them. In photographs these infants have a haunting, faraway look in their eyes. The turned-off or "paralytic" states occur when children give up all hope of finding their lost parent again. But how could Mr. L., who entered similar states, have registered such early experiences in his memory?

Neuroscientists recognize two major memory systems. Both are plastically altered in psychotherapy.

The well-developed memory system in twenty-six-month-old children is called "procedural" or "implicit" memory. These terms are often used interchangeably by Kandel. Procedural/implicit memory functions when we learn a procedure or group of automatic actions, occurring outside our focused attention, in which words are generally not required. Our nonverbal interactions with people and many of our emotional memories are part of our procedural memory system. As Kandel says, "During the first 2–3 years of life, when an infant's interaction with its mother is particularly important, the infant relies primarily on its procedural memory systems." Procedural memories are generally unconscious. Riding a bike depends on procedural memory, and most people who ride easily would have trouble consciously explaining precisely how they do it. The procedural memory system confirms that we can have unconscious memories, as Freud proposed.

The other form of memory is called "explicit" or "declarative" memory, which is just beginning to develop in the twenty-six-month-old. Explicit memory consciously recollects specific facts, events, and episodes. It is the memory we use when we describe and make explicit what we did on the weekend, and with whom, and for how long. It helps us to organize our memories by time and place. Explicit memory is supported by language and becomes more important once children can talk.

People who have been traumatized in their first three years can be expected to have few if any explicit memories of their traumas. (Mr. L. said he had not a single memory of his first four years.) But procedural/implicit memories for these traumas exist and are commonly *evoked* or triggered when people get into situations that are similar to the trauma. Such memories often seem to come at us "out of the blue" and do not seem to be classified by time, place, and context, the way most explicit memories are. Procedural memories of emotional interactions often get repeated in transference, or in life.

Explicit memory was discovered through observation of the most famous memory case in neuroscience—a young man named H.M., who had had severe epilepsy. To treat it, his doctors cut out a part of his brain the size of the human thumb, the hippocampus. (There are actually two "hippocampi," one in each hemisphere, and both were removed.) After surgery H.M. at first seemed normal. He recognized his family and could conduct conversations. But it was soon apparent that since his operation, he could not learn any new facts. When his doctors visited him, chatted, left, and then returned again, he had no memory whatsoever of the previous meeting. We learn from H.M.'s case that the hippocampus turns our short-term explicit memories into long-term explicit memories for people, places, and things—the memories to which we have conscious access.

Analysis helps patients put their unconscious procedural memories and actions into words and into context, so they can better understand them. In the process they plastically retranscribe these procedural memories, so that they become conscious explicit memories, sometimes for the first time, and patients no longer need to "relive" or "reenact" them, especially if they were traumatic.

Mr. L. took quickly to analysis and free association and started to find, as many patients do, that dreams from the night before often came to mind. Soon he began reporting his recurring dream about searching for an unidentified object, but added new details—the "object" might be a person:

> The lost object may be part of me, maybe not, maybe a toy, possession, or a person. I absolutely must have it. I will know it when I find it. Yet sometimes I am not sure it existed at all, and hence I am unsure anything was lost.

I pointed out to him that a pattern was emerging. He reported not only these dreams but also his depressions and feelings of paralysis after holidays that interrupted our work. At first he didn't believe me, but the depressions and the dreams of loss—possibly of a person—continued to appear at breaks. Then he remembered that interruptions at work also led to mysterious depressions.

The thoughts in his dream of *desperately searching* were associated, in his memory, with *interruptions of his care,* and the neurons encoding these memories were presumably wired together early in his development. But he was no longer consciously aware—if he ever was—of this past link. The "lost toy" in the dream was the clue that his current suffering was colored by childhood losses. But the dream implied that the loss was happening *now*. Past and present were being mixed up together, and a transference was being activated. At this point I, as an analyst, did what an attuned mother does, when she develops the orbitofrontal system, by pointing out emotional "basics"—helping him name his emotions, their triggers, and how they influenced his mental and bodily states. Soon, he was able to spot the triggers and emotions himself.

The interruptions evoked three different types of procedural memories: an anxious state, in which he was pining and searching for his lost mother and family; a depressed state, in which he despaired of finding what he sought; and a paralyzed state, when he turned off and time stood still, probably because he was totally overwhelmed.

By talking about these experiences, he was able, for the first time in his adult life, to connect his desperate searching with its true trigger, the loss of a person, and to realize that his mind and brain still fused the idea of separation with the idea of his mother's death. Making these connections, and also realizing that he was no longer a helpless child, he felt less overwhelmed.

In neuroplastic terms, activating and *paying close attention* to the link between everyday separations and his catastrophic response to them allowed him to unwire the connection and alter the pattern.

As Mr. L. became aware that he was reacting to our short separations as though they were major losses, he had the following dream:

> I'm with a man moving a big wooden box with a weight in it.

When he freely associated to the dream, several thoughts came to mind. The box reminded him of his toy box but also of a coffin. The dream seemed to be saying in symbolic images that he was carrying around the weight of his mother's death. Then the man in the dream said:

> "Look at what you paid for this box." I start disrobing, and my leg is in bad condition, scarred, covered with scabs, and healed with a protuberance that is a dead part of me. I didn't know the price would be this high.

The words "I didn't know the price would be this high" were linked in his mind to a growing realization that he was still influenced by his mother's death. He had been wounded and was still "scarred." Right after articulating that thought, he grew silent and had one of the major epiphanies of his life.

"Whenever I am with a woman," he said, "I soon think that she is not the one for me, and I imagine that some other ideal woman is out there somewhere, waiting." Then, sounding utterly shocked, he said, "I just realized that that other woman seems to be some vague sense of my mother that I had as a child, and it is *she* that I must be faithful to, but whom I never find. The woman I am with becomes my adoptive mother, and loving her is betraying my real mother."

He suddenly realized that his urge to cheat had occurred just as he was getting closer to his wife, threatening his buried tie to his mother. His infidelity was always in the service of a "higher" but unconscious fidelity. This revelation was also the first hint that he had registered some kind of attachment to his mother.

When I next wondered aloud whether he might be experiencing me as the man (in his dream) who pointed out how damaged he felt, Mr. L. burst into tears for the first time in his adult life.

Mr. L. did not get better all at once. He had first to experience cycles of separations, dreams, depressions, and insights—the repetition, or "working through," required for long-term neuroplastic change. New ways of relating had to be learned, wiring new neurons together, and old ways of responding had to be unlearned, weakening neuronal links. Because Mr. L. had linked the ideas of separation and death, they were wired together in his neuronal networks. Now that he was conscious of his association, he could unlearn it.

We all have defense mechanisms, really reaction patterns, that hide unbearably painful ideas, feelings, and memories from conscious awareness. One of these defenses is called dissociation, which keeps threatening ideas or feelings separated from the rest of the psyche. In analysis Mr. L. began to have opportunities to reexperience painful autobiographical memories of searching for his mother that had been frozen in time and dissociated from his conscious memories. Each time he did so he felt more whole as neuronal groups encoding his memories that had been disconnected were connected.

Psychoanalysts since Freud have noted that some patients in analysis develop powerful feelings toward the analyst. This happened in Mr. L.'s case. A certain warmth and positive sense of closeness developed between us. Freud thought that these powerful, positive transference feelings became one of the many engines that promoted the cure. In neuroscientific terms, this probably helps because emotions and the patterns we display in relationships are part of the procedural memory system. When such patterns are triggered in therapy, it gives the patient a chance to look at them and change them, for as we saw in chapter 4, "Acquiring Tastes and Loves," positive bonds appear to facilitate neuroplastic change by triggering unlearning and dissolving existing neuronal networks, so the patient can alter his existing intentions.

"There is no longer any doubt," writes Kandel, "that psychotherapy can result in detectable changes in the brain." Recent brain scans done before and after psychotherapy show both that the brain plastically reorganizes itself in treatment and that the more successful the treatment the greater the change. When patients relive their traumas and have flashbacks and uncontrollable emotions, the flow of blood to the prefrontal and frontal lobes, which help regulate our behavior, decreases, indicating that these areas are less active. According to neuropsychoanalyst Mark Solms and neuroscientist Oliver Turnbull, "The aim of the talking cure . . . from the neurobiological point of view [is] to extend the functional sphere of influence of the prefrontal lobes."

A study of depressed patients treated with interpersonal psychotherapy—a short-term treatment that is partially based on the theoretical work of two psychoanalysts, John Bowlby and Harry Stack Sullivan—showed that prefrontal brain activity normalized with treatment. (The right orbitofrontal system, which is so important in recognizing and regulating emotions and relationships—a function that was disturbed in Mr. L.—is part of the prefrontal cortex.) A more recent fMRI brain scan study of anxious patients with panic disorder found that the tendency of their limbic systems to be abnormally activated by potentially threatening stimuli was reduced following psychoanalytic psychotherapy.

As Mr. L. began to understand his post-traumatic symptoms, he began "regulating" his emotions better. He reported that outside the analysis he had more self-control. His mysterious paralytic states decreased. When he had painful feelings, he didn't resort to drinking nearly as often. Now Mr. L. began to let down his guard and became less defensive. He was more comfortable expressing anger, when called for, and felt closer to his children. Increasingly he used his sessions to face his pain instead of turning it off completely. Now Mr. L. slipped into long silences that had a profoundly resolute quality. His facial expression showed that he was in extraordinary pain, feeling an awful sadness that he wouldn't discuss.

Because his feelings about his mother's loss were not talked about while he was growing up, and the family dealt with its pain by getting on with chores, and because he had been silent for so long, I took a risk and tried to put what he was conveying nonverbally into words. I said, "It's as though you are saying to me, as perhaps you once wanted to say to your family, 'Can't you see, after this terrible loss, I have to be depressed right now?' "

He burst into tears for the second time in the analysis. He started involuntarily and rhythmically to protrude his tongue between weeping spells, making him look like a baby from whom the breast had been withdrawn and who was protruding his tongue to find it. Then he covered his face, put his hand in his mouth like a two-year-old, and broke out into loud, primitive sobbing. He said, "I want to be consoled for my pains and losses, yet don't come too close to console me. I want to be alone in my sullen misery. Which you can't understand, because I can't understand. It is a grief that is too big."

Hearing this, we both became aware that he often took the stance of "rejecting consolation" and that it contributed to the "remoteness" of his character. He was working through a defense mechanism that had been in place since childhood and that helped him block off the immensity of his loss. That defense, by being repeated many thousands of times, had been plastically reinforced. This most pronounced of his character traits, his remoteness, wasn't genetically predetermined but plastically learned, and now it was being unlearned.

It may seem unusual that Mr. L. wept and stuck out his tongue like a baby, but it was the first of several such "infantile" experiences that he would have on the couch. Freud observed that patients who have had early trauma will often, at key moments, "regress" (to use his term) and not only remember early memories but briefly experience them in childlike ways. This makes perfect sense from a neuroplastic point of view. Mr. L. had just given up a defense that he had used since childhood—the denial of the emotional impact of his loss—and it exposed the memories and emotional pain that the defense had hidden. Recall that Bach-y-Rita described something very similar happening in patients undergoing brain reorganization. If an established brain network is blocked, then older networks, in place long before the established one, must be used. He called this the "unmasking" of older neuronal paths and thought it one of the chief ways the brain reorganizes itself. Regression in analysis at a neuronal level is, I believe, an instance of unmasking, which often precedes psychological reorganization. That is what followed for Mr. L.

In his next session he reported that his recurring dream had changed. This time he went to visit his old house, looking for "adult possessions." The dream signaled that the part of him that had been deadened was coming alive again:

> I go to visit an old house. I don't know whose it is, yet it's mine. I am searching for something—not toys now, but

> adult possessions. There is a spring thaw, the end of winter. I enter the house, and it is the house where I was born. I had thought the house was empty, but my ex-wife—whom I felt was like a good mother to me—appeared from the back room, which was flooding. She welcomed me and was pleased to see me, and I felt elation.

He was emerging from a sense of isolation, of being cut off from people and from parts of himself. The dream was about his emotional "spring thaw" and a motherlike person being present with him in the house where he spent his earliest childhood. It was not empty, after all. Similar dreams followed in which he reclaimed his past, his sense of himself, and the sense that he had had a mother.

One day he mentioned a poem about a starving Indian mother who gave her child her last morsel of food before dying herself. He couldn't understand why the poem moved him so. Then he paused and burst into an ear-splitting wail, "My mommy sacrificed her life for me!" He wailed, his whole body shaking, fell silent, and then yelled, "I want my mommy!"

Mr. L., not given to hysterics, was now experiencing all the emotional pain that his defenses had pushed away, reliving thoughts and feelings he had had as a child—he was regressing and unmasking older memory networks, even ways of talking. But again this was followed by psychological reorganization at a higher level.

After acknowledging his great sense of missing his mother, he went to visit her grave for the first time. It was as though a part of his mind had held on to the magical idea that she was alive. Now he was able to accept, at the core of his being, that she was dead.

The next year Mr. L. fell profoundly in love for the first time in his adult life. He also became possessive of his lover and suffered normal jealousy, also for the first time. He now understood why women had been infuriated by his aloofness and lack of commitment and felt sad and guilty. He felt too that he discovered a part of himself that had been linked to his mother and lost when she died. Finding that part of himself that had once loved a woman allowed him to fall in love again.

Then he had the last dream of his analysis:

> I saw my mother playing the piano, and then I go to get someone, and when I return, she is in a coffin.

As he associated to the dream, he was startled by the image of his being held up to see his mother in her open coffin, reaching out to her, and being overwhelmed by the dreadful, terrifying realization that she did not respond. He let out a loud wail, and overcome with primitive grieving, his whole body convulsed for ten minutes. When he settled down, he said, "I believe this was a memory of my mother's wake, which was conducted with an open coffin."

Mr. L. was feeling better, and different. He was in a stable, loving relationship with a woman, his connection to his children had deepened significantly, and he was no longer remote. In his final session he reported that he had spoken to an older sibling, who confirmed that there had been an open coffin at his mother's funeral and that he had been present. When we parted, Mr. L. was consciously sad but no longer depressed or paralyzed at the thought of a permanent separation. Ten years have passed since he completed his analysis, and he remains free of his deep depressions and says his analysis "changed my life and gave me control of it."

Many of us, because of our own infantile amnesia, may doubt adults can remember as far back as Mr. L. ultimately did. This doubt was once so widespread that no research was conducted to investigate the matter, but new studies show that infants in the first and second years can store such facts and events, including traumatic ones. While the explicit memory system is not robust in the first few years, research by Carolyn Rovee-Collier and others shows it exists, even in preverbal or barely verbal infants. Infants can remember events from the first few years of life if they are reminded. Older children can remember events that occurred before they could talk and, once they learn to speak, can put those memories into words. At times Mr. L. was doing just this, putting events he experienced into words for the first time. Other times he unblocked events that had been in his explicit memory all along, such as the thought *My mommy sacrificed her life for me*, or his memory of being at his mother's wake, which was independently verified. And still other times, he "retranscribed" experiences from his procedural memory system to his explicit system. And interestingly, his core dream seemed to register that he had a major problem with his memory—he was searching for something but couldn't recall what—though he sensed he'd recognize it if he found it.

Why are dreams so important in analysis, and what is their relationship to plastic change? Patients are often haunted by recurring dreams of their traumas and awaken in terror. As long as they remain ill, these dreams don't change their basic structure. The neural network that represents the trauma—such as Mr. L.'s dream that he was missing something—is persistently reactivated, without being retranscribed. Should these traumatized patients get better, these nightmares gradually become less frightening, until ultimately the patient dreams something like *At first I think the trauma is recurring, but it isn't; it's over now, I've survived.* This kind of progressive dream series shows the mind and brain slowly changing, as the patient learns that he is safe now. For this to happen, neural networks must unlearn certain associations—as Mr. L. unlearned his association between separation and death—and change existing synaptic connections to make way for new learning.

What physical evidence exists that dreams show our brains in the process of plastic change, altering hitherto buried, emotionally meaningful memories, as in Mr. L.'s case?

The newest brain scans show that when we dream, that part of the brain that processes emotion, and our sexual, survival, and aggressive instincts, is quite active. At the same time the prefrontal cortex system, which is responsible for inhibiting our emotions and instincts, shows lower activity. With instincts turned up and inhibitions turned down, the dreaming brain can reveal impulses that are normally blocked from awareness.

Scores of studies show that sleep affects plastic change by allowing us to consolidate learning and memory. When we learn a skill during the day, we will be better at it the next day if we have a good night's sleep. "Sleeping on a problem" often does make sense.

A team led by Marcos Frank has also shown that sleep enhances neuroplasticity during the critical period when most plastic change takes place. Recall that Hubel and Wiesel blocked one eye of a kitten in the critical period and showed that the brain map for the blocked eye was taken over by the good eye—a case of use it or lose it. Frank's team did the same experiment with two groups of kittens, one group that it deprived of sleep, and another group that got a full amount of it. They found that the more sleep the kittens got, the greater the plastic change in their brain map.

The dream state also facilitates plastic change. Sleep is divided into two stages, and most of our dreaming occurs during one of them, called rapid-eye-movement sleep, or REM sleep. Infants spend many more hours in REM sleep than adults, and it is during infancy that neuroplastic change occurs most rapidly. In fact, REM sleep is required for the plastic development of the brain in infancy. A team led by Gerald Marks did a study similar to Frank's that looked at the effects of REM sleep on kittens and on their brain structure. Marks found that in kittens deprived of REM sleep, the neurons in their visual cortex were actually smaller, so REM sleep seems necessary for neurons to grow normally. REM sleep has also been shown to be particularly important for enhancing our ability to retain emotional memories and for allowing the hippocampus to turn short-term memories of the day before into long-term ones (i.e., it helps make memories more permanent, leading to structural change in the brain).

Each day, in analysis, Mr. L. worked on his core conflicts, memories, and traumas, and at night there was dream evidence not only of his buried emotions but of his brain reinforcing the learning and unlearning he had done.

We understand why Mr. L., at the outset of his analysis, had no conscious memories of the first four years of his life: most of his memories of the period were unconscious procedural memories—automatic sequences of emotional interactions—and the few explicit memories he had were so painful, they were repressed. In treatment he gained access to both procedural and explicit memories from his first four years. But why was he unable to recall his adolescent memories? One possibility is that he repressed some of his adolescence; often when we repress one thing, such as a catastrophic early loss, we repress other events loosely associated with it, to block access to the original.

But there is another possible cause. It has recently been discovered that early childhood trauma causes massive plastic change in the hippocampus, shrinking it so that new, long-term explicit memories cannot form. Animals removed from their mothers let out desperate cries, then enter a turned-off state—as Spitz's infants did—and release a stress hormone called "glucocorticoid." Glucocorticoids kill cells in the hippocampus so that it cannot make the synaptic connections in neural networks that make learning and explicit long-term memory possible. These early stresses predispose these motherless animals to stress-related illness for the rest of their lives. When they undergo long separations, the gene to initiate production of glucocorticoids gets turned on and stays on for extended periods. Trauma in infancy appears to lead to a supersensitization—a plastic alteration—of the brain neurons that regulate glucocorticoids. Recent research in humans shows that adult survivors of childhood abuse also show signs of glucocorticoid supersensitivity lasting into adulthood.

That the hippocampus shrinks is an important neuroplastic discovery and may help explain why Mr. L. had so few explicit memories from adolescence. Depression, high stress, and childhood trauma all release glucocorticoids and kill cells in the hippocampus, leading to memory loss. The longer people are depressed, the smaller their hippocampus gets. The hippocampus of depressed adults who suffered prepubertal childhood trauma is 18 percent smaller than that of depressed adults without childhood trauma—a downside of the plastic brain: we literally lose essential cortical real estate in response to illness.

If the stress is brief, this decrease in size is temporary. If it is too prolonged, the damage is permanent. As people recover from depression, their memories return, and research suggests their hippocampi can grow back. In fact, the hippocampus is one of two areas where new neurons are created from our own stem cells as part of normal functioning. If Mr. L. had hippocampal damage, he had recovered from it by his early twenties when he began forming explicit memories again.

Antidepressant medications increase the number of stem cells that become new neurons in the hippocampus. Rats given Prozac for three weeks had a 70 percent increase in the number of cells in their hippocampi. It usually takes three to six weeks for antidepressants to work in humans—perhaps coincidentally, the same amount of time it takes for newly born neurons in the hippocampus to mature, extend their projections, and connect with other neurons. So we may, without knowing it, have been helping people get out of depression by using medications that foster brain plasticity. Since people who improve in psychotherapy also find that their memories improve, it may be that it also stimulates neuronal growth in their hippocampi.

The many changes that Mr. L. made might even have surprised Freud, given Mr. L.'s age at the time of his analysis. Freud used the term "mental plasticity" to describe people's capacity for change and recognized that people's overall ability to change seemed to vary. He also observed that a "depletion of the plasticity" tended to occur in many older people, leading them to become "unchangeable, fixed, and rigid." He attributed this to "force of habit" and wrote, "There are some people, however, who retain this mental plasticity far beyond the usual age-limit, and others who lose it very prematurely." Such people, he observed, have great difficulties getting rid of their neuroses in psychoanalytic treatment. They can activate transferences but have difficulty changing them. Mr. L. had certainly had a fixed character structure for over fifty years. How then was he able to change?

The answer is part of a larger riddle that I call the "plastic paradox" and that I consider one of the most important lessons of this book. The plastic paradox is that the same neuroplastic properties that allow us to change our brains and produce more flexible behaviors can also allow us to produce more rigid ones. All people start out with plastic potential. Some of us develop into increasingly flexible children and stay that way through our adult lives. For others of us, the spontaneity, creativity, and unpredictability of childhood gives way to a routinized existence that repeats the same behavior and turns us into rigid caricatures of ourselves. Anything that involves unvaried repetition—our careers, cultural activities, skills, and neuroses—can lead to rigidity. Indeed, it is because we have a neuroplastic brain that we can develop these rigid behaviors in the first place. As Pascual-Leone's metaphor illustrates, neuroplasticity is like pliable snow on a hill. When we go down the hill on a sled, we can be flexible because we have the option of taking different paths through the soft snow each time. But should we choose the same path a second or third time, tracks will start to develop, and soon we will tend to get stuck in a rut—our route will now be quite rigid, as neural circuits, once established, tend to become *self-sustaining*.

Because our neuroplasticity can give rise to both mental flexibility and mental rigidity, we tend to underestimate our own potential for flexibility, which most of us experience only in flashes.

Freud was right when he said that the absence of plasticity seemed related to force of habit. Neuroses are prone to being entrenched by force of habit because they involve repeating patterns of which we are not conscious, making them almost impossible to interrupt and redirect without special techniques. Once Mr. L. was able to understand the causes of his often defensive habits, and his view of himself and the world, he could make use of his innate plasticity, despite his age.

When Mr. L. started analysis, he experienced his mother as a ghost he could not see; a presence both alive and dead; someone he was faithful to yet was never sure existed. By accepting that she had really died, he lost his sense of her as a ghost and instead gained a feeling that he really had had a substantial mother, a good person, who had loved him as long as she was alive. Only when his ghost was turned into a loving ancestor was he freed to form a close relationship with a living woman.

Psychoanalysis is often about turning our ghosts into ancestors, even for patients who have not lost loved ones to death. We are often haunted by important relationships from the past that influence us unconsciously in the present. As we work them through, they go from haunting us to becoming simply part of our history. We can turn our ghosts into ancestors because we can transform implicit memories—which we are often not aware exist until they are evoked and thus seem to come at us "out of the blue"—into declarative memories that now have a clear context, which makes them easier to recollect and experience as part of the past.

Today H.M., the most famous case in neuropsychology, is still alive, in his seventies, his mind locked in the 1940s, in the moment before he had his surgery and lost both of his hippocampi, the gateways through which memories must pass if they are to be preserved and long-term plastic change is to be achieved. Unable to convert short-term memories into long, the structure of his brain and memory, and his mental and physical images of himself, are frozen where they were when he had his surgery. Sadly, he cannot even recognize himself in the mirror. Eric Kandel, who was born at roughly the same time, continues to probe the hippocampus, and the plasticity of memory, down to alterations in individual molecules. He has further dealt with his painful memories of the 1930s by writing a poignant, informative memoir, *In Search of Memory*. Mr. L.—now also in his seventies—is no longer emotionally locked in the 1930s because he was able to bring to consciousness events that happened almost sixty years before, retranscribe them, and in the process rewire his plastic brain.

10

Rejuvenation

The Discovery of the Neuronal Stem Cell and Lessons for Preserving Our Brains

Ninety-year-old Dr. Stanley Karansky seems unable to believe that just because he is old, his life must wind down. He has nineteen descendants—five children, eight grandchildren, and six great-grandchildren. His wife of fifty-three years died of cancer in 1995, and he now lives in California, with his second wife, Helen.

Born in New York City in 1916, he went to Duke University medical school, did his internship in 1942, and in World War II was a medic in the D-day invasion. He served as a medical officer in the infantry, in the European theater, for almost four years, then was shipped to Hawaii, where he eventually settled. He practiced as an anesthesiologist until he retired at seventy. But retirement didn't suit him, so he retrained himself as a family doctor and practiced in a small clinic for ten more years, until he was eighty.

I talked with him shortly after he completed the series of brain exercises Merzenich's team developed with Posit Science. Dr. Karansky hadn't seen cognitive decline, though he adds, "My handwriting was good but not as good as it was before." He simply hoped to keep his brain fit.

He began the auditory memory program in August 2005 by inserting a CD into his computer, and found the exercises "sophisticated and entertaining." They required him to determine if sounds were sweeping up in frequency or down, to pick the order in which he heard certain syllables, identify similar sounds, and listen to stories and answer questions about them—all in order to sharpen brain maps and stimulate the mechanisms that regulate brain plasticity. He worked on the exercises for an hour and a quarter, three times a week, for three months.

"I didn't notice anything for the first six weeks. At about week seven I began to notice that I was more alert than I had been before. And I could tell from the program itself, from the way I was monitoring my progress, that I was getting better at getting correct answers, and I felt better about everything. My driving alertness, both during the day and at night, also improved. I was talking to people more, and talking came more easily. In the last few weeks I think my handwriting has improved. When I sign my name, I think I'm writing the way I did twenty years ago. My wife, Helen, told me, 'I think you are more alert, more active, more responsive.'" He intends to wait a number of months, then redo the exercises to stay in shape. Even though the exercises are for auditory memory, he's been getting general benefits, as did the children who did *Fast ForWord,* because he is stimulating not only his auditory memory but also the brain centers that regulate plasticity.

He also does physical exercise. "My wife and I do muscle exercises three times a week on the CYBEX machines, followed by a thirty- to thirty-five-minute workout on an exercise bicycle."

Dr. Karansky describes himself as a lifelong self-educator. He reads serious mathematics and loves games, word puzzles, double acrostics, and Sudoku.

"I like to read about history," he says. "I tend to get interested in a period, for whatever reason, and I get started, and I mine that period for a while, until I feel I've learned enough about it to learn something else." What might be thought of as dilettantism has the effect of keeping him constantly exposed to novelty and new subjects, which keeps the regulatory system for plasticity and dopamine from atrophying.

Each new interest becomes an engaging passion. "I became interested in astronomy five years ago and became an amateur astronomer. I bought a telescope because we were living in Arizona at the time, and the natural viewing conditions were so good." He is also a serious rock collector and has spent much of what many would call old age crawling in mines looking for specimens.

"Is there longevity in your family?" I ask. "No," he says. "My mother died in her late forties. My father died in his sixties—he had some hypertension."

"How has your health been?"

"Well, I died once." He laughs. "You have to forgive me for being the type of person who likes to astound people. I used to do some long-distance running, and in 1982, I was sixty-five years old and had an episode of ventricular fibrillation"—an often fatal arrhythmia of the heart—"on a practice run in Honolulu, and I literally died on the sidewalk. The guy I was running with was wise enough to give me streetside CPR, and some of the runners called the fire department paramedics, and they got to me quickly enough, and zapped me, and returned me to normal sinus rhythm and took me to Straub Hospital." After that he underwent bypass surgery. He got actively engaged in rehab and recovered rapidly. "I did not do competitive running after that, but I ran about twenty-five miles per week at a slower pace." He then had another heart attack in 2000, when he was eighty-three.

He is social, but not in large groups. "I don't go readily to cocktail parties, where people just come together and talk. I don't tend to like that kind of thing. I'd rather sit down with somebody and find a mutual topic of interest, and explore it in depth with that person, or maybe two or three people. Not a conversation that says how do you feel."

He says that he and his wife are not strong travelers, but that is a matter of opinion. When he was eighty-one, he learned some Russian and then went on a Russian scientific vessel to visit Antarctica.

"What for?" I ask.

"Because it was there."

In the last few years he's been to the Yucatán, England, France, Switzerland, and Italy, spent six weeks in South America, visited his daughter in the United Arab Emirates, and traveled to Oman, Australia, New Zealand, Thailand, and Hong Kong.

He is always looking for novel things to do, and once he's engaged in something, he turns his full attention to it—the necessary condition for plastic change. He says, "I'm willing to put pretty intense concentration and attention into something that interests me at the moment. Then after I feel I've gotten to a higher level at it, I don't pay quite as much attention to that activity, and I start sending interest tentacles to something else."

His philosophical attitude also protects his brain because he doesn't get worked up about little things—no small matter, since stress releases glucocorticoids, which can kill cells in the hippocampus.

"You seem less anxious and nervous than most people," I say.

"I've seen it to be very beneficial for people."

"Are you an optimistic person?"

"Not so much, but I think I understand what random events are. There are many things that go on that can affect me that are beyond my control. I can't control them, only how I react to them. I've spent my time worrying about things I can control and can affect the outcome of, and I've managed to develop a philosophy that enables me to deal with those."

At the beginning of the twentieth century the world's most outstanding neuroanatomist, Nobel Prize winner Santiago Ramón y Cajal, who laid the groundwork for our understanding of how neurons are structured, turned his attention to one of the most vexing problems of human brain anatomy. Unlike the brains of simpler animals, such as lizards, the human brain seemed unable to regenerate itself after an injury. This helplessness is not typical of all human organs. Our skin, when cut, can heal itself, by producing new skin cells; our fractured bones can mend themselves; our liver and intestinal lining can repair themselves; lost blood can replenish itself because cells in our marrow can become red or white blood cells. But our brains seemed to be a disturbing exception. It was known that millions of neurons die as we age. Whereas other organs make new tissues from stem cells, none could be found in the brain. The main explanation for the absence was that the human brain, as it evolved, must have become so complex and specialized that it lost the power to produce replacement cells. Besides, scientists asked, how could a new neuron enter a complex, existing neuronal network and create a thousand synaptic connections without causing chaos in that network? The human brain was assumed to be a closed system.

Ramón y Cajal devoted the later part of his career to searching for any sign that either the brain or spinal cord could change, regenerate, or reorganize its structure. He failed.

In his 1913 masterpiece, *Degeneration and Regeneration of the Nervous System,* he wrote, "In adult [brain] centers the nerve paths are something fixed, ended, immutable. Everything may die, nothing may be regenerated. It is for the science of the future to change, if possible, this harsh decree."

There matters stood.

I am staring down a microscope in the most advanced lab I have ever visited, at the Salk Laboratories in La Jolla, California, looking at living, human neuronal stem cells in a petri dish in the lab of Frederick "Rusty" Gage. He and Peter Eriksson of Sweden discovered these cells in 1998, in the hippocampus.

The neuronal stem cells I see are vibrating with life. They are called "neuronal" stem cells because they can divide and differentiate to become neurons or glial cells, which support neurons in the brain. The ones I am looking at have yet to differentiate into either neurons or glia and have yet to "specialize," so they all look identical. Yet what stem cells lack in personality, they make up for in immortality. For stem cells don't have to specialize but can continue to divide, producing exact replicas of themselves, and they can go on doing this endlessly without any signs of aging. For this reason stem cells are often described as the eternally young, baby cells of the brain. This rejuvenating process is called "neurogenesis," and it goes on until the day that we die.

Neuronal stem cells were long overlooked, in part, because they went against the theory that the brain was like a complex machine or computer, and machines don't grow new parts. When, in 1965, Joseph Altman and Gopal D. Das of the Massachusetts Institute of Technology discovered them in rats, their work was disbelieved.

Then in the 1980s Fernando Nottebohm, a bird specialist, was struck by the fact that songbirds sing new songs each season. He examined their brains and found that every year, during the season when the birds do the most singing, they grow new brain cells in the area of the brain responsible for song learning. Inspired by Nottebohm's discovery, scientists began examining animals that were more like human beings. Elizabeth Gould of Princeton University was the first to discover neuronal stem cells in primates. Next, Eriksson and

Gage found an ingenious way to stain brain cells with a marker, called BrdU, that gets taken into neurons only at the moment they are created and that lights up under the microscope. Eriksson and Gage asked terminally ill patients for permission to inject them with the marker. When these patients died, Eriksson and Gage examined their brains and found new, recently formed baby neurons in their hippocampi. Thus we learned from these dying patients that living neurons form in us until the very end of our lives.

The search continues for neuronal stem cells in other parts of the human brain. So far they've also been found active in the olfactory bulb (a processing area for smell) and dormant and inactive in the septum (which processes emotion), the striatum (which processes movement), and the spinal cord. Gage and others are working on treatments that might activate dormant stem cells with drugs and be useful if an area where they are dormant suffers damage. They are also trying to find out whether stem cells can be transplanted into injured brain areas, or even induced to move to those areas.

To find out if neurogenesis can strengthen mental capacity, Gage's team has set out to understand how to increase the production of neuronal stem cells. Gage's colleague Gerd Kempermann raised aging mice in enriched environments, filled with mice toys such as balls, tubes, and running wheels, for only forty-five days. When Kempermann sacrificed the mice and examined their brains, he found they had a 15 percent increase in the volume of their hippocampi and forty thousand new neurons, also a 15 percent increase, compared with mice raised in standard cages.

Mice live to about two years. When the team tested older mice raised in the enriched environment for ten months in the second half of their lives, there was a fivefold increase in the number of neurons in the hippocampus. These mice were better at tests of learning, exploration, movement, and other measures of mouse intelligence than those raised in unenriched conditions. They developed new neurons, though not quite as quickly as younger mice, proving that long-term enrichment had an immense effect on promoting neurogenesis in an aging brain.

Next the team looked at which activities caused cell increases in the mice, and they found that there are two ways to increase the overall number of neurons in the brain: by creating new neurons, and by extending the life of existing neurons.

Gage's colleague Henriette van Praag showed that the most effective contributor to increased proliferation of *new* neurons was the running wheel. After a month on the wheel, the mice had doubled the number of new neurons in the hippocampus. Mice don't really run on running wheels, Gage told me; it only looks like they do, because the wheel provides so little resistance. Rather, they walk quickly.

Gage's theory is that in a natural setting, long-term fast walking would take the animal into a new, different environment that would require new learning, sparking what he calls "anticipatory proliferation."

"If we lived in this room only," he told me, "and this was our entire experience, we would not need neurogenesis. We would know everything about this environment and could function with all the basic knowledge we have."

This theory, that novel environments may trigger neurogenesis, is consistent with Merzenich's discovery that in order to keep the brain fit, we must learn something new, rather than simply replaying already-mastered skills.

But as we've said, there is a second way to increase the number of neurons in the hippocampus: by extending the life of neurons already there. Studying the mice, the team found that learning how to use the other toys, balls, and tubes didn't make new neurons, but it did cause the new neurons in the area to live longer. Elizabeth Gould also found that learning, even in a nonenriched environment, enhances survival of stem cells. Thus physical exercise and learning work in complementary ways: the first to make new stem cells, the second to prolong their survival.

Though the discovery of neuronal stem cells was momentous, it is only one of the ways the aging brain can rejuvenate and improve itself. Paradoxically, sometimes losing neurons can improve brain function, as happens in the massive "pruning back" that occurs during adolescence when synaptic connections and neurons that have not been extensively used die off, in perhaps the most dramatic case of use it or lose it. Keeping unused neurons supplied with blood, oxygen, and energy is wasteful, and getting rid of them keeps the brain more focused and efficient.

That we still have some neurogenesis in old age is not to deny that our brains, like our other organs, gradually decline. But even in the midst of this deterioration, the brain undergoes massive plastic reorganization, possibly to adjust for the brain's losses. Researchers Mellanie Springer and Cheryl Grady of the University of Toronto have shown that as we age, we tend to perform cognitive activities in different lobes of the brain from those we use when we are young. When Springer and Grady's young subjects, aged fourteen to thirty years, did a variety of cognitive tests, brain scans showed that they performed them largely in their temporal lobes, on the sides of the head, and that the more education they'd had, the more they used these lobes.

Subjects over sixty-five years had a different pattern. Brain scans showed that they performed these same cognitive tasks largely in their frontal lobes, and again, the more education they'd had, the more they used the frontal lobes.

This shift within the brain is another sign of plasticity—shifting processing areas from one lobe to another is about as large a migration as a function can make. No one knows for sure why this shift happens, or why so many studies suggest that people with more education seem better protected from mental decline. The most popular theory is that years of education create a "cognitive reserve"—many more networks devoted to mental activity—that we can call upon as our brains decline.

Another major reorganization of the brain occurs as we age. As we have seen, many brain activities are "lateralized." Much of speech is a left-hemispheric function, while visual-spatial processing is a right-hemispheric function, a phenomenon called "hemispheric asymmetry." But recent research by Duke University's Roberto Cabeza and others shows that some lateralization is lost as we age. Prefrontal activities that took place in one hemisphere now take place in both. While we don't know for sure why this happens, one theory is that as we age and one of our hemispheres starts to become less effective, the other hemisphere compensates—suggesting that the brain restructures itself in response to its own weaknesses.

We now know that exercise and mental activity in animals generate and sustain more brain cells, and we have many studies confirming that humans who lead mentally active lives have better brain function. The more education we have, the more socially and physically active we are, and the more we participate in mentally stimulating activities, the less likely we are to get Alzheimer's disease or dementia.

Not all activities are equal in this regard. Those that involve genuine concentration—studying a musical instrument, playing board games, reading, and dancing—are associated with a lower risk for dementia. Dancing, which requires learning new moves, is both physically and mentally challenging and requires much concentration. Less intense activities, such as bowling, babysitting, and golfing, are not associated with a reduced incidence of Alzheimer's.

These studies are suggestive but stop short of proving that we can prevent Alzheimer's disease with brain exercises. These activities are associated with or correlated with less Alzheimer's, but correlations don't prove causality. It is possible that people with very early onset but undetectable Alzheimer's begin slowing down early in life and so stop being active. The most we can say about the relationship between brain exercises and Alzheimer's at the moment is that it seems very promising.

As Merzenich's work has shown, however, a condition often confused with Alzheimer's disease, and much more common—age-related memory loss, a typical decline in memory that occurs in advanced years—seems almost certainly reversible with the right mental exercises. Though Dr. Karansky didn't complain of general cognitive decline, he did experience some "senior moments," which was part of age-related memory loss, and the benefits he got from the exercises certainly showed he had other reversible cognitive deficits that he hadn't even been aware of.

Dr. Karansky, it turns out, was doing everything right to fight off age-related memory loss, making him an exemplary model for the common practices we should all be pursuing.

Physical activity is helpful not only because it creates new neurons but because the mind is based in the brain, and the brain needs oxygen. Walking, cycling, or cardiovascular exercise strengthens the heart and the blood vessels that supply the brain and helps people who engage in these activities feel mentally sharper—as pointed out by the Roman philosopher Seneca two thousand years ago. Recent research shows that exercise stimulates the production and release of the neuronal growth factor BDNF, which, as we saw in chapter 3, "Redesigning the Brain," plays a crucial role in effecting plastic change. In fact, whatever keeps the heart and blood vessels fit invigorates the brain, including a healthy diet. A brutal workout is not necessary—consistent natural movement of the limbs will do. As van Praag and Gage discovered, simply walking, at a good pace, stimulates the growth of new neurons.

Exercise stimulates your sensory and motor cortices and maintains your brain's balance system. These functions begin to deteriorate as we age, making us prone to falling and becoming housebound.

Nothing speeds brain atrophy more than being immobilized in the same environment; the monotony undermines our dopamine and attentional systems crucial to maintaining brain plasticity. A cognitively rich physical activity such as learning new dances will probably help ward off balance problems and have the added benefit of being social, which also preserves brain health. Tai chi, though it hasn't been studied, requires intense concentration on motor movements and stimulates the brain's balance system. It also has a meditative aspect, which has been proven very effective in lowering stress and so is likely to preserve memory and the hippocampal neurons.

Dr. Karansky is always learning new things, which plays a role in being happy and healthy in old age, according to Dr. George Vaillant, a Harvard psychiatrist who heads up the largest, longest ongoing study of the human life cycle, the Harvard Study of Adult Development. He studied 824 people from their late teens through to old age from three groups: Harvard graduates, poor Bostonians, and women with extremely high IQs. Some of these people, now in their eighties, have been tracked for over six decades. Vaillant concluded that old age is not simply a process of decline and decay, as many younger people think. Older people often develop new skills and are often wiser and more socially adept than they were as younger adults. These elderly people are actually less prone to depression than younger people and usually do not suffer from incapacitating disease until they get their final illness.

Of course, challenging mental activities will increase the likelihood that our hippocampal neurons will survive. One approach is to use tested brain exercises, such as those Merzenich developed. But life is for living and not only for doing exercises, so it is best that people also choose to do something they've always wanted to do, because they will be highly motivated, which is crucial. Mary Fasano, at age eighty-nine, earned her undergraduate degree from Harvard. David Ben-Gurion, the first prime minister of Israel, taught himself ancient Greek in old age to master the classics in the original. We might think, "What for? Who am I fooling? I'm at the end of the road." But that thinking is a self-fulfilling prophecy, which hastens the mental decline of the use-it-or-lose-it brain.

At ninety, the architect Frank Lloyd Wright designed the Guggenheim Museum. At seventy-eight, Benjamin Franklin invented bifocal spectacles. In studies of creativity, H. C. Lehman and Dean Keith Simonton found that while the ages thirty-five to fifty-five are the peak of creativity in most fields, people in their sixties and seventies, though they work at a slower speed, are as productive as they were in their twenties.

When Pablo Casals, the cellist, was ninety-one years old, he was approached by a student who asked, "Master, why do you continue to practice?" Casals replied, "Because I am making progress."

11

More than the Sum of Her Parts

A Woman Shows Us How Radically Plastic the Brain Can Be

The woman joking with me across the table was born with only half her brain. Something catastrophic happened while she was in her mother's womb, though no one knows what for sure. It wasn't a stroke, because stroke destroys healthy tissue, and Michelle Mack's left hemisphere simply never developed. Her doctors speculated that her left carotid artery, which supplies blood to the left hemisphere, may have become blocked while Michelle was still a fetus, preventing that hemisphere from forming. At birth the doctors gave her the usual tests and told her mother, Carol, that she was a normal baby. Even today a neurologist would not likely guess, without a brain scan, that a whole hemisphere is missing. I find myself wondering how many others have lived out their lives with half a brain, without themselves or anyone knowing it.

I am visiting Michelle to discover how much neuroplastic change is possible in a human being whose brain has undergone such a challenge, but a doctrinaire localizationism, which posits that each hemisphere is genetically hardwired to have its own specialized functions, is itself seriously challenged if Michelle can function with only one. It's hard to imagine a better illustration or indeed a greater test of human neuroplasticity.

Though she has only a right hemisphere, Michelle is not a desperate creature barely surviving on life support. She is twenty-nine years old. Her blue eyes peer through thick glasses. She wears blue jeans, sleeps in a blue bedroom, and speaks fairly normally. She holds a part-time job, reads, and enjoys movies and her family. She can do all this because her right hemisphere took over for her left, and such essential mental functions as speech and language moved to her right. Her development makes it clear that neuroplasticity is no minor phenomenon operating at the margins; it has allowed her to achieve massive brain reorganization.

Michelle's right hemisphere must not only carry out the key functions of the left but also economize on its "own" functions. In a normal brain each hemisphere helps refine the development of the other by sending electrical signals informing its partner of its activities, so the two will function in a coordinated way. In Michelle, the right hemisphere had to evolve without input from the left and learn to live and function on its own.

Michelle has some extraordinary calculating skills—savant skills—that she employs at lightning speed. She also has special needs and disabilities. She doesn't like to travel and gets easily lost in unfamiliar surroundings. She has trouble understanding certain kinds of abstract thought. But her inner life is alive, and she reads, prays, and loves. She speaks normally, except when frustrated. She adores Carol Burnett comedies. She follows the news and basketball and votes in elections. Her life is a demonstration that the whole is more than the sum of its parts and that half a brain does not make for half a mind.

One hundred and forty years ago Paul Broca opened the era of localizationism, saying, "One speaks with the left hemisphere," and initiated not only localizationism but the related theory of "laterality," which explored the difference between our left and right hemispheres. The left came to be seen as the verbal domain, where such symbolic activities as language and arithmetic calculation took place; the right housed many of our "nonverbal" functions including visual-spatial activities (as when we look at a map or navigate through space), and more "imaginative" and "artistic" activities.

Michelle's experience reminds us how ignorant we are about some of the most basic aspects of human brain functions. What happens when the functions of both hemispheres must compete for the same space? What, if anything, must be sacrificed? How much brain is needed for survival? How much brain is required to develop wit, empathy, personal taste, spiritual longing, and subtlety? If we can survive and live without half our brain tissue, why is it there in the first place?

And then there is the question, what is it like to be her?

I am in Michelle's family's living room, in their middle-class house, in Falls Church, Virginia, looking at the film of her MRI, illustrating the anatomy of her brain. On the right I can see the gray convolutions of a normal right hemisphere. On the left, except for a thin, wayward peninsula of gray brain tissue—the minuscule amount of left hemisphere that developed—there is only the deep black that denotes emptiness. Michelle has never looked at the film.

She calls this emptiness "my cyst," and when she speaks of "my cyst" or "the cyst," it sounds as though it has become substantial for her, an eerie character in a science fiction movie. And indeed, peering at her scan is an eerie experience. When I look at Michelle, I see her whole face, her eyes and smile, and cannot help but project that symmetry backward into the brain behind. The scan is a rude awakening.

Michelle's body does show some signs of her missing hemisphere. Her right wrist is bent and a bit twisted, but she can use it—though normally almost all the instructions for the right side of the body come from the left hemisphere. Probably she's developed a very thin strand of nerve fibers from the right hemisphere to her right hand. Her left hand is normal, and she's a lefty. When she gets up to walk, I see that a brace supports her right leg.

The localizationists showed that everything we see on our right—our "right visual field"—is processed on the left side of the brain. But because Michelle has no left hemisphere, she has trouble seeing things coming from her right and is blind in the right visual field. Her brothers used to steal her french fries from her right side, but she'd catch them because what she lacks in vision, she has made up for with supercharged hearing. Her hearing is so acute that she can clearly hear her parents talking in the kitchen when she is upstairs at the other end of the house. This hyperdevelopment of hearing, so common in the totally blind, is another sign of the brain's ability to adjust to a changed situation. But this sensitivity has a cost. In traffic, when a horn blares, she puts her hands over her ears, to avoid sensory overload. At church she escapes the sound of the organ pipes by slipping out the door. School fire drills frightened her because of the noise and confusion.

She is also supersensitive to touch. Carol cuts the tags off Michelle's clothing so she won't feel them. It's as though her brain lacks a filter to keep out excess sensation, so Carol often "filters" for her, protecting her. If Michelle has a second hemisphere, it is her mother.

"You know," said Carol, "I was never supposed to have children, so we adopted two," Michelle's older siblings, Bill and Sharon. As often happens, Carol then found she was pregnant with a son, Steve, who was born a healthy child. Carol and her husband, Wally, wanted more children but again had trouble conceiving.

One day, feeling ill with what seemed to be a bout of morning sickness, she ran a pregnancy test, but it was negative. Not quite believing the result, she ran more tests, with a strange result each time. A test strip that changes color within two minutes indicates a pregnancy. Each of Carol's tests was negative until two minutes and ten seconds, and then turned positive.

In the meantime Carol was having intermittent spotting and bleeding. She told me, "I went back to the doctor three weeks after the pregnancy tests, at which point the doctor said, 'I don't care what the tests have been saying, you are three months pregnant.' At the time we didn't think anything of it. But in hindsight I am convinced that because of the damage that Michelle had suffered in utero, my body was trying to have a miscarriage. It did not happen."

"Thank God it didn't!" said Michelle.

"Thank goodness you're right," said Carol.

Michelle was born November 9, 1973. The first days of her life are a blur for Carol. The day she brought Michelle home from the hospital, Carol's mother, who was living with them, had a stroke. The house was in chaos.

As time passed, Carol began noticing problems. Michelle didn't gain weight. She wasn't active and hardly made sounds. She also didn't seem to be tracking moving objects with her eyes. So Carol began what would become an endless series of visits to doctors. The first hint that there might be some kind of brain damage came when Michelle was six months old. Carol, thinking Michelle had a problem with her eye muscles, took her to an eye specialist who discovered that both her optic nerves were damaged and very pale, though not totally white as in people who are blind. He told Carol that Michelle's vision would never be normal. Glasses wouldn't help, because her optic nerves, not her lenses, were damaged. Even more upsetting were hints of a serious problem originating in Michelle's brain and causing her optic nerves to waste away.

At around the same time, Carol observed that Michelle wasn't turning over and that her right hand was clenched. Tests established that she was "hemiplegic," meaning the right half of her body was partially paralyzed. Her twisted right hand resembled that of a person who's had a stroke in the left hemisphere. Most kids start crawling at about seven months. But Michelle would sit on her bottom and get around by grabbing things with her good arm.

Though she didn't fit a clear category, her doctor assigned her a diagnosis of Behr syndrome, so she could get medical care and disability help. Indeed, she did have some symptoms consistent with Behr syndrome: optic atrophy and her neurologically based coordination problems. But Carol and Wally knew the diagnosis was absurd because Behr syndrome is a rare genetic condition, and neither of them showed a trace of it in their families. At three, Michelle was sent to a facility that treated cerebral palsy, though she didn't have that diagnosis either.

When Michelle was in her infancy, the computerized axial tomograph, or CAT scan, had just become available. This sophisticated X-ray takes numerous pictures of the head in cross section and feeds the images into a computer. Bone is white, brain tissue gray, and body cavities pitch black. Michelle had a CAT scan when she was six months old, but early scans had such poor resolution that hers showed only a mush of gray, from which the doctors could draw no conclusions.

Carol was devastated by the prospect that her child would never see properly. Then one day Wally was walking around the dining room while Carol was feeding Michelle breakfast, and Carol noticed that she was tracking him with her eyes.

"That cereal hit the ceiling, I was so elated," she says, "because it meant Michelle wasn't totally blind, that she had some vision."

A few weeks later, when Carol was sitting on the porch with Michelle, a motorcycle came up the street, and Michelle followed it with her eyes.

Then one day, when Michelle was about a year old, her clenched right arm, which she had always held close to her heart, opened out.

When she was about two, this girl who barely talked started to get interested in language.

"I would come home," said Wally, "and she would say, 'ABCs! ABCs!'" Sitting on his lap, she would put her fingers on his lips to feel the vibrations as he spoke. The doctors told Carol that Michelle did not have a learning disability and in fact seemed to have normal intelligence.

But at two she still couldn't crawl, so Wally, who knew that she loved music, would play her favorite record, and when the song was over, Michelle would cry, "Hmmm, hmmm, hmmm, want it again!" Then Wally would insist she crawl to the record player before he'd play it again. Michelle's overall learning pattern was becoming clear—a significant delay in development; a message from the clinicians to her parents to get used to it; and then somehow Michelle would pull herself out of it. Carol and Wally became more hopeful.

In 1977, when Carol was pregnant for the third time with Michelle's brother Jeff, one of her doctors persuaded Carol to arrange another CAT scan for Michelle. He said Carol owed it to her unborn child to try to determine what had happened to Michelle in the womb in order to prevent it from happening again.

By now the resolution of CAT scans had improved radically, and when Carol looked at the new scan, "the pictures showed—like night and day: brain and no brain." She was in shock. She told me, "If they had shown me those pictures when we had the CAT scan done at six months, I don't think I could have handled it." But at three and a half Michelle had already shown that her brain could adjust and change, so Carol felt there might be hope.

Michelle knows that researchers at the National Institutes of Health (NIH) under the direction of Dr. Jordan Grafman are studying her. Carol brought Michelle to the NIH because she read an article in the press about neuroplasticity in which Dr. Grafman contradicted many of the things that she had been told about brain problems. Grafman believed that with help the brain could often develop and change throughout life, even after injuries. Doctors had told Carol that Michelle would develop mentally only until she was about twelve, but she was now already twenty-five. If Dr. Grafman was right, Michelle had lost many years when other treatments might have been tried, a realization that stirred guilt in Carol but also hope.

One of the things Carol and Dr. Grafman worked on together was helping Michelle better understand her condition and better control her feelings.

Michelle is disarmingly honest about her emotions. "For many years," she said, "ever since I was little, whenever I did not get my way, I threw a fit. Last year I got tired of people always thinking that I needed to have my own way, otherwise my cyst would take over." But she adds, "Ever since last year I have tried to tell my parents that my cyst can handle changes."

Though she can repeat Dr. Grafman's explanation that her right hemisphere now handles such left-brain activities as speaking, reading, and math, she sometimes speaks of the cyst as though it has substance, as though it were a kind of alien being with a personality and will, rather than an emptiness inside her skull, where a left hemisphere should have been. This paradox displays two tendencies in her thought. She has a superior memory for concrete details but difficulty with abstract thought. Being concrete has some advantages. Michelle is a great speller and can remember the arrangement of letters on the page, because like many concrete thinkers, she can record events in memory and keep them as *fresh* and *vivid* as the moment when she first perceived them. But she can find it difficult to understand a story illustrating an underlying moral, theme, or main point that is not explicitly spelled out, because that involves abstraction.

Over and over I encountered examples of Michelle's interpreting symbols concretely. When Carol was talking about how shocked she was to see that second CAT scan with no left hemisphere, I heard a noise. Michelle, who had been listening, started sucking and blowing into the bottle she had been drinking from.

"What are you doing?" Carol asked her.

"Oh well, see, um, I am getting my feelings out in the bottle," said Michelle. It seemed as though she felt that her feelings could be almost literally breathed out into the bottle.

I asked Michelle whether her mother's describing the CAT scan was upsetting.

"No, no, no, um, um, see, it's important that this is said, and I'm just keeping my right side in control"—an example of Michelle's belief that when she gets upset, her cyst is "taking over."

At times she uses nonsense words, not so much to communicate as to discharge feelings. She mentioned in passing that she loved doing crosswords and word searches, even while watching TV.

"Is it because you want to improve your vocabulary?" I asked.

She answered, "Actually—ACTING BEES! ACTING BEES!—I do that while I'm watching sitcoms on television so as to not let my mind get bored."

She sang "ACTING BEES!" out loud, a bit of music inserted into her answer. I asked her to explain.

"Utter nonsense, when, when, when, when, when I am asked things that frustrate me," said Michelle.

She often chooses words not so much for their abstract meaning as for their physical quality, their similar rhyming sound—a sign of her concreteness. Once, while scooting out of the car, she erupted singing, "TOOPERS IN YOUR POOPERS." She often sings her exclamations aloud in restaurants, and people look at her. Before she started to sing, she would clench her jaw so hard when she was frustrated that she broke her two front teeth, then broke the bridge that replaced them several times. Somehow singing nonsense helped her break the biting habit. I asked her if her singing nonsense words soothed her.

"I KNOW YOUR PEEPERS!" she sang. "When I sing, my right side is controlling my cyst."

"Does it soothe you?" I persisted.

"I think so," she said.

The nonsense often has a joking quality, as though she is getting one up on the situation, by using funny words. But it typically occurs when she senses that her mind is failing her and she cannot understand why.

"My right side," she says, "cannot do some of the things that other people's right side can do. I can make simple decisions, but not decisions that require a lot of subjective thinking."

That's why she not only likes but loves repetitive activities that might drive others crazy, like data entry. She currently enters and maintains all the data for the roster of five thousand parishioners at the church where her mother works. On her computer she shows me one of her favorite pastimes—solitaire. As I watch her, I am amazed at how quickly she can play. At this task, where no "subjective" assessments are required, she is *extremely* decisive.

"Oh! Oh! And look, oh, oh, look here!" As she squeals with delight, calling out the names of the cards and placing them, she starts singing. I realize that she visualizes the *entire* deck in her head. She knows the position and identity of each card she has seen, whether it is currently turned over or not.

The other repetitive task she enjoys is folding. Each week, with a smile on her face, she folds, at lightning speed, one thousand sheets of church flyers in a half hour—using only one hand.

Her abstraction problem may well be the dearest price she's paid for having an overcrowded right hemisphere. To get a better sense of her ability with abstractions, I asked her to explain some proverbs.

What does "Don't cry over spilt milk" mean?

"It means don't spend your time worrying about one thing."

I asked her to tell me more, hoping she might add that it's no use focusing on misfortunes about which nothing can be done.

She started breathing very heavily and singing, in an upset voice, "DON'T LIKE PARTIES, PARTIES, OOOOO."

Then she said that she knew one symbolic phrase: "That's the way the ball bounces." She said it meant "That's the way things are."

Next I asked her to interpret a proverb she hadn't heard: "People in glass houses shouldn't throw stones."

Again she started breathing heavily.

Because she goes to church, I asked her about Jesus saying, "Let he who has not sinned cast the first stone," recalling for her the story in which he said it.

She sighed and breathed heavily. "I AM FINDING YOUR PEAS! This is something I really have to think about."

I went on to ask her about similarities and differences between objects, a test of abstraction that is not as challenging as interpreting proverbs or allegories, which involves longer sequences of symbols. Similarities and differences work much more closely with the details.

Here, she performed much faster than most people. What is similar about a chair and a horse? Without missing a beat she said, "They both have four legs and you can sit on them." "And a difference?" "A horse is alive, and a chair isn't. And a horse is able to move by itself." I went through a number of these, and she answered all perfectly and at lightning speed. This time there was no nonsense singing. I gave her some arithmetic problems and memory problems, and she answered them perfectly too. She told me that in school arithmetic was always very easy, and she was so good at it they took her out of her special education class and put her into a regular class. But in eighth grade, when algebra, which is more abstract, was introduced, she found it very hard. The same thing happened in history as well. At first she shone, but when historical concepts were introduced in eighth grade, she found them hard to grasp. A consistent picture emerged: her memory for details was excellent; abstract thinking was a challenge.

I began to suspect Michelle was a savant with some extraordinary mental abilities when, in our conversations, almost as an aside, she would unobtrusively, but with uncommon accuracy and confidence, correct her mother about the date of a particular event. Her mother mentioned a trip to Ireland and asked Michelle when that was.

"May of '87," Michelle said immediately.

I asked her how she did that. "I remember most things . . . I think it's more vivid or something." She said her vivid memory goes back eighteen years, to the mid-1980s. I asked her if she had a formula or rules for figuring out dates, as many savants do. She said she usually remembers the day and the event without calculation but also knows that the calendar follows a pattern for six years and then moves to a five-year pattern, depending on where leap years occur. "Like today is Wednesday, June 4. Six years ago June 4 was also on a Wednesday."

"Are there other rules?" I asked. "What was June 4 three years ago?"

"That was a Sunday then."

"Did you use a rule?" I asked.

"No, I didn't. I just went back in my memory."

Amazed, I asked her if she had ever been fascinated with calendars. She said no, flatly. I asked if she enjoyed remembering things.

"It's just something I do."

I asked her a number of dates rapid fire that I would check later.

"March 2, 1985?"

"That was a Saturday." Her answer was *immediate* and correct.

"July 17, 1985?"

"A Wednesday." Immediate and correct. I realized it was harder for me to think of random dates than for her to answer.

Because she said that she often could remember days back into the mid-1980s without using a formula, I tried to push her past her recall and asked her the day of the week for August 22, 1983.

This time she took half a minute and was clearly calculating, whispering to herself, instead of remembering.

"August 22, 1983, um, that was a Tuesday."

"That was harder because?"

"Because in my mind I only go back to the fall of 1984. That's when I remember things well." She explained she had a clear memory of each day and what happened on it during the period she was in school, and that she used those days as an anchor.

"August 1985 started on a Thursday. So what I did was go back two years. August of '84 started on a Wednesday."

Then she said, "I made a boo-boo," and laughed. "I said August 22, 1983, was on a Tuesday. It was actually on a Monday." I checked it, and her correction was right.

Her calculating speed was dazzling, but more impressive was the vivid way in which she remembered events that had happened throughout the previous eighteen years.

Sometimes savants have unusual ways of representing experiences. The Russian neuropsychologist Aleksandr Luria worked with a mnemonist, or memory artist, "S," who could memorize long tables of random numbers, and he made his living performing these skills. S had a photographic memory, going all the way back to infancy, and was also a "synesthete," so that certain senses, not normally connected, were "cross-wired." High-level synesthetes can experience concepts, such as the days of the week, as having colors, which allows them to have particularly vivid experiences and memories. S associated certain numbers with colors and, like Michelle, often could not get the main point.

"There are certain people," I said to Michelle, "who, when they imagine a day of the week, see a color—which makes it more vivid. They might think of Wednesdays as red, Thursdays as blue, Fridays as black—"

"Ooh, ooh!" she said. I asked her if she had that ability.

"Well, not a color code like that." She had *scenes* for the days of the week. "For Monday I picture my classroom at the Child Development Center. For the word 'hello' I picture the little room off to the right of the lobby of Belle Willard."

"Holy cow!" Carol erupted. She explained that Michelle went to Belle Willard, a special education center, from the time she was fourteen months old until she was two years and ten months.

I went through the days of the week with her. Each was attached to a scene. *Saturday.* She explained that she sees a toy merry-go-round with a light green bottom and a yellow top with holes in it, near where she lives. She imagines having "sat" on a merry-go-round toy as a child and "sat is the first syllable of Saturday," which she guesses is why she experiences Saturday connected to the scene. *Sunday* has a scene with sunshine, and the "sun" sound is the link. But other days have scenes she couldn't explain. *Friday.* "A bird's-eye view of the pancake griddle that was used in our old kitchen," which she had last seen about eighteen years ago, before the kitchen was remodeled. (Perhaps she associated *Fri*-day with the griddle because it is used to *fry* foods.)

Jordan Grafman is the research scientist trying to figure out how Michelle's brain works. After Carol read his article on plasticity, she contacted him, and he said she could bring Michelle in for a visit. Ever since, Michelle has gone for testing, and he has used what he discovers to help her adapt to her situation and to better understand how her brain developed.

Grafman has a warm smile, a musical voice, and fair hair, and his broad white-coated, six-foot frame fills his small book-lined office at the National Institutes of Health. He is chief of the Cognitive Neurosciences Section, National Institute of Neurological Disorders and Stroke. He has two main interests: understanding the frontal lobes and neuroplasticity—the two subjects, taken together, that help explain Michelle's extraordinary strengths and her cognitive difficulties.

For twenty years Grafman served as a captain in the United States Air Force, Biomedical Sciences Command. He received a Defense Meritorious Service Medal for his work as head of the Vietnam Head Injury Study. He has probably seen more people with frontal lobe injuries than anyone else in the world.

His own life is an impressive story of transformation. When Jordan was in elementary school, his father had a devastating stroke that caused a type of brain damage, then poorly understood by physicians, that changed his personality. He had emotional outbursts and what is called, euphemistically, in neurology "social disinhibition"—meaning the release of the aggressive and sexual instincts normally repressed or inhibited. Nor could he seem to grasp the main point of what people were saying. Jordan did not understand what was causing his father's behavior. Jordan's mother divorced her husband, who lived the rest of his life in a transient hotel in Chicago, where he died of a second stroke alone in a back alley.

Jordan, in deep pain, stopped attending elementary school and became a juvenile delinquent. Yet something in him longed for more, and he started spending his mornings in the public library reading, discovering Dostoyevksy and other great novelists. In the afternoons he visited the Art Institute until he learned it was a cruising spot where young boys were targets. He spent evenings in Old Town's jazz and blues clubs. On the streets he got a real psychological education, learning by trial and error what makes people tick. To avoid being sent to the St. Charles reformatory, essentially a jail for kids under sixteen, he spent four years in a boys' home and reform school, where he saw a social worker for the psychotherapy, which he felt rescued him and "prepared me for the rest of my life." He graduated from high school and fled what had become for him a brown-gray Chicago, to a pastel California. He fell in love with Yosemite and decided to become a geologist. But by chance he took a course in the psychology of dreams and found it so fascinating he changed his concentration to psychology.

His first encounter with neuroplasticity was in 1977, when he was in graduate school at the University of Wisconsin, working with a brain-damaged African-American woman who made an unexpected recovery. "Renata," as he calls her, had been strangled in an assault in Central Park in New York City and left for dead. The attack cut off oxygen to her brain long enough to cause an anoxic injury—neuronal death from the lack of oxygen. Grafman first saw her more than five years after the attack, after the doctors had given up on her. Her motor cortex had been so severely damaged that she had great trouble moving, was disabled and wheelchair bound, her muscles wasted away. The team believed she probably had damage to her hippocampus; she had severe memory problems and could barely read. Since the assault, her life had been one downward spiral. She couldn't work and lost her friends. Patients like Renata were assumed to be beyond help, since anoxic injury leaves behind vast amounts of dead brain tissue, and most clinicians believed that when brain tissue dies, the brain cannot recover.

Nevertheless, the team Grafman was working on began giving Renata intensive training—the kinds of physical rehab usually given to patients only in the first weeks after their injuries. Grafman had been doing memory research, knew about rehabilitation, and wondered what would happen if the two fields were integrated. He suggested that Renata begin memory, reading, and thinking exercises. Grafman had no idea that Paul Bach-y-Rita's father had actually benefited from a similar program twenty years before.

She started to move more and became more communicative and more able to concentrate and think and to remember day-to-day events. Ultimately she was able to go back to school, get a job, and reenter the world. Though she never completely recovered, Grafman was amazed by her progress, saying these interventions "had so improved the quality of her life that it was stunning."

The U.S. Air Force put Grafman through graduate school. In return, he was commissioned a captain and made director of the neuropsychological component of the Vietnam Head Injury Study, where he had his second exposure to brain plasticity. Since soldiers face toward the battlefield, its torrent of flying metal often enters and damages the tissue in the front of their brains, the frontal lobes, which coordinate other parts of the brain and help the mind focus on the main point of a situation, form goals, and make lasting decisions.

Grafman wanted to understand what factors most affected recovery from frontal lobe injuries, so he began to examine how a soldier's health, genetics, social status, and intelligence prior to his injury might predict his chance of recovery. Since everyone in the service has to take the Armed Forces Qualifications Test (roughly equivalent to an IQ test), Grafman could study the relationship of preinjury intelligence to that after recovery. He found that aside from the size of the wounds and the location of the injury, a soldier's IQ was a very important predictor of how well he would recover his lost brain functions. Having more cognitive ability—intelligence to spare—enabled the brain to respond better to severe trauma. Grafman's data suggested that highly intelligent soldiers seemed better able to reorganize their cognitive abilities to support the areas that had been injured.

As we have seen, according to strict localizationism, each cognitive function is processed in a different genetically predetermined location. If that location is wiped out by a bullet, so should its functions be—forever—unless the brain is plastic and capable of adapting and creating new structures to replace the damaged ones.

Grafman wanted to explore plasticity's limits and potential, to discover how long structural reorganization takes, and to understand whether there are different types of plasticity. He reasoned that because each person with a brain injury has uniquely affected areas, paying close attention to individual cases was often more productive than were large group studies.

Grafman's view of the brain integrates a nondoctrinaire version of localizationism with plasticity.

The brain is divided into sectors, and in the course of development each acquires a primary responsibility for a particular kind of mental activity. In complex activities several sectors must interact. When we read, the meaning of a word is stored or "mapped" in one sector of the brain; the visual appearance of the letters is stored in another, and its sound in yet another. Each sector is bound together in a network, so that when we encounter the word, we can see it, hear it, and understand it. Neurons from each sector have to be activated at the same time—coactivated—for us to see, hear, and understand at once.

The rules for storing all this information reflect the use-it-or-lose-it principle. The more frequently we use a word, the more easily we'll find it. Even patients with brain damage to the word sector are better able to retrieve words they used frequently before their injury than those they used infrequently.

Grafman believes that in any area of the brain that performs an activity, such as storing words, it is the neurons in the center of that area that are most committed to the task. Those on the border are far less committed, so adjacent brain areas compete with each other to recruit these border neurons. Daily activities determine which brain area wins this competition. For a postal worker who looks at addresses on envelopes without thinking about their meaning, the neurons on the boundary between the visual area and the meaning area will become committed to representing the "look" of the word. For a philosopher, interested in the meaning of words, those boundary neurons will become committed to representing meaning. Grafman believes that everything we know from brain scans about these boundary areas tells us that they can expand quickly, within minutes, to respond to our moment-by-moment needs.

From his research Grafman has identified four kinds of plasticity.

The first is "map expansion," described above, which occurs largely at the boundaries between brain areas as a result of daily activity.

The second is "sensory reassignment," which occurs when one sense is blocked, as in the blind. When the visual cortex is deprived of its normal inputs, it can receive new inputs from another sense, such as touch.

The third is "compensatory masquerade," which takes advantage of the fact that there's more than one way for your brain to approach a task. Some people use visual landmarks to get from place to place. Others with "a good sense of direction" have a strong spatial sense, so if they lose their spatial sense in a brain injury, they can fall back on landmarks. Until neuroplasticity was recognized, compensatory masquerade—also called compensation or "alternative strategies," such as switching people with reading problems to audio tapes—was the chief method used to help children with learning disabilities.

The fourth kind of plasticity is "mirror region takeover." When part of one hemisphere fails, the mirror region in the opposite hemisphere adapts, taking over its mental function as best it can.

This last idea grew out of work Grafman and his colleague Harvey Levin did with a boy I shall call Paul, who was in a car accident when he was seven months old. A blow to his head pushed the bones of his fractured skull into his *right* parietal lobe, the top central part of the brain, behind the frontal lobes. Grafman's team first saw Paul when he was seventeen.

Surprisingly, he was having problems with calculation and number processing. People with *right* parietal injuries are expected to have problems processing visual-spatial information. Grafman and others had established that it is the *left* parietal lobe of the brain that normally stores mathematical facts and performs calculations involved in simple arithmetic, yet Paul's left lobe had not been injured.

A CAT scan showed that Paul had a cyst on his injured right side. Then Grafman and Levin did an fMRI scan (functional magnetic resonance imaging), and, while Paul's brain was being scanned, gave him simple arithmetic problems. The scan showed there was *a very weak* activation of the left parietal area.

They concluded from these odd results that the left area was weakly activated during arithmetic because it was now processing the visual-spatial information that could no longer be processed by the right parietal lobe.

The car accident occurred before seven-month-old Paul was required to learn arithmetic, therefore *before* the left parietal lobe was committed to becoming a specialized processing area for calculation. During the time between seven months and six years, when he started learning arithmetic, it had been far more important for him to navigate, for which he required visual-spatial processing. So visual-spatial activity found its home in the part of the brain that most closely approximated the right parietal lobe—the left parietal lobe. Paul could now navigate through the world, but at a cost. When he had to learn arithmetic, the central part of the left parietal sector was already committed to visual-spatial processing.

Grafman's theory provides an explanation of how Michelle's brain evolved. Michelle's loss of brain tissue occurred before there could have been any significant commitment of her right hemisphere. Since plasticity is at its height in the earliest years, what probably saved Michelle from certain death was that her damage occurred so early. When her brain was still forming, her right hemisphere had time to adjust in the womb, and Carol was there to care for her.

It is possible that her right hemisphere, which normally processes visual-spatial activities, was able to process speech because, being partially blind and barely able to crawl, Michelle learned to speak before she learned to see and walk. Speech would have trumped visual-spatial needs in Michelle, just as visual-spatial needs had trumped arithmetical needs in Paul.

Migration of a mental function to the opposite hemisphere can happen because early in development our hemispheres are quite similar, and only later do they gradually specialize. Brain scans of babies in their first year show that they process new sounds in both hemispheres. By age two they usually process these new sounds in the left hemisphere, which has begun to specialize in speech. Grafman wonders whether visual-spatial ability, like language in babies, is initially present in both hemispheres and then inhibited in the left as the brain specializes. In other words, each hemisphere *tends* to specialize in certain functions but is not hardwired to do so. The age at which we learn a mental skill strongly influences the area in which it gets processed. As infants, we are *slowly* exposed to the world around us, and as we learn new skills, the most suitable processing sectors of our brains that are as yet uncommitted are the ones used to process those skills.

"Which means," says Grafman, "that if you take a million people, and you look at the same areas of their brains, you will see those areas more or less committed to performing the same functions or processes." But he adds, "They may not be in the exact same place. And they shouldn't be, because each of us will have different life experiences."

The riddle of the relationship between Michelle's extraordinary abilities and her difficulties is explained by Grafman's work on the frontal lobe. Specifically, his work on the prefrontal cortex helps explain the price Michelle has had to pay to survive. The prefrontal lobes are the part of the brain that is most uniquely human, as they are most developed in human beings, relative to other animals.

Grafman's theory is that over the course of evolution the prefrontal cortex developed the ability to capture and retain information over longer and longer periods of time, allowing human beings to develop both foresight and memory. The left-frontal lobe became specialized in storing memories of *individual events* and the right in *extracting a theme* or the main point from a series of events or from a story.

Foresight involves extracting the theme from a series of events before they completely unfold, and it is a great advantage in life: to know that when a tiger crouches it is preparing to attack may help you survive. The person with foresight doesn't have to experience the entire series of events to know what is likely coming.

People with right-prefrontal lesions have impaired foresight. They can watch a movie, but they can't get the main point or see where the plot is going. They don't plan well, since planning involves ordering a series of events so that they lead toward a desired outcome, goal, or main point. Nor do people with right frontal lesions execute their plans well. Unable to keep to the main point, they are easily distracted. They are often socially inappropriate because they don't get the main point of social interactions, which are also a series of events, and they have difficulty understanding metaphors and similes, which require extracting the main point or theme from various details. If a poet says, "A marriage is a battle zone," it is important to know that the poet doesn't mean the marriage consists of actual explosions and dead bodies; rather, it is a husband and wife fighting intensely.

All the areas Michelle has difficulty with—getting the main point, understanding proverbs, metaphors, concepts, and abstract thought—are right prefrontal activities. Grafman's standardized psychological testing confirmed that she has difficulty planning, sorting out social situations, understanding motives (a version of getting the main theme, applied to social life), and also some problems empathizing with and forecasting the behavior of others. Her relative absence of foresight, Grafman thinks, increases her level of anxiety and makes it harder for her to control her impulses. On the other hand, she has a savant's ability to remember individual events and the exact dates they occurred—a left-prefrontal function.

Grafman believes Michelle has the same kind of mirror area adaptation as Paul but that her mirror sites are her prefrontal lobes. Because one usually masters registering the occurrence of events before one learns to extract their main theme, event registration—most often a left prefrontal function—so occupied her right-prefrontal lobe that theme extraction never had a chance to fully develop.

When I met with Grafman after I saw Michelle, I asked him why she remembers events so much better than the rest of us. Why not just a normal ability?

Grafman thinks that her superior ability to remember events may be related to the fact that she has only one hemisphere. Normally the two hemispheres are in constant communication. Each not only informs the other of its own activities but also corrects its mate, at times restraining it and balancing the other's eccentricities. What happens when that hemisphere is stricken and can no longer inhibit its partner?

A dramatic example has been described by Dr. Bruce Miller, a professor of neurology at the University of California, San Francisco, who has shown that some people who develop frontotemporal lobe dementia in the *left* side of their brain lose their ability to understand the meaning of words but *spontaneously* develop unusual artistic, musical, and rhyming skills—skills usually processed in the right temporal and parietal lobes. Artistically, they become particularly good at drawing details. Miller argues that the left hemisphere normally acts like a bully, inhibiting and suppressing the right. As the left hemisphere falters, the right's uninhibited potential can emerge.

In fact, people without disabilities can benefit from liberating one hemisphere from another. Betty Edwards's popular book *Drawing on the Right Side of the Brain,* written in 1979, years before

Miller's discovery, taught people to draw by developing ways to stop the verbal, analytical left hemisphere from inhibiting the right hemisphere's artistic tendencies. Inspired by the neuroscientific research of Richard Sperry, Edwards taught that the "verbal," "logical," and "analytical" left hemisphere perceives in ways that actually interfere with drawing and tends to overpower the right hemisphere, which is better at drawing. Edwards's primary tactic was to deactivate the left hemisphere's inhibition of the right by giving the student a task the left hemisphere would be unable to understand and so "turn down." For instance, she had students draw a picture of a Picasso sketch while looking at it upside down and found they did a far better job than when doing it right-side up. Students would develop a sudden knack for drawing rather than acquiring the skill gradually.

In Grafman's view, Michelle's superior registration of events might have developed because, once event registration developed in her right hemisphere, there was no left hemisphere to inhibit it, as usually happens after the main point has been extracted and the details are often no longer important.

Since there are many thousands of brain activities going on at once, we need forces to inhibit, control, and regulate our brains in order to keep us sane, organized and in control of ourselves so we don't "ride off in all directions at once." It would seem that the most frightening thing about brain disease is that it might erase certain mental functions. But just as devastating is a brain disease that leads us to express parts of ourselves we wish didn't exist. Much of the brain is inhibitory, and when we lose that inhibition, unwanted drives and instincts emerge full force, shaming us and devastating our relationships and families.

A few years ago Jordan Grafman was able to get the records from the hospital where his own father had been diagnosed with the stroke that led to his loss of inhibition and his ultimate deterioration. He discovered that his father's stroke had been in the right-frontal cortex, the area Grafman had spent the past quarter century studying.

Before I leave, I am to get the tour of Michelle's inner sanctum. "This is my bedroom," she says proudly. It is painted blue and crammed with her collection of stuffed bears, Mickey and Minnie Mouse, and Bugs Bunny. On her bookshelves are hundreds of the Baby-Sitters Club books, a series that often appeals to girls in the years just before puberty. She has a collection of Carol Burnett tapes and loves easy rock from the 1960s and 1970s. Seeing the room, I wonder about her social life. Carol explains that she was a loner growing up; she loved books instead.

"You didn't seem to want to have others around," she says to Michelle. One doctor thought she demonstrated some autistic behaviors but that she was not autistic, and I can see she's not. She's courteous, recognizes people's comings and goings, and is warm and connected to her parents. She longs for a connection to people and feels hurt when they don't look her in the eye, as so often happens when "normal people" encounter those with disabilities.

Hearing the autism comment, Michelle pipes up, "My theory is that I always liked to be alone because that way I would not cause any trouble." She has many painful memories of trying to play with other kids, and of their not knowing how to play with someone with her disabilities—particularly her hypersensitivity to sounds. I ask her if she has any friends from the past whom she keeps in touch with now.

"No," she says.

"Nope, nobody," whispers Carol solemnly.

I ask Michelle whether during eighth or ninth grades, when boys and girls become more social, she got interested in dating.

"No, no, I didn't." She says she's never had a crush on anyone. She's never really been interested.

"Did you ever dream you'd get married?"

"I don't think so."

There is a theme to her preferences, tastes, and longings. The Baby-Sitters Club, Carol Burnett's harmless humor, the toy bear collection, and everything else I saw in Michelle's blue room were part of that phase of development called "latency," the relatively calm period that precedes the storm of puberty, with its erupting instincts. Michelle, it seemed to me, showed many latency passions, and I found myself wondering whether the absence of her left lobe had affected her hormonal development even though she was a fully developed woman. Perhaps these tastes were the result of her protected upbringing, or perhaps her difficulty understanding the motives of others led her to a world in which the instincts are quieted and where humor is gentle.

Carol and Wally, loving parents of a child with a disability, believe they must make preparations for Michelle after they are gone. Carol is doing her best to line up Michelle's siblings to help out, so that Michelle isn't left alone. She's hoping Michelle will be able to get a job in the local funeral home when the woman who does data entry retires, sparing Michelle the travel she dreads.

The Macks have had other anxieties and near tragedies to endure. Carol has had cancer. Michelle's brother Bill, whom Carol describes as a thrill seeker, has had many incidents. The day he was voted head of the rugby team, his mates flipped him into the air to celebrate, and he landed on his head, breaking his neck. Fortunately, a top-notch surgical team saved him from a life of paralysis. As Carol started to tell me how she went to the hospital to tell Bill that God was trying to get his attention, I looked at Michelle. She seemed serenely quiet, and a smile was on her face.

"What are you thinking, Michelle?" I asked.

"I'm fine," she said.

"But you are smiling—are you finding this interesting?"

"Yeah," she said.

"I bet I know what she's thinking," said Carol.

"What?" said Michelle.

"About heaven," said Carol.

"I think so, yeah."

"Michelle," said Carol, "has very deep faith. In many ways it is a very simple faith." Michelle has an idea of what heaven is going to be like, and whenever she thinks about it, "you see this smile."

"Do you ever dream at night?" I asked.

"Yes," she answered, "in little snatches. But no nightmares. Mostly daydreams."

"What about?" I asked.

"Mostly about upstairs. Heaven."

I asked her to tell me about it, and she got excited.

"Okay, sure!" she said. "There are some people for whom I have a very high regard, and my wish is that these people would live together, unisexually, nearby, the women in one place, and the men in another. And two of the men would agree with each other that I should be given an offer to live with the women." Her mother and father are also there. They all live in a high-rise apartment building, but her parents are on a lower floor, and Michelle lives with the women.

"She broke it to me one day," said Carol. "She said, 'I hope you don't mind, but when we all get to heaven, I don't want to live with you.' I said, 'Okay.'"

I asked Michelle what the people will do for entertainment, and she answered, "Things that they would normally do on vacation here. You know, like play miniature golf. Not work-type stuff."

"Would the men and women ever date?"

"I don't know. I know they would get together. But for fun stuff."

"Do you see heaven as having material things, like trees and birds?"

"Oh yeah! yeah! And another thing about heaven is that all the food up there is fat-free and calorie-free, so that we would be able to have all the food we want. And we wouldn't have to use money to pay for things." And then she added something her mother had always told her about heaven. "There is always happiness in heaven. There aren't any medical problems at all. Just happiness."

I see the smile—an overflow of inner peace. In Michelle's heaven are all the things she's striving for—more human contact, vague hints of increased but safely circumscribed relations between men and women, all that has given her pleasure. Yet all this occurs in an afterworld where, though she is more independent, she can find the parents she so loves not too far away. She has no medical problems, nor does she wish for the other half of her brain. She's fine there just as she is.

Appendix 1

The Culturally Modified Brain

Not Only Does the Brain Shape Culture, Culture Shapes the Brain

What is the relationship between the brain and culture?

The conventional answer of scientists has been that the human brain, from which all thought and action emanate, produces culture. Based on what we have learned about neuroplasticity, this answer is no longer adequate.

Culture is not just produced by the brain; it is also by definition a series of activities that shape the mind. *The Oxford English Dictionary* gives one important definition of "culture": "the cultivating or development . . . of the mind, faculties, manners, etc. . . . improvement or refinement by education and training . . . the training, development and refinement of the mind, tastes and manners." We become cultured through training in various activities, such as customs, arts, ways of interacting with people, and the use of technologies, and the learning of ideas, beliefs, shared philosophies, and religion.

Neuroplastic research has shown us that every sustained activity ever mapped—including physical activities, sensory activities, learning, thinking, and imagining—changes the brain as well as the mind. Cultural ideas and activities are no exception. Our brains are modified by the cultural activities we do—be they reading, studying music, or learning new languages. We all have what might be called a culturally modified brain, and as cultures evolve, they continually lead to new changes in the brain. As Merzenich puts it, "Our brains are vastly different, in fine detail, from the brains of our ancestors . . . In each stage of cultural development . . . the average human had to learn complex new skills and abilities that all involve massive brain change . . . Each one of us can actually learn an incredibly elaborate set of ancestrally developed skills and abilities in our lifetimes, in a sense generating a re-creation of this history of cultural evolution via brain plasticity."

So a neuroplastically informed view of culture and the brain implies a two-way street: the brain and genetics produce culture, but culture also shapes the brain. Sometimes these changes can be dramatic.

The Sea Gypsies

The Sea Gypsies are nomadic people who live in a cluster of tropical islands in the Burmese archipelago and off the west coast of Thailand. A wandering water tribe, they learn to swim before they learn to walk, and live over half their lives in boats on the open sea, where they are often born and die. They survive by harvesting clams and sea cucumbers. Their children dive down, often thirty feet beneath the water's surface, and pluck up their food, including small morsels of marine life, and have done so for centuries. By learning to lower their heart rate, they can stay under water twice as long as most swimmers. They do this without any diving equipment. One tribe, the Sulu, dive over seventy-five feet for pearls.

But what distinguishes these children, for our purposes, is that they can see clearly at these great depths, without goggles. Most human beings cannot see clearly under water because as sunlight passes through water, it is bent, or "refracted," so that light doesn't land where it should on the retina.

Anna Gislén, a Swedish researcher, studied the Sea Gypsies' ability to read placards under water and found that they were more than twice as skillful as European children. The Gypsies learned to control the shape of their lenses and, more significantly, to control the size of their pupils, constricting them 22 percent. This is a remarkable finding, because human pupils reflexively get larger under water, and pupil adjustment has been thought to be a fixed, innate reflex, controlled by the brain and nervous system.

This ability of the Sea Gypsies to see under water isn't the product of a unique genetic endowment. Gislén has since taught Swedish children to constrict their pupils to see under water—one more instance of the brain and nervous system showing unexpected training effects that alter what was thought to be a hardwired, unchangeable circuit.

Cultural Activities Change Brain Structure

The Sea Gypsies' underwater sight is just one example of how cultural activities can change brain circuits, in this case leading to a new and seemingly impossible change in perception. Though the Gypsies' brains have yet to be scanned, we do have studies that show cultural activities changing brain structure. Music makes extraordinary demands on the brain. A pianist performing the eleventh variation of the Sixth Paganini Etude by Franz Liszt must play a staggering eighteen hundred notes per minute. Studies by Taub and others of musicians who play stringed instruments have shown that the more these musicians practice, the larger the brain maps for their active left hands become, and the neurons and maps that respond to string timbres increase; in trumpeters the neurons and maps that respond to "brassy" sounds enlarge. Brain imaging shows that musicians have several areas of their brains—the motor cortex and the cerebellum, among others—that differ from those of nonmusicians. Imaging also shows that musicians who begin playing before the age of seven have larger brain areas connecting the two hemispheres.

Giorgio Vasari, the art historian, tells us that when Michelangelo painted the Sistine Chapel, he built a scaffold almost to the ceiling and painted for twenty months. As Vasari writes, "The work was executed in great discomfort, as Michelangelo had to stand with his head thrown back, and he so injured his eyesight that for several months he could only read and look at designs in that posture." This may have been a case of his brain rewiring itself, to see only in the odd position that it had adapted itself to. Vasari's claim might seem incredible, but studies show that when people wear prism inversion glasses, which turn the world upside down, they find that, after a short while, their brain changes and their perceptual centers "flip," so that they perceive the world right side up and even read books held upside down. When they take the glasses off, they see the world as though it were upside down, until they readapt, as Michelangelo did.

It is not just "highly cultured" activities that rewire the brain. Brain scans of London taxi drivers show that the more years a cabbie spends navigating London streets, the larger the volume of his hippocampus, that part of the brain that stores spatial representations. Even leisure activities change our brain; meditators and meditation teachers have a thicker insula, a part of the cortex activated by paying close attention.

Unlike musicians, taxi drivers, and meditation teachers, the Sea Gypsies are an entire culture of hunter-gatherers on the open sea, all of whom share underwater sight.

In all cultures members tend to share certain common activities, the "signature activities of a culture." For Sea Gypsies it is seeing under water. For those of us living in the information age, signature activities include reading, writing, computer literacy, and using electronic media. Signature activities differ from such universal human activities as seeing, hearing, and walking, which develop with minimal prompting and are shared by all humanity, even those rare people who have been raised outside culture. Signature activities require training and cultural experience and lead to the development of a new, specially wired brain. Human beings did not evolve to see clearly under water—we left our "aquatic eyes" behind with scales and fins, when our ancestors emerged from the sea and evolved to see on land. Underwater sight is not the gift of evolution; the gift is brain plasticity, which allows us to adapt to a vast range of environments.

Are Our Brains Stuck in the Pleistocene Age?

A popular explanation of how our brain comes to perform cultural activities is proposed by evolutionary psychologists, a group of researchers who argue that all human beings share the same basic brain modules (departments in the brain), or brain hardware, and these modules developed to do specific cultural tasks, some for language, some for mating, some for classifying the world, and so on. These modules evolved in the Pleistocene age, from about 1.8 million to ten thousand years ago, when humanity lived as hunter-gatherers, and the modules have been passed on, essentially unchanged genetically. Because we all share these modules, key aspects of human nature and psychology are fairly universal. Then, in an addendum, these psychologists note that the adult human brain is therefore anatomically unchanged since the Pleistocene. This addendum goes too far, because it doesn't take plasticity, also part of our genetic heritage, into account.

The hunter-gatherer brain was as plastic as our own, and it was not "stuck" in the Pleistocene at all but rather was able to reorganize its structure and functions in order to respond to changing conditions. In fact, it was that ability to modify itself that enabled us to emerge from the Pleistocene, a process that has been called "cognitive fluidity" by the archaeologist Steven Mithen and that, I would argue, probably has its basis in brain plasticity. All our brain modules are plastic to some degree and can be combined and differentiated over the course of our individual lives to perform a number of functions—as in Pascual-Leone's experiment in which he blindfolded people and demonstrated that their occipital lobe, which normally processes vision, could process sound and touch. Modular change is necessary for adaptation to the modern world, which exposes us to things our hunter-gatherer ancestors never had to contend with. An fMRI study shows that we recognize cars and trucks with the same brain module we use to recognize faces. Clearly, the hunter-gatherer brain did not evolve to recognize cars and trucks. It is likely that the face module was most competitively suited to process these shapes—headlights are sufficiently like eyes, the hood like a nose, the grill like a mouth—so that the plastic brain, with a little training and structural alteration, could process a car with the facial recognition system.

The many brain modules a child must use for reading, writing, and computer work evolved millennia before literacy, which is only several thousand years old. Literacy's spread has been so rapid that the brain could not have evolved a genetically based module specifically for reading. Literacy, after all, can be taught to illiterate hunter-gatherer tribes in a single generation, and there is no way the whole tribe could develop a gene for a reading module in that time. A child today, when it learns to read, recapitulates the stages humanity went through. Thirty thousand years ago humanity learned to draw on cave walls, which required forming and strengthening links between the visual functions (which process images) and the motor functions (which move the hand). This stage was followed in about 3000 B.C. by the invention of hieroglyphics, where simple standardized images were used to represent objects—not a big change. Next, these hieroglyphic images were converted into letters, and the first phonetic alphabet was developed to represent sounds instead of visual images. This change required strengthening neuronal connections between different functions that process the images of letters, their sound, and their meaning, as well as motor functions that move the eyes across the page.

As Merzenich and Tallal learned, it is possible to see reading circuits on brain scans. Thus signature cultural activities give rise to signature brain circuits that did not exist in our ancestors. According to Merzenich, "Our brains are different from those of all humans before us . . . Our brain is modified on a substantial scale, physically and functionally, each time we learn a new skill or develop a new ability. Massive changes are associated with our modern cultural specializations." And though not everyone uses the same brain areas to read, because the brain is plastic, there are typical circuits for reading—physical evidence that cultural activity leads to modified brain structures.

Why Human Beings Became the Preeminent Bearers of Culture

One could rightly ask, why is it that human beings, and not other animals, which also have plastic brains, developed culture? True, other animals, such as chimpanzees, have rudimentary forms of culture and can both make tools and teach their descendants to use them, or perform rudimentary operations with symbols. But these are very limited. As the neuroscientist Robert Sapolsky points out, the answer lies in a very slight genetic variation between us and chimpanzees. We share 98 percent of our DNA with chimpanzees. The human genome project enabled scientists to determine precisely which genes differed, and it turns out that one of them is a gene that determines how many neurons we will make. Our neurons are basically identical to those of chimps and even of marine snails. In the embryo, all our neurons start from a single cell, which divides and makes two, then four, and so on. A regulatory gene determines when that process of division will stop, and it is this gene that differs between humans and chimps. That process goes on for enough rounds until human beings have about 100 billion neurons. It stops a few rounds earlier in chimps, so they have a brain one-third the size of our own. Chimpanzee brains are plastic, but the sheer quantitative difference between ours and theirs leads to "an exponentially greater number of interactions between them," because each neuron can be connected to thousands of cells.

As the scientist Gerald Edelman has pointed out, the human cortex alone has 30 billion neurons and is capable of making 1 million billion synaptic connections. Edelman writes, "If we considered the number of possible neural circuits, we would be dealing with hyper-astronomical numbers: 10 followed by at least a million zeros. (There are 10 followed by 79 zeros, give or take a few, of particles in the known universe.)" These staggering numbers explain why the human brain can be described as the most complex known object in the universe, and why it is capable of ongoing, massive microstructural change, and capable of performing so many different mental functions and behaviors, including our different cultural activities.

A Non-Darwinian Way to Alter Biological Structures

Up until the discovery of neuroplasticity, scientists believed that the only way that the brain changes its structure is through evolution of the species, which in most cases takes many thousands of years. According to modern Darwinian evolutionary theory, new biological brain structures develop in a species when genetic mutations arise, creating variation in the gene pool. If these variations have survival value, they are more likely to be passed on to the next generation.

But plasticity creates a new way—beyond genetic mutation and variation—of introducing new biological brain structures in individuals by non-Darwinian means. When a parent reads, the microscopic structure of his or her own brain is changed. Reading can be taught to children, and it changes the biological structure of their brains.

The brain is changed in two ways. The fine details of the circuits connecting the modules are altered—no small matter. But so are the original hunter-gatherer brain modules themselves, because, in the plastic brain, change in one area or brain function "flows" through the brain, typically altering the modules that are connected to it.

Merzenich demonstrated that change in the auditory cortex—increasing firing rates—leads to changes in the frontal lobe connected to it, and says, "You can't change the primary auditory cortex without changing what is happening in the frontal cortex. It's an absolute impossibility." The brain doesn't have one set of plastic rules for one part and another set for another part. (If that were the case, the different parts of the brain would not be able to interact.) When two modules are linked in a new way in a cultural activity—as when reading links visual and auditory modules as never before—the modules for both functions are changed by the interaction, creating a new whole, greater than the sum of the parts. A view of the brain that takes plasticity and localizationism into account sees the brain as a complex system in which, as Gerald Edelman argues, "smaller parts form a heterogeneous set of components which are more or less independent. But as these parts connect with each other in larger and larger aggregates, their functions tend to become integrated, yielding new functions that depend on such higher order integration."

Similarly, when one module fails, others connected to it are altered. When we lose a sense—hearing, for example—other senses become more active and more acute to make up for the loss. But they increase not only the *quantity* of their processing but also the *quality*, becoming more like the lost sense. The plasticity researchers Helen Neville and Donald Lawson (measuring neuronal firing rates to determine which sectors of the brain are active) found that deaf people intensify their peripheral vision to make up for the fact that they can't hear things coming at them from a distance. People who can hear use their parietal cortex, near the top of the brain, to process peripheral vision, whereas the deaf use their visual cortex, at the back of the brain. Change in one brain module—here a decrease in output—leads to structural and functional change in another brain module, so that the eyes of the deaf come to behave much more like ears, more able to sense the periphery.

Plasticity and Sublimation: How We Civilize Our Animal Instincts

This principle that modules working together modify each other may even help explain how it is possible for us to mix together brute predatory and dominance instincts (processed by instinctual modules) with our more cognitive-cerebral tendencies (processed by intelligence modules), as we do in sports or competitive games such as chess, or in artistic competitions, to come up with activities that express both the instinctual and the intellectual in one activity.

An activity of this kind is called a "sublimation," a hitherto mysterious process by which brutish animal instincts are "civilized." How sublimation occurs has always been a riddle. Clearly, much of parenting involves "civilizing" children by teaching them to restrain or channel these instincts into acceptable expressions, such as in contact sports, board and computer games, theater, literature, and art. In aggressive sports such as football, hockey, boxing, and soccer, fans often express these brute wishes ("Kill him! Flatten him! Eat him alive! "and so on), but the civilizing rules modify the expression of the instinct, so the fans leave satisfied if their team wins enough points.

For over a century, thinkers influenced by Darwin conceded that we had within us brutish animal instincts, but they were unable to explain how sublimation of these instincts might occur. Nineteenth-century neurologists, such as John Hughlings Jackson and the young

Freud, following Darwin, divided the brain into "lower" parts that we share with animals, and that process our brute animal instincts, and "higher" parts that are uniquely human, and that can inhibit the expression of our brutishness. Indeed, Freud believed that civilization rests on the partial inhibition of sexual and aggressive instincts. He also believed we could go too far in repressing our instincts, leading us to develop neuroses. The ideal solution was to express these instincts in ways that were acceptable and even rewarded by our fellow humans, which was possible because the instincts, being plastic, could change their aim. He called this process sublimation, yet as he conceded, he never really explained exactly how an instinct might be transformed into something more cerebral.

The plastic brain solves the riddle of sublimation. Areas that evolved to perform hunter-gatherer tasks such as stalking prey can, because they are plastic, be sublimated into competitive games, since our brains evolved to link different neuronal groups and modules in novel ways. There is no reason why neurons from the instinctual parts of our brains cannot be linked to our more cognitive-cerebral ones and to our pleasure centers, so that they literally get wired together to form new wholes.

These wholes are more than, and different from, the sum of their parts. Recall that Merzenich and Pascual-Leone argued that a fundamental rule of brain plasticity is that when two areas begin to interact, *they influence each other and form a new whole*. When an instinct, such as stalking prey, is linked up to a civilized activity, such as cornering the opponent's king on the chessboard, and the neuronal networks for the instinct and the intellectual activity are also linked, the two activities appear to temper each other—playing chess is no longer about bloodthirsty stalking, though it still has some of the exciting emotions of the hunt. The dichotomy between "low" instinctual and "high" cerebral begins to disappear. Whenever the low and the high transform each other to create a new whole, we can call it a sublimation.

Civilization is a series of techniques in which the hunter-gatherer brain teaches itself to rewire itself. And the sad proof that civilization is a composite of the higher and lower brain functions is seen when civilization breaks down in civil wars, and brutal instincts emerge full-force, and theft, rape, destruction, and murder become commonplace. Because the plastic brain can always allow brain functions that it has brought together to separate, a regression to barbarism is always possible, and civilization will always be a tenuous affair that must be taught in each generation and is always, at most, one generation deep.

When the Brain Is Caught Between Two Cultures

The culturally modified brain is subject to the plastic paradox (discussed in chapter 9, "Turning Our Ghosts into Ancestors"), which can make us either more flexible or more rigid—a major problem when changing cultures, in a multicultural world.

Immigration is hard on the plastic brain. The process of learning a culture—acculturation—is an "additive" experience, of learning new things and making new neuronal connections as we "acquire" culture. Additive plasticity occurs when brain change involves growth. But plasticity is also "subtractive" and can involve "taking things away," as occurs when the adolescent brain prunes away neurons, and when neuronal connections not being used are lost. Each time the plastic brain acquires culture and uses it repeatedly, there is an opportunity cost: the brain loses some neural structure in the process, because plasticity is competitive.

Patricia Kuhl, of the University of Washington in Seattle, has done brain-wave studies that show that human infants are capable of hearing *any* sound distinction in all the thousands of languages of our species. But once the critical period of auditory cortex development closes, an infant reared in a single culture loses the capacity to hear many of those sounds, and unused neurons are pruned away, until the brain map is dominated by the language of its culture. Now its brain filters out thousands of sounds. A Japanese six-month-old can hear the English *r-l* distinction as well as an American infant. At one year she no longer can. Should that child later immigrate, she will have difficulty hearing and speaking new sounds properly.

Immigration is usually an unending, brutal workout for the adult brain, requiring a massive rewiring of vast amounts of our cortical real estate. It is a far more difficult matter than simply learning new things, because the new culture is in plastic competition with neural networks that had their critical period of development in the native land. Successful assimilation, with few exceptions, requires at least a generation. Only immigrant children who pass through their critical periods in the new culture can hope to find immigration less disorienting and traumatizing. For most, culture shock is brain shock.

Cultural differences are so persistent because when our native culture is learned and wired into our brains, it becomes "second nature," seemingly as "natural" as many of the instincts we were born with. The tastes our culture creates—in foods, in type of family, in love, in music—often seem "natural," even though they may be acquired tastes. The ways we conduct nonverbal communication—how close we stand to other people, the rhythms and volume of our speech, how long we wait before interrupting a conversation—all seem "natural" to us, because they are so deeply wired into our brains. When we change cultures, we are shocked to learn that these customs are not natural at all. Indeed, even when we make a modest change, such as moving to a new house, we discover that something as basic as our sense of space, which seems so natural to us, and numerous routines we were not even aware we had, must slowly be altered while the brain rewires itself.

Sensing and Perceiving Are Plastic

"Perceptual learning" is the kind of learning that occurs whenever the brain learns how to perceive with more acuteness or, as occurs in the Sea Gypsies, in a new way and in the process develops new brain maps and structures. Perceptual learning is also involved in the plasticity-based structural change that occurs when Merzenich's *Fast ForWord* helps children with auditory discrimination problems develop more refined brain maps, so they can hear normal speech for the first time.

It has long been assumed that we absorb culture through universally shared, standard-issue, human perceptual equipment, but perceptual learning shows that this assumption is not completely accurate. *To a larger degree than we suspected, culture determines what we can and cannot perceive.*

One of the first people to begin thinking about how plasticity must change the way we think about culture was the Canadian cognitive neuroscientist Merlin Donald, who argued in 2000 that culture changes our *functional* cognitive architecture, meaning that, as with learning to read and write, mental functions are reorganized. We now know that for this to happen, anatomical structures must change too. Donald also argued that complex cultural activities like literacy and language change brain functions, but our most basic brain functions such as vision and memory are not altered. As he put it, "No one suggests that culture determines anything fundamental about vision or basic memory capacity. However, this is obviously not true of the functional architecture of literacy and probably not of language."

Yet in the few years since that statement, it has become clear that even such brain fundamentals as visual processing and memory capacity are to some extent neuroplastic. The idea that culture may change such fundamental brain activities as sight and perception is a radical one. While almost all social scientists—anthropologists, sociologists, psychologists—concede that different cultures interpret the world differently, most scientists and lay people assumed for several thousand years—as the University of Michigan social psychologist Richard E. Nisbett puts it—that "where people in one culture differ from those in another in their beliefs, it can't be because they have different cognitive processes. Rather, they must have been exposed to different aspects of the world, or taught different things." The most famous European psychologist of the mid-twentieth century, Jean Piaget, believed he showed, in a series of brilliant experiments on European children, that perceiving and reasoning unfold in development in the same way for all human beings, and that these processes are universal. True, scholars, travelers, and anthropologists had long observed that the peoples of the East (those Asian peoples influenced by Chinese traditions) and those of the West (the heirs to the traditions of the ancient Greeks) perceive in different ways, but scientists assumed these differences were based on different *interpretations* of what was seen, not on microscopic differences in their perceptual equipment and structures.

For instance, it was often observed that Westerners approach the world "analytically," dividing what they observe into individual parts. Easterners tend to approach the world more "holistically," perceiving by looking at "the whole," and emphasizing the interrelatedness of all things. It was also observed that the differing cognitive styles of the analytic West and the holistic East parallel differences between the brain's two hemispheres. The left hemisphere tends to perform more sequential and analytical processing, while the right hemisphere is often engaged in simultaneous and holistic processing. Were these different ways of seeing the world based on different interpretations of what was seen, or were Easterners and Westerners actually seeing different things?

The answer was unclear because almost all the studies of perception had been done by Western academics on Westerners—typically, on their own American college students—until Nisbett designed experiments to compare perception in the East and the West, working with colleagues in the United States, China, Korea, and Japan. He did so reluctantly because he believed that we all perceive and reason in the same way.

In a typical experiment, Nisbett's Japanese student, Take Masuda, showed students in the United States and Japan eight color animations of fish swimming under water. Each scene had one "focal fish" that moved faster or was bigger, brighter, or more prominent than the other smaller fish it swam among.

When asked to describe the scene, the Americans usually referred to the focal fish. The Japanese referred to the less prominent fish, background rocks, plants, and animals 70 percent more often than did the Americans. Subjects were then shown some of these objects by themselves, not as part of the original scene. The Americans recognized the objects whether they were shown in the original scene or not. The Japanese were better able to recognize an object if it was shown in the original scene. They perceived the object in terms of what it had been "bound" to. Nisbett and Masuda also measured how quickly subjects recognized objects—a test of how *automatic* their perceptual processing was. When objects were put against a new background, the Japanese made mistakes. The Americans did not. These aspects of perception are not under our conscious control and are dependent on trained neuronal circuits and brain maps.

These experiments and many others like them confirm that Easterners perceive holistically, viewing objects as they are related to each other or in a context, whereas Westerners perceive them in isolation. Easterners see through a wide-angle lens; Westerners use a narrow one with a sharper focus. Everything we know about plasticity suggests that these different ways of perceiving, repeated hundreds of times a day, in massed practice, must lead to changes in neural networks responsible for sensing and perceiving. High-resolution brain scans of Easterners and Westerners sensing and perceiving could likely settle the matter.

Further experiments by Nisbett's team confirm that when people change cultures, they learn to perceive in a new way. After several years in America the Japanese begin to perceive in a way indistinguishable from Americans, so clearly the perceptual differences are not based on genetics. The children of Asian-American immigrants perceive in a way that reflects both cultures. Because they are subject to Eastern influences at home and Western influences at school and elsewhere, they sometimes process scenes holistically, and sometimes they focus on prominent objects. Other studies show that people raised in a bicultural situation actually alternate between Western and Eastern perception. Hong Kongers, having lived under both a British and a Chinese influence, can be "primed" to perceive in either Eastern or Western fashion by experiments showing them a Western image of Mickey Mouse or the U.S. Capitol, or an Eastern image of a temple or a dragon. Nisbett and his colleagues are thus doing the first experiments that demonstrate cross-cultural "perceptual learning."

Culture can influence the development of perceptual learning because perception is not (as many assume) a passive, "bottom up" process that begins when energy in the outside world strikes the sense receptors, then passes signals to the "higher" perceptual centers in the brain. The perceiving brain is active and always adjusting itself. Seeing is as active as touching, when we run our fingers over an object to discover its texture and shape. Indeed, the stationary eye is virtually incapable of perceiving a complex object. Both our sensory *and* our motor cortices are always involved in perceiving. The neuroscientists Manfred Fahle and Tomaso Poggio have shown experimentally that "higher" levels of perception affect how neuroplastic change in the "lower," sensory parts of the brain develops.

The fact that cultures differ in perception is not proof that one perceptual act is as good as the next, or that "everything is relative" when it comes to perception. Clearly some contexts call for a more narrow angle of view, and some for more wide-angle, holistic perception. The Sea Gypsies have survived using a combination of their experience of the sea and holistic perception. So attuned are they to the moods of the sea that when the tsunami of December 26, 2004, hit the Indian Ocean, killing hundreds of thousands, they all survived. They saw that the sea had begun to recede in a strange way, and this drawing back was followed by an unusually small wave; they saw dolphins begin to swim for deeper water, while the elephants started stampeding to higher ground, and they heard the cicadas fall silent.

The Sea Gypsies began telling each other their ancient story about "The Wave That Eats People," saying it had come again. Long before modern science put this all together, they had either fled the sea to the shore, seeking the highest ground, or gone into very deep waters, where they also survived. What they were able to do, as more modern people under the influence of analytical science were not, was put all these unusual events together and see the whole, using an exceptionally wide-angle lens, exceptional even by Eastern standards. Indeed, Burmese boatmen were also at sea when these preternatural events were occurring, but they did not survive. A Sea Gypsy was asked how it was that the Burmese, who also knew the sea, all perished.

He replied, "They were looking at squid. They were not looking at anything. They saw nothing, they looked at nothing. They don't know how to look."

Neuroplasticity and Social Rigidity

Bruce Wexler, a psychiatrist and researcher from Yale University, argues, in his book *Brain and Culture,* that the relative decline in neuroplasticity as we age explains many social phenomena. In childhood our brains readily shape themselves in response to the world, developing neuropsychological structures, which include our pictures or representations of the world. These structures form the neuronal basis for all our perceptual habits and beliefs, all the way up to complex ideologies. Like all plastic phenomena, these structures tend to get reinforced early on, if repeated, and become self-sustaining.

As we age and plasticity declines, it becomes increasingly difficult for us to change in response to the world, even if we want to. We find familiar types of stimulation pleasurable; we seek out like-minded individuals to associate with, and research shows we tend to ignore or forget, or attempt to discredit, information that does not match our beliefs, or perception of the world, because it is very distressing and difficult to think and perceive in unfamiliar ways. Increasingly, the aging individual acts to preserve the structures within, and when there is a mismatch between his internal neurocognitive structures and the world, he seeks to change the world. In small ways he begins to micromanage his environment, to control it and make it familiar. But this process, writ large, often leads whole cultural groups to try to impose their view of the world on other cultures, and they often become violent, especially in the modern world, where globalization has brought different cultures closer together, exacerbating the problem. Wexler's point, then, is that much of the cross-cultural conflict we see is a product of the relative decrease in plasticity.

One could add that totalitarian regimes seem to have an intuitive awareness that it becomes hard for people to change after a certain age, which is why so much effort is made to indoctrinate the young from an early age. For instance, North Korea, the most thoroughgoing totalitarian regime in existence, places children in school from ages two and a half to four years; they spend almost every waking hour being immersed in a cult of adoration for dictator Kim Jong Il and his father, Kim Il Sung. They can see their parents only on weekends. Practically every story read to them is about the leader. Forty percent of the primary school textbooks are devoted wholly to describing the two Kims. This continues all the way through school. Hatred of the enemy is drilled in with massed practice as well, so that a brain circuit forms linking the perception of "the enemy" with negative emotions automatically. A typical math quiz asks, "Three soldiers from the Korean People's Army killed thirty American soldiers. How many American soldiers were killed by each of them, if they all killed an equal number of enemy soldiers?" Such perceptual emotional networks, once established in an indoctrinated people, do not lead only to mere "differences of opinion" between them and their adversaries, but to plasticity-based anatomical differences, which are much harder to bridge or overcome with ordinary persuasion.

Wexler's emphasis is on the relative tapering-off of plasticity as we age, but it must be said that certain practices used by cults, or in brainwashing, which obey the laws of neuroplasticity, demonstrate that sometimes individual identities can be changed in adulthood, even against a person's will. Human beings can be broken down and then develop, or at least "add on," neurocognitive structures, if their daily lives can be totally controlled, and they can be conditioned by reward and severe punishment and subjected to massed practice, where they are forced to repeat or mentally rehearse various ideological statements. In some cases, this process can actually lead them to "unlearn" their preexisting mental structures, as Walter Freeman has observed. These unpleasant outcomes would not be possible if the adult brain were not plastic.

A Vulnerable Brain—How the Media Reorganize It

> *The Internet is just one of those things that contemporary humans can spend millions of "practice" events at, that the average human a thousand years ago had absolutely no exposure to. Our brains are massively remodeled by this exposure—but so, too, by reading, by television, by video games, by modern electronics, by contemporary music, by contemporary "tools," etc.*
>
> MICHAEL MERZENICH, 2005

We have discussed several reasons plasticity was not discovered sooner—such as the lack of a window into the living brain, and the more simplistic versions of localizationism. But there is another reason we did not recognize it, one that is particularly relevant to the culturally modified brain. Almost all neuroscientists, as Merlin Donald writes, had a view of the brain as an isolated organ, almost as though it were contained in a box, and they believed that "the mind exists and develops entirely in the head, and that its basic structure is a biological given." The behaviorists and many biologists championed this view. Among those who rejected it were developmental psychologists, because they have generally been sensitive to how outside influences might harm brain development.

Television watching, one of the signature activities of our culture, correlates with brain problems. A recent study of more than twenty-six hundred toddlers shows that early exposure to television between the ages of one and three correlates with problems paying attention and controlling impulses later in childhood. For every hour of TV the toddlers watched each day, their chances of developing serious attentional difficulties at age seven increased by 10 percent. This study, as psychologist Joel T. Nigg argues, did not perfectly control for other possible factors influencing the correlation between TV watching and later attentional problems. It might be argued that parents of children with more attentional difficulties deal with them by putting them in front of television sets. Still, the study's findings are extremely suggestive and, given the rise in television watching, demand further investigation. Forty-three percent of U.S. children two years or younger watch television daily, and a quarter have TVs in their bedrooms. About twenty years after the spread of TV, teachers of young children began to notice that their students had become more restless and had increasing difficulty paying attention. The educator Jane Healy documented these changes in her book *Endangered Minds,* speculating they were the product of plastic changes in the children's brains. When those children entered college, professors complained of having to "dumb down" their courses each new year, for students who were increasingly interested in "sound bites" and intimidated by reading of any length. Meanwhile, the problem was buried by "grade inflation" and accelerated by pushes for "computers in every classroom," which aimed to increase the RAM and gigabytes in the class computers rather than the attention spans and memories of the students. The Harvard psychiatrist Edward Hallowell, an expert on attention deficit disorder (ADD), which is genetic, has linked the electronic media to the rise of attention deficit traits, which are not genetic, in much of the population.

Ian H. Robertson and Redmond O'Connell have had promising results using brain exercises to treat attention deficit disorder, and if that can be done, we have reason to hope that mere traits can be treated as well.

Most people think that the dangers created by the media are a result of content. But Marshall McLuhan, the Canadian who founded media studies in the 1950s and predicted the Internet twenty years before it was invented, was the first to intuit that the media change our brains irrespective of content, and he famously said, "The medium is the message." McLuhan was arguing that each medium reorganizes our mind and brain in its own unique way and that the consequences of these reorganizations are far more significant than the effects of the content or "message."

Erica Michael and Marcel Just of Carnegie Mellon University did a brain scan study to test whether the medium is indeed the message. They showed that different brain areas are involved in hearing speech and reading it, and *different comprehension centers* in hearing words and reading them. As Just put it, "The brain constructs the message . . . differently for reading and listening. The pragmatic implication is that the medium is part of the message. Listening to an audio book leaves a different set of memories than reading does. A newscast heard on the radio is processed differently from the same words read in a newspaper." This finding refutes the conventional theory of comprehension, which argues that a single center in the brain understands words, and it doesn't really matter how (by what sense or medium) information enters the brain, because it will be processed in the same way and place. Michael and Just's experiment shows that each medium creates a different sensory and semantic experience—and, we might add, develops different circuits in the brain.

Each medium leads to a change in the balance of our individual senses, increasing some at the expense of others. According to McLuhan, preliterate man lived with a "natural" balance of hearing, seeing, feeling, smelling, and tasting. The written word moved preliterate man from a world of sound to a visual world, by switching from speech to reading; type and the printing press hastened that process. Now the electronic media are bringing sound back and, in some ways, restoring the original balance. Each new medium creates a unique form of awareness, in which some senses are "stepped up" and others "stepped down." McLuhan said, "The ratio among our senses is altered." We know from Pascual-Leone's work with blindfolded people (stepping down sight) how quickly sensory reorganizations can take place.

To say that a cultural medium, such as television, radio, or the Internet, alters the balance of senses does not prove it is harmful. Much of the harm from television and other electronic media, such as music videos and computer games, comes from their effect on attention. Children and teenagers who sit in front of fighting games are engaged in massed practice and are incrementally rewarded. Video games, like Internet porn, meet all the conditions for plastic brain map changes. A team at the Hammersmith Hospital in London designed a typical video game in which a tank commander shoots the enemy and dodges enemy fire. The experiment showed that dopamine—the reward neurotransmitter, also triggered by addictive drugs—is released in the brain during these games. People who are addicted to computer games show all the signs of other addictions: cravings when they stop, neglect of other activities, euphoria when on the computer, and a tendency to deny or minimize their actual involvement.

Television, music videos, and video games, all of which use television techniques, unfold at a much faster pace than real life, and they are getting faster, which causes people to develop an increased appetite for high-speed transitions in those media. It is the *form* of the television medium—cuts, edits, zooms, pans, and sudden noises—that alters the brain, by activating what Pavlov called the "orienting response," which occurs whenever we sense a sudden change in the world around us, especially a sudden movement. We instinctively interrupt whatever we are doing to turn, pay attention, and get our bearings. The orientation response evolved, no doubt, because our forebears were both predators and prey and needed to react to situations that could be dangerous or could provide sudden opportunities for such things as food or sex, or simply to novel situations. The response is physiological: the heart rate decreases for four to six seconds. Television triggers this response at a far more rapid rate than we experience it in life, which is why we can't keep our eyes off the TV screen, even in the middle of an intimate conversation, and why people watch TV a lot longer than they intend. Because typical music videos, action sequences, and commercials trigger orienting responses at a rate of one per second, watching them puts us into continuous orienting response with no recovery. No wonder people report feeling drained from watching TV. Yet we acquire a taste for it and find slower changes boring. The cost is that such activities as reading, complex conversation, and listening to lectures become more difficult.

McLuhan's insight was that the communications media both extend our range and implode into us. His first law of media is that all the media are extensions of aspects of man. Writing extends memory, when we use a paper and pen to record our thoughts; the car extends the foot, clothing the skin. Electronic media are extensions of our nervous systems: the telegraph, radio, and telephone extend the range of the human ear, the television camera extends the eye and sight, the computer extends the processing capacities of our central nervous system. He argued that the process of extending our nervous system also alters it.

The implosion of the media into us, affecting our brains, is less obvious, but we have seen many examples already. When Merzenich and colleagues devised the cochlear implant, a medium that translates sound waves into electrical impulses, the brain of an implant patient rewired itself to read those impulses.

Fast ForWord is a medium that, like radio or an interactive computer game, conveys language, sounds, and images and radically rewires the brain in the process. When Bach-y-Rita attached blind people to a camera, and they were able to perceive shapes, faces, and perspective, he demonstrated that the nervous system can become part of a larger electronic system. All electronic devices rewire the brain. People who write on a computer are often at a loss when they have to write by hand or dictate, because their brains are not wired to translate thoughts into cursive writing or speech at high speed. When computers crash and people have mini–nervous breakdowns, there is more than a little truth in their cry, "I feel like I've lost my mind!" As we use an electronic medium, our nervous system extends outward, and the medium extends inward.

Electronic media are so effective at altering the nervous system because they both work in similar ways and are basically compatible and thus easily linked. Both involve the instantaneous transmission of electric signals to make linkages. Because our nervous system is plastic, it can take advantage of this compatibility and merge with the electronic media, making a single, larger system. Indeed, it is the nature of such systems to merge whether they are biological or man-made. The nervous system is an internal medium, communicating messages from one area of the body to another, and it evolved to do, for multicelled organisms such as ourselves, what the electronic media do for humanity—connect disparate parts. McLuhan expressed this electronic extension of the nervous system and the self in comic terms: "Now man is beginning to wear his brain outside his skull, and his nerves outside his skin." In a famous formulation, he said, "Today, after more than a century of electric technology, we have extended our central nervous system itself in a global embrace, abolishing both space and time as far as our planet is concerned." Space and time are abolished because electronic media link faraway places instantaneously, giving rise to what he called the "global village." This extension is possible because our plastic nervous system can integrate itself with an electronic system.

Appendix 2

Plasticity and the Idea of Progress

The idea of the brain as plastic has appeared in previous times, in flashes, then disappeared. But even though it is only now being established as a fact of mainstream science, these earlier appearances left their traces and made possible a receptivity to the idea, in spite of the enormous opposition each of the neuroplasticians faced from fellow scientists.

As early as 1762 the Swiss philosopher Jean-Jacques Rousseau (1712–1778), who faulted the mechanistic view of nature of his time, argued that nature was alive and had a history and was changing over time; our nervous systems are not like machines, he said, but are alive and able to change. In his book *Émile, or On Education*—the first detailed book on child development ever written—he proposed that the "organization of the brain" was affected by our experience, and that we need to "exercise" our senses and mental abilities the way we exercise our muscles. Rousseau maintained that even our emotions and passions are, to a great extent, learned early in childhood. He imagined radically transforming human education and culture, based on the premise that many aspects of our nature that we think are fixed are, in fact, changeable and that this malleability is a defining human trait. He wrote, "To understand a man, look to men; and to understand men, look to the animals." When he compared us with other species, he saw what he called human "perfectibility"—and brought the French word *perfectibilité* into vogue—using it to describe a specifically human plasticity or malleability, which distinguishes us in degree from animals. Several months after an animal's birth, he observed, it is for the most part what it will be for the rest of its life. But human beings change throughout life because of their "perfectibility."

It was our "perfectibility," he argued, that allowed us to develop different kinds of mental faculties and to change the balance among our existing mental faculties and senses, but this could also be problematic because it disrupted the natural balance of our senses. Because our brains were so sensitive to experience, they were also more vulnerable to being shaped by it. Educational schools such as the Montessori School, with its emphasis on the education of the senses, grew out of Rousseau's observations. He was also the precursor to McLuhan, who would argue centuries later that certain technologies and media alter the ratio or balance of the senses. When we say that the instantaneous electronic media, television sound bites, and a shift away from literacy have created overly intense, "wired" people with short attention spans, we are speaking Rousseau's language, about a new kind of environmental problem that interferes with our cognition. Rousseau was also concerned that the balance between our senses and our imagination can be disturbed by the wrong kinds of experience.

In 1783 Rousseau's contemporary Charles Bonnet (1720–1793), also a Swiss philosopher and a naturalist familiar with Rousseau's writings, wrote to an Italian scientist, Michele Vincenzo Malacarne (1744–1816), proposing that neural tissue might respond to exercise as do muscles. Malacarne set out to test Bonnet's hypothesis experimentally. He took pairs of birds that came from the same clutch of eggs and raised half of them under enriched circumstances, stimulated by intensive training for several years. The other half received no training. He did the same experiment with two litter-mate dogs. When Malacarne sacrificed the animals and compared their brain size, he found that the animals that received training had larger brains, particularly in a part of the brain called the cerebellum, demonstrating the influence of "enriched circumstances" and "training" on the development of an individual's brain. Malacarne's work was all but forgotten, until revived and mastered by Rosenzweig and others in the twentieth century.

Perfectibilité—the Mixed Blessing

Though Rousseau, who died in 1778, could not have known Malacarne's results, he showed an uncanny ability to anticipate what *perfectibilité* meant for humanity. It provided hope but was not always a blessing. Because we could change, we did not always know what was natural in us and what was acquired from our culture. Because we could change, we could be overly shaped by culture and society, to a point where we drifted too far from our true nature and became alienated from ourselves.

While we may rejoice at the thought that the brain and human nature may be "improved," the idea of human perfectibility or plasticity stirs up a hornet's nest of moral problems.

Earlier thinkers, going back to Aristotle, who did not speak of a plastic brain, argued that there was an obvious ideal or "perfect" mental development. Our mental and emotional faculties were provided by nature, and a healthy mental development was achieved by using those faculties and perfecting them. Rousseau understood that if human mental and emotional life and the brain are malleable, we can no longer be so certain what a normal or perfect mental development would look like; there could be many different kinds of development. Perfectibility meant that we could no longer be so certain about what it meant to perfect ourselves. Realizing this moral problem, Rousseau used the term "perfectibility" in an ironic sense.

From Perfectibility to the Idea of Progress

Any change in how we understand the brain ultimately affects how we understand human nature. After Rousseau the idea of perfectibility quickly got tied to the idea of "progress." Condorcet (1743–1794), the French philosopher and mathematician, who was a major participant in the French Revolution, argued that human history was the story of progress and linked it to our perfectibility. He wrote, "Nature has set no term to the perfection of human faculties; . . . the perfectibility of man is truly indefinite, and . . . the progress of this perfectibility . . . has no other limit than the duration of the globe upon which nature has cast us." Human nature was continually improvable, in intellectual and moral terms, and humans should not give themselves fixed limits to their possible perfection. (This view was somewhat less ambitious than seeking ultimate perfection, but still naïvely utopian.)

The twin ideas of progress and perfectibility came to America through the thought of Thomas Jefferson, who appears to have been introduced to Condorcet by Benjamin Franklin. Among the American founders, Jefferson was most open to the idea and wrote, "I am among those who think well of the human character generally . . . I believe also, with Condorcet . . . that his mind is perfectible to a degree of which we cannot as yet form any conception." Not all the founders agreed with Jefferson, but Alexis de Tocqueville, visiting America from France in 1830, remarked that Americans, in contrast to others, seemed to believe in the "indefinite perfectibility of man." It is the idea of scientific and political progress—and its constant ally, the idea of individual perfectibility—that may well make Americans so interested in self-improvement, self-transformation, and self-help books, as well as in solving problems and in having a can-do attitude.

As hopeful as this all sounds, the idea of human perfectibility in theory has also had a dark side in practice. When utopian revolutionaries in France and Russia, smitten with the idea of progress and embracing a naïve belief in the plasticity of human beings, looked around them and saw an imperfect society, they tended to blame individuals for "standing in the way of progress." A Reign of Terror and the Gulag followed. We must be careful clinically too, as we speak of brain plasticity, not to fall into blaming those who, despite this new science, cannot benefit or change. Clearly neuroplasticity teaches that the brain is more malleable than some have thought, but to move from calling it malleable to calling it perfectible raises expectations to a dangerous level. The plastic paradox teaches that neuroplasticity can also be responsible for many rigid behaviors, and even some pathologies, along with all the potential flexibility that is within us. As the idea of plasticity becomes the focus of human attention in our time, we would be wise to remember that it is a phenomenon that produces effects we think of as both bad and good—rigidity and flexibility, vulnerability, and an unexpected resourcefulness.

The economist and scholar Thomas Sowell has observed, "While the use of the word 'perfectibility' has faded away over the centuries, the concept has survived, largely intact, to the present time. The notion that 'the human being is highly plastic material' is still central among many contemporary thinkers . . ." Sowell's detailed study *A Conflict of Visions* shows that many major Western political philosophers can be classified, and better understood, by taking into account the extent to which they reject or accept this human plasticity and have a more or less constrained view of human nature. While it has often been the case that more "conservative" or "right-leaning" thinkers such as Adam Smith or Edmund Burke seemed to champion the constrained view of human nature, while "liberal" or "left-leaning" thinkers such as Condorcet or William Godwin have tended to believe that it is less constrained, there are times or issues about which conservatives appear to have the more plastic view and liberals the more constrained view. For instance, in recent times, a number of conservative commentators have argued that sexual orientation is a matter of choice and have spoken as though it might be changed by effort or experience—i.e., that it is a plastic phenomenon—whereas, by and large, liberal commentators have tended to argue that it is "hardwired" and "all in the genes." But not all thinkers offer a strictly constrained or unconstrained vision of human nature, and there are those who have had a mixed view of human changeability, perfectibility, and progress.

What we have learned by looking closely at neuroplasticity and the plastic paradox is that human neuroplasticity contributes to both the constrained and the unconstrained aspects of our nature. Thus, while it is true that the history of Western political thought turns in large part upon the attitudes that various ages and thinkers have held toward the question of human plasticity broadly understood, the elucidation of human neuroplasticity in our time, if carefully thought through, shows that plasticity is far too subtle a phenomenon to unambiguously support a more constrained or unconstrained view of human nature, because in fact it contributes to both human rigidity and flexibility, depending upon how it is cultivated.

Acknowledgments

My debts are great and widely spread. First and foremost, they are to two people.

Karen Lipton-Doidge, my wife, gave daily guidance and assistance with this book, discussed the ideas with me as they were formed, helped tirelessly with research, read every draft countless times, and provided every conceivable kind of intellectual and emotional support.

My editor James H. Silberman immediately intuited the importance of neuroplasticity and worked with me for over three years, spurring me on from the earliest parts of this project, taking updates on my travels, watching me—possibly horrified—as I forgot how to write while I internalized the language of neuroplasticity, hoping to grasp the subject on its own terms, and then helping me to slowly reemerge speaking English. He was more attentive, hardworking, plainspoken, and unrelentingly devoted to this project than I ever imagined an editor would be, and his presence, counsel, and craft are on each page. It has been an honor to work with him.

I thank all the neuroplasticians and their colleagues, assistants, subjects, and patients who shared with me the stories told in these chapters. They gave of their time, and I hope I have been able to convey the excitement they feel about the birth of their new field. It is with very great sadness that I learned, as this book was about to go to press, that Paul Bach-y-Rita, the gentle, ingenious iconoclast who was in many ways the father of the idea of neuroplasticity in our time, passed away after a several-year battle with cancer. Amazingly, he worked until three days before he died. In our meetings I found him to be uniquely open, without airs, adventurous, and an altogether endearing, compassionate human being with a panoramic mind. Several of my own patients' stories of neuroplastic change are told in this book, and I am immensely grateful to them. There are also many other patients who spoke with me over the years, and whose changes helped deepen my understanding of human neuroplasticity's potential and limits.

The generosity of spirit of the following people provided great encouragement, and none of them should underestimate how helpful they have been. Arthur Fish championed this project from the very beginning. Geoffrey Clarfield, Jacqueline Newell, Cyril Levitt, Corrine Levitt, Philip Kyriacou, Jordan Peterson, Gerald Owen, Neil Hrab, Margaret-Ann Fitzpatrick-Hanly, and Charles Hanly each read drafts and made extremely helpful comments. Waller Newell, Peter Gellman, George Jonas, Maya Jonas, Mark Doidge, Elizabeth Yanowski, Donna Orwin, David Ellman, Stephen Connell, Kenneth Green, and Sharon Green each provided moral support.

I thank my medical colleagues and teachers at the Center for Psychoanalytic Training and Research, Columbia University, Department of Psychiatry, where this line of thought began: Drs. Meriamne Singer, Mark Sorensen, Eric Marcus, Stan Bone, Robert Glick, Lila Kalinich, Donald Meyers, Roger MacKinnon, and Yoram Yovell.

Though I did not work with him, Eric Kandel, through his publications and towering influence at Columbia, drew me there to better understand the project he championed—a wish to integrate biology, psychiatry, and psychoanalysis.

Dianne de Fenoyl, Hugo Gurdon, John O'Sullivan, Dianna Symonds, Mark Stevenson, and Kenneth Whyte, of the *National Post*, *Saturday Night*, and *Macleans*, supported my writing about neuroscience and neuroplasticity for a general audience. Some of the ideas on neuroplasticity in this book were first discussed in those publications. Chapter 2 appeared in a slightly different version in *Saturday Night*.

Jay Grossman, Dan Kiesel, James Fitzpatrick, and Yaz Yamaguchi were very helpful to me during this period and charitable with their time and conversation.

Among those I interviewed who are not mentioned in chapters, or are mentioned only in passing, I thank Martha Burns for spending much time with me on brain exercises, and Steve Miller and William Jenkins of Scientific Learning, Jeff Zimman and Henry Mahncke of Posit Science, and Gitendra Uswatte of the Taub Therapy Clinic.

Gerald Edelman, Nobel laureate, who has developed the most ambitious theory of consciousness that gives neuroplasticity a central role, was generous with his time when I visited him. Though there is no single chapter in this book devoted to his work—because I chose to describe plasticity by matching the work of a scientist or clinician with a patient whenever possible, and his work is theoretical—Dr. Edelman's theory looms large for me behind these stories, and shows how far a plastic theory of the brain can go. I thank V. S. Ramachandran, not only for our time spent together, but for arranging a memorable lunch with Francis Crick, codiscoverer of DNA, and philosopher Patricia Churchland, at which we had an animated discussion of Dr. Edelman's work, allowing me to see the extraordinary neuroscience community of San Diego at work.

A number of academics and other learned people responded to e-mail questions, including Walter J. Freeman, Mriganka Sur, Richard

C. Friedman, Thomas Pangle, Ian Robertson, Nancy Byl, Orlando Figes, Anna Gislén, Cheryl Grady, Adrian Morrison, Eric Nestler, Clifford Orwin, Allan N. Schore, Myrna Weissman, and Yuri Danilov. A number of agencies gave me grants and awards over the years that allowed me to further my scientific development and writing, including the National Institute of Mental Health, Washington, DC, and the National Health Research and Development Program of Health Canada.

I am immensely grateful to Chris Calhoun, my agent at Sterling Lord, for his wit, enthusiasm, intellectual interest, and spirited guidance throughout. At Viking, editor Hilary Redmon did an extraordinary job as she went over the manuscript, giving numerous helpful suggestions to unify it. I thank Janet Biehl and Bruce Giffords for their astute and erudite copyediting and editorial production (and Bruce for being so levelheaded, supportive, patient, and meticulous throughout the process), and Holly Lindem and Jaya Miceli for the magical cover for the hardcover edition, which captures, in a single image, what this book is about and even the mood I hope the book creates. I also thank the ever-helpful Jacqueline Powers in editorial and Spring Hoteling, the book's designer.

Finally, I would like to thank my daughter, Brauna Doidge, for help with transcripts and my son, Joshua Doidge, who tried different kinds of brain exercises with me, and showed me that they really work.

Despite this mountain of support, to err is as human as the wish to evade responsibility for errors. Still, the responsibility for any oversights or mistakes is mine.

Notes and References

A Note to the Reader About These Notes

These notes are of two kinds. First, there are comments about interesting details, exceptions, historical notes, and more scholarly matters, and these are all preceded by a black dot (•). Second, there are references to articles upon which studies mentioned in the book are based. All notes are preceded by the page number and a phrase from the text to which they refer. There are notes both for the main chapters and the appendices. These phrases are full enough to contain the main relevant idea from the text, so that the reader will only occasionally have to flip back to the text to get the context for the note.

Chapter 1
A Woman Perpetually Falling . . .

Page

7 ***falls in the elderly:*** N. R. Kleinfeld. 2003. For elderly, fear of falling is a risk in itself. *New York Times,* March 5.

10 ***The article described a device that enabled people who had been blind from birth to see:*** P. Bach-y-Rita, C. C. Collins, F. A. Saunders, B. White, and L. Scadden. 1969. Vision substitution by tactile image projection. *Nature,* 221(5184): 963–64.

12 ***•the two-thousand-year-old Greek idea that viewed all nature as a vast living organism:*** The Greeks, who invented the idea of nature, saw all nature as a vast *living organism.* All things, insofar as they took up space, were made of matter; insofar as they moved, were alive; and insofar as they were orderly, partook of intelligence. This was the first great idea of nature that humanity developed. In effect, the Greeks had projected themselves onto the macrocosm, and said it was alive and a reflection of themselves. Because nature was alive, they would not have been opposed to the idea of plasticity in principle, or the idea that the organ of thought could grow. Socrates, in the *Republic,* argued that a person could train his mind the way gymnasts trained their muscles.

After the discoveries of Galileo, the second great idea of nature emerged, that of *nature as mechanism.* The mechanists projected an image of a machine onto the cosmos, describing the universe as a vast "cosmic clock." Then they internalized that image and applied it to human beings. For instance, the physician Julien Offray de La Mettrie (1709–1751) wrote *Man a Machine (L'Homme-machine),* reducing human beings to mechanisms.

But then a new, third, grander idea of nature emerged, inspired by Buffon and others, which restored life to it; this was the idea of nature as an unfolding historical process, or *nature as history.* In this view, the universe is not a mechanism but an evolving historical process that is changing over time. The idea of natural history laid the groundwork for Darwin's theory of evolution. But the key point for our purposes is that this view was not opposed to the notion of plastic change in principle. This is discussed in more detail in appendix 2 and in the first note for that appendix. See R. G. Collingwood. 1945. *The idea of nature.* Oxford: Oxford University Press; R. S. Westfall. 1977. *The construction of modern science: Mechanisms and mechanics.* Cambridge: Cambridge University Press, 90.

13 •***Like a machine, the brain:*** The machine metaphor was not without major accomplishments; it made possible a more sober study of the brain based on observation, freed from mysticism. But it was always nonetheless an impoverished way to view the living brain, and the mechanists themselves knew it. Harvey was as interested in vital forces as in mechanisms, and Descartes famously argued that the complex cerebral contraption he depicted was animated and moved by the soul, though he never could explain how. The cost was dear, for he "dissected" us into a living, immaterial soul that was alive and could change, and a material brain that could not. In other words, he put, as a witty philosopher once said, "a ghost in the machine." Incidentally, Descartes's model of the nervous system was inspired by the hydraulic fountains of Saint-Germain-en-Laye, where pumping water animated moving statues of mythological figures.

13 •***Localizationism was applied to the senses as well, theorizing that each . . . specializes in detecting one of the various forms of energy:*** Starting in the early nineteenth century scientists labored to understand what makes each of our senses different, and a great debate began. Some argued that our nerves all carried the *same* kind of energy and that the only difference between vision and touch was quantitative: the eye could pick up the impingement of light because it was far more delicate, and sensitive, than the sense of touch. Others argued that the nerves of each sense carried a *different* form of energy, specific to that sense, and that the nerves from one sense could not replace or perform the function of

nerves of another sense. This point of view won out and was enshrined as "the law of the specific energy of nerves," proposed by Johannes Müller, in 1826. He wrote, "The nerve of each sense seems to be capable of one determinate kind of sensation only, and not of those proper to the other organs of sense; hence one nerve of sense cannot take the place and perform the function of the nerve of another sense." J. Müller. 1838. *Handbuch der Physiologie des Menschen,* bk. 5, Coblenz, reprinted in R. J. Herrnstein and E. G. Boring, eds. 1965. *A source book in the history of psychology.* Cambridge, MA: Harvard University Press, 26–33, especially 32.

Müller qualified his law, though, and conceded that he wasn't certain whether the specific energy of a particular nerve was caused by the nerve itself or by the brain or the spinal cord. His qualification was often forgotten.

Emil du Bois-Reymond (1818–1896), Müller's student and successor, speculated that if it were somehow possible to cross-connect the optic and auditory nerves, we would be able to see sounds and hear light impressions. E. G. Boring. 1929. *A history of experimental psychology.* New York: D. Appleton-Century Co., 91. See also S. Finger. 1994. *Origins of neuroscience: A history of explorations into brain function.* New York: Oxford University Press, 135.

16 •***Bach-y-Rita determined that skin . . . could substitute for a retina:*** Technically, a picture can form on the two-dimensional surfaces of both skin and retina because both can detect information *simultaneously.* Because they can both detect information *serially,* over time, they can both form moving pictures.

17 ***"one function, one location"***: S. Finger and D. Stein. 1982. *Brain damage and recovery: Research and clinical perspectives.* New York: Academic Press, 45.

17 ***Yet these children could still speak:*** A. Benton and D. Tranel. 2000. Historical notes on reorganization of function and neuroplasticity. In H. S. Levin and J. Grafman, eds., *Cerebral reorganization of function after brain damage.* New York: Oxford University Press.

17 ***Otto Soltmann removed the motor cortex . . . yet found they were still able to move:*** O. Soltmann. 1876. Experimentelle studien über die functionen des grosshirns der neugeborenen. *Jahrbuch für kinderheilkunde und physische Erzeihung,* 9:106–48.

17 ***But when the cat's paw was . . . stroked, the visual area . . . fired:*** K. Murata, H. Cramer, and P. Bach-y-Rita. 1965. Neuronal convergence of noxious, acoustic and visual stimuli in the visual cortex of the cat. *Journal of Neurophysiology,* 28(6): 1223–39; P. Bach-y-Rita. 1972. *Brain mechanisms in sensory substitution.* New York: Academic Press, 43–45, 54.

18 •***Bach-y-Rita realized that the areas . . . are far more homogeneous:*** The relative homogeneity of the cortex is demonstrated by the fact that scientists working with rats can transplant bits of "visual" cortex to the part of the brain that usually processes touch, and these transplants will start processing touch. See J. Hawkins and S. Blakeslee. 2004. *On intelligence.* New York: Times Books, Henry Holt & Co., 54.

18 •***Bach-y-Rita began to study all the exceptions to localizationism:*** In 1977, a new technique showed that (contrary to Broca's assertion that one speaks with the left hemisphere) 95 percent of healthy right-handers have language processed in their left hemisphere, and the remaining 5 percent have it processed in their right. Seventy percent of left-handers have language processed in the left hemisphere, but 15 percent have it processed in the right, and 15 percent have it processed bilaterally. S. P. Springer and G. Deutsch, G. 1999. *Left brain right brain: Perspectives from cognitive neuroscience.* New York: W. H. Freeman and Company, 22.

18 •***He discovered the work of Marie-Jean-Pierre Flourens:*** Flourens showed that if he removed large parts of a bird's brain, mental functions were lost. But because he observed his animals for a whole year, he also discovered that the lost functions often returned. He concluded that the brains had reorganized themselves, since the remaining parts were able to take over lost functions.

Flourens argued that the nervous system and brain had to be understood as a dynamic whole, more than the sum of its parts, and that it was premature to assume that mental functions had an invariant location in the brain. M.-J.-P. Flourens. 1824/1842. *Recherches expérimentales sur les propriétés et les fonctions du système nerveux dans les animaux vertébrés.* Paris: Ballière. Bach-y-Rita also drew inspiration from scientists Karl Lashley, Paul Weiss, and Charles Sherrington, all of whom showed that the brain and nervous system could, if parts were removed or disconnected, reacquire lost functions.

19 •***"a large body of evidence indicates that the brain demonstrates both motor and sensory plasticity":*** This paper was ultimately published as P. Bach-y-Rita. 1967. Sensory plasticity: Applications to a vision substitution system. *Acta Neurologica Scandinavica,* 43:417–26.

19 •***began to lay out . . . the evidence for brain plasticity:*** P. Bach-y-Rita. 1972. *Brain mechanisms and sensory substitution.* New York: Academic Press. This paper was his first sustained discussion in print.

23 ***"She wanted me to . . . coauthor . . . the paper":*** M. J. Aguilar. 1969. Recovery of motor function after unilateral infarction of the basis pontis. *American Journal of Physical Medicine,* 48:279–88; P. Bach-y-Rita. 1980. Brain plasticity as a basis for therapeutic procedures. In P. Bach-y-Rita, ed., *Recovery of function: Theoretical considerations for brain injury rehabilitation.* Bern: Hans Huber Publishers, 239–41.

23 ***Shepherd Ivory Franz:*** S. I. Franz. 1916. The function of the cerebrum. *Psychological Bulletin,* 13:149–73; S. I. Franz. 1912. New phrenology. *Science,* 35(896): 321–28; see 322.

24 •***the consolidation stage:*** We now suspect that during the consolidation stage of learning, neurons are making new proteins and changing their structure. See E. R. Kandel. 2006. *In search of memory.* New York: W. W. Norton & Co., 262.

24 ***scientists . . . have put patients under brain scans and confirmed:*** Maurice Ptito of Canada, in collaboration with Ron Kupers at the Université of Århus, Denmark.

25 ***Mriganka Sur, a neuroscientist, surgically rewired:*** M. Sur. 2003. *How experience rewires the brain.* Presentation at "Reprogramming the Human Brain" Conference, Center for Brain Health, University of Texas at Dallas, April 11.

26 ***"natural-born cyborgs":*** A. Clark. 2003. *Natural-born cyborgs: Minds, technologies, and the future of human intelligence.* Oxford: Oxford University Press.

Chapter 2
Building Herself a Better Brain

32 ***Luria . . . was deeply interested in psychoanalysis:*** K. Kaplan-Solms and M. Solms. 2000. *Clinical studies in neuro-psychoanalysis: Introduction to a depth neuropsychology.* Madison, CT: International Universities Press, 26–43; O. Sacks. 1998. The other road: Freud as neurologist. In M. S. Roth, ed., *Freud: Conflict and culture.* New York: Alfred A. Knopf, 221–34.

42 ***in an immature brain the number of . . . synapses, is 50 percent greater:*** D. Bavelier and H. Neville. 2002. Neuroplasticity, developmental. In V. S. Ramachandran, ed., *Encyclopedia of the human brain,* vol. 3. Amsterdam: Academic Press, 561.

43 ***Acetylcholine . . . is higher in rats trained:*** M. J. Renner and M. R. Rosenzweig. 1987. *Enriched and impoverished environments.* New York: Springer-Verlag.

43 ***Mental training . . . increases brain weight by 5 percent:*** M. R. Rosenzweig, D. Krech, E. L. Bennet, and M. C. Diamond. 1962. Effects of environmental complexity and training on brain chemistry and anatomy: A replication and extension. *Journal of Comparative and Physiological Psychology,* 55:429–37; M. J. Renner and M. R. Rosenzweig, 1987, 13.

43 ***9 percent in areas that the training directly stimulates:*** M. J. Renner and M. R. Rosenzweig, 1987, 13–15.

43 ***Trained . . . neurons develop 25 percent more branches:*** W. T. Greenough and F. R. Volkmar. 1973. Pattern of dendritic branching in occipital cortex of rats reared in complex environments. *Experimental Neurology,* 40:491–504; R. L. Hollaway. 1966. Dendritic branching in the rat visual cortex. Effects of extra environmental complexity and training. *Brain Research,* 2(4): 393–96.

43 ***Trained . . . neurons . . . increase their size:*** M. C. Diamond, B. Lindner, and A. Raymond. 1967. Extensive cortical depth measurements and neuron size increases in the cortex of environmentally enriched rats. *Journal of Comparative Neurology,* 131(3): 357–64.

43 ***increase . . . the number of connections per neuron:*** A. M. Turner and W. T. Greenough. 1985. Differential rearing effects on rat visual cortex synapses. I. Synaptic and neuronal density and synapses per neuron. *Brain Research,* 329:195–203.

43 ***increase . . . their blood supply:*** M. C. Diamond. 1988. *Enriching heredity: The impact of the environment on the anatomy of the brain.* New York: Free Press.

43 ***do not develop as rapidly in older animals as in younger:*** M. R. Rosenzweig. 1996. Aspects of the search for neural mechanisms of memory. *Annual Review of Psychology,* 47:1–32.

43 ***Similar effects . . . seen in all types of animals:*** M. J. Renner and M. R. Rosenzweig, 1987, 54–59.

43 ***For people . . . education increases the number of branches among neurons:*** B. Jacobs, M. Schall, and A. B. Scheibel. 1993. A quantitative dendritic analysis of Wernicke's area in humans. II. Gender, hemispheric, and environmental factors. *Journal of Comparative Neurology,* 327(1): 97–111.

43 ***increase in the volume and thickness of the brain:*** M. J. Renner and M. R. Rosenzweig, 1987, 44–48; M. R. Rosenzweig, 1996; M. C. Diamond, D. Krech, and M. R. Rosenzweig. 1964. The effects of an enriched environment on the histology of rat cerebral cortex. *Journal of Comparative Neurology,* 123:111–19.

Chapter 3
Redesigning the Brain

47 ***Merzenich argues that practicing a new skill . . . can change hundreds of millions . . . of the connections:*** M. M. Merzenich, P. Tallal, B. Peterson, S. Miller, and W. M. Jenkins. 1999. Some neurological principles relevant to the origins of—and the cortical plasticity-based remediation of—developmental language impairments. In J. Grafman and Y. Christen, eds., *Neuronal plasticity: Building a bridge from the laboratory to the clinic.* Berlin: Springer-Verlag, 169–87.

47 ***it is always "learning how to learn":*** M. M. Merzenich. 2001. Cortical plasticity contributing to childhood development. In J. L. McClelland and R. S. Siegler, eds., *Mechanisms of cognitive development: Behavioral and neural perspectives.* Mahwah, NJ: Lawrence Erlbaum Associates, 68.

48 •***brain maps . . . were first made vivid in human beings:*** The somatosensory cortex was first mapped by Wade Marshall in cats and monkeys.

49 ***By touching different parts of this map, he could trigger movements:*** W. Penfield and T. Rasmussen. 1950. *The cerebral cortex of man.* New York: Macmillan.

49 ***The Penfield maps shaped several generations':*** J. N. Sanes and J. P. Donoghue. 2000. Plasticity and primary motor cortex. *Annual Review of Neuroscience,* 23:393–415, especially 394; G. D. Schott. 1993. Penfield's homunculus: A note on cerebral cartography. *Journal of Neurology, Neurosurgery and Psychiatry,* 56:329–33.

49 •***the maps were fixed, immutable, and universal:*** Nobel laureate Eric Kandel writes, "When I was a medical student in the 1950s, we were taught that the map of the somatosensory cortex . . . was fixed and immutable throughout life." See E. R. Kandel. 2006. *In search of memory.* New York: W. W. Norton & Co., 216.

50 ***approximately 100 billion:*** G. M. Edelman and G. Tononi. 2000. *A universe of consciousness.* New York: Basic Books, 38.

50 ***•miss an extraordinary amount of information:*** Brain scans, such as fMRIs, can measure the activity in a 1-millimeter brain area. But a neuron is typically *a thousandth* of a millimeter across. S. P. Springer and G. Deutsch. 1999. *Left brain right brain: Perspectives from cognitive neuroscience.* New York: W. H. Freeman & Co., 65.

52 ***In fact, second languages . . . are not processed in the same part:*** P. R. Huttenlocher. 2002. *Neural plasticity: The effects of environment on the development of the cerebral cortex.* Cambridge, MA: Harvard University Press, 141, 149, 153.

55 ***Graham Brown and Charles Sherrington had shown that stimulating* one point:** T. Graham Brown and C. S. Sherrington. 1912. On the instability of a cortical point. *Proceedings of the Royal Society of London, Series B, Containing Papers of a Biological Character,* 85(579): 250–77.

55 ***the movement produced often changed:*** D. O. Hebb. 1963, commenting in the introduction to K. S. Lashley, *Brain mechanisms and intelligence: A quantitative study of the injuries to the brain.* New York: Dover Publications, xii. (Original edition, University of Chicago Press, 1929.)

56 ***When the paper was published, no mention was made of plasticity:*** R. L. Paul, H. Goodman, and M. M. Merzenich. 1972. Alterations in mechanoreceptor input to Brodmann's areas 1 and 3 of the postcentral hand area of *Macaca mulatta* after nerve section and regeneration. *Brain Research,* 39(1): 1–19. See also R. L. Paul, M. M. Merzenich, and H. Goodman. 1972. Representation of slowly and rapidly adapting cutaneous mechanoreceptors of the hand in Brodmann's areas 3 and 1 of *Macaca mulatta. Brain Research,* 36(2): 229–49.

58 ***Merzenich's contribution was . . . to determine the kind of input patients needed:*** R. P. Michelson. 1985. Cochlear implants: Personal perspectives. In R. A. Schindler and M. M. Merzenich, eds., *Cochlear implants.* New York: Raven Press, 10.

59 ***This time he and Kaas . . . called the changes "spectacular":*** M. M. Merzenich, J. H. Kaas, J. Wall, R. J. Nelson, M. Sur, and

D. Felleman. 1983. Topographic reorganization of somatosensory cortical areas 3b and 1 in adult monkeys following restricted deafferentation. *Neuroscience,* 8(1): 33–55.

60 ***He mapped a monkey's hand map . . . Then he amputated the monkey's middle finger:*** M. M. Merzenich, R. J. Nelson, M. P. Stryker, M. S. Cynader, A. Schoppmann, and J. M. Zook. 1984. Somatosensory cortical map changes following digit amputation in adult monkeys. *Journal of Comparative Neurology,* 224(4): 591–605.

61 ***Wiesel . . . has gracefully acknowledged in print:*** T. N. Wiesel. 1999. Early explorations of the development and plasticity of the visual cortex: A personal view. *Journal of Neurobiology,* 41(1): 7–9.

62 •***"nobody paid any attention":*** Jon Kaas attempted to deal with the early anti-adult plasticity bias in visual neuroscience head-on. He mapped the adult visual cortex, then cut the retinal input into it. He was able to show with remapping that in a matter of weeks new receptive fields moved into the cortical map space of the lesioned area. A reviewer at *Science* dismissed it as an impossible finding. It was eventually published in J. H. Kaas, L. A. Krubitzer, Y. M. Chino, A. L. Langston, E. H. Polley, and N. Blair. 1990. Reorganization of retinotopic cortical maps in adult mammals after lesions of the retina. *Science,* 248(4952): 229–31. Merzenich assembled the scientific evidence for plasticity in D. V. Buonomano and M. M. Merzenich. 1998. Cortical plasticity: From synapses to maps. *Annual Review of Neuroscience,* 21:149–86.

62 ***He cut a monkey's median nerve and then did multiple mappings:*** M. M. Merzenich, J. H. Kaas, J. T. Wall, M. Sur, R. J. Nelson, and D. Felleman. 1983. Progression of change following median nerve section in the cortical representation of the hand in areas 3b and 1 in adult owl and squirrel monkeys. *Neuroscience,* 10(3): 639–65.

62 •***These maps sprang up so quickly, it was as though they had been hidden:*** Recall that Bach-y-Rita thought that one way the brain rewired itself was by "unmasking" older paths, and that if one neuronal path in the brain is cut off, preexisting paths are

used instead, the way drivers rediscover old country back roads when their superhighway is blocked off. And like old country roads, these older maps were more primitive than the map they replaced, perhaps for lack of use.

62 ***The radial and ulnar maps . . . expanded to occupy almost the entire median nerve map:*** M. M. Merzenich, J. H. Kaas, J. T. Wall, M. Sur, R. J. Nelson, and D. Felleman. 1983. Progression of change following median nerve section in the cortical representation of the hand in areas 3b and 1 in adult owl and squirrel monkeys. *Neuroscience,* 10(3): 649.

63 ***Hebb . . . proposed that when two neurons fire at the same time repeatedly:*** D. O. Hebb. 1949. *The organization of behavior: A neuropsychological theory.* New York: John Wiley & Sons, 62.

63 ***•Hebb's concept—actually proposed by Freud sixty years before:*** Freud stated that when two neurons fire *simultaneously,* this firing facilitates their ongoing *association.* In 1888 he called it the law of association by simultaneity. Freud emphasized that what linked neurons was their firing together *in time.* See P. Amacher. 1965. *Freud's neurological education and its influence on psychoanalytic theory.* New York: International Universities Press, 57–59; K. H. Pribram and M. Gill. 1976. *Freud's "Project" re-assessed: Preface to contemporary cognitive theory and neuropsychology.* New York: Basic Books, 62–66; S. Freud, 1895. Project for a Scientific Psychology. Translated by J. Strachey. In *Standard edition of the complete psychological works of Sigmund Freud,* vol. 1. London: Hogarth Press, 281–397.

63 ***Following Hebb, Merzenich's new theory was that neurons in brain maps develop strong connections:*** M. M. Merzenich, W. M. Jenkins, and J. C. Middlebrooks. 1984. Observations and hypotheses on special organizational features of the central auditory nervous system. In G. Edelman, W. Einar Gall, and W. M. Cowan, eds., *Dynamic aspects of neocortical function.* New York: Wiley, 397–424; M. M. Merzenich, T. Allard, and W. M. Jenkins. 1991. Neural ontogeny of higher brain function: Implications of some

recent neurophysiological findings. In O. Franzén and J. Westman, eds., *Information processing in the somatosensory system.* London: Macmillan, 193–209.

63 ***In one ingenious experiment . . . sewed together two of the monkey's fingers:*** S. A. Clark, T. Allard, W. M. Jenkins, and M. Merzenich. 1988. Receptive fields in the body-surface map in adult cortex defined by temporally correlated inputs. *Nature*, 332(6163): 444–45; T. Allard, S. A. Clark, W. M. Jenkins, and M. M. Merzenich. 1991. Reorganization of somatosensory area 3b representations in adult owl monkeys after digital syndactyly. *Journal of Neurophysiology*, 66(3): 1048–58.

64 ***•each had one large map for their fused fingers instead of two separate ones:*** The scan technique used is called the magnetoencephalograph (MEG). Neuronal activity generates both electrical activity and magnetic fields. A magnetoencephalograph detects these magnetic fields and can tell us where the activity is occurring. A. Mogilner, J. A. Grossman, U. Ribary, M. Joliot, J. Volkmann, D. Rapaport, R. W. Beasley, and R. Llinás. 1993. Somatosensory cortical plasticity in adult humans revealed by magnetoencephalography. *Proceedings of the National Academy of Sciences, USA*, 90(8): 3593–97.

64 ***In the next experiment . . . created a map for . . . a nonexistent finger:*** X. Wang, M. M. Merzenich, K. Sameshima, and W. M. Jenkins. 1995. Remodelling of hand representation in adult cortex determined by timing of tactile stimulation. *Nature*, 378(6552): 71–75.

64 ***In the final and most brilliant demonstration, Merzenich . . . proved that maps cannot be anatomically based:*** S. A. Clark, T. Allard, W. M. Jenkins, and M. M. Merzenich. 1986. Cortical map reorganization following neurovascular island skin transfers on the hand of adult owl monkeys. *Neuroscience Abstracts*, 12:391.

65 ***•how does this topographic order emerge:*** In making topographical maps, nature performs two ingenious translations: a spatial organization (of the fingers on the hand) turns into an organized time sequence, which then turns into a spatial organization (of

the fingers on the brain map). The power of the brain to create its topographical order anew was demonstrated in a most remarkable way in France. A man from Lyon had both hands amputated in 1996 and then got two new hands transplanted to replace his lost ones. While he was still an amputee, his French doctors performed an fMRI scan to map his motor cortex, which showed, as might be expected, that he had developed an abnormally organized topography in the map in response to the total loss of nervous input from his hands. In 2000, after his bilateral hand transplant, they mapped him after two, four, and six months and found that the grafted hands came to be "recognized and activated normally by the sensory cortex" and the map developed a normal topography. P. Giraux, A. Sirigu, F. Schneider, and J-M. Dubernard. 2001. Cortical reorganization in motor cortex after graft of both hands. *Nature Neuroscience,* 4(7): 691–92.

65 •***A topographic order emerges because . . . everyday activities involve repeating sequences in a fixed order:*** By realizing that our maps are formed by the timing of the input to them, Merzenich thus solved the mystery of his first experiment, when he cut nerves to a monkey's hand, and they got shuffled—the "wires were crossed"—and yet that monkey still had a normally organized topographical map. Even after the nerves were shuffled, signals from the fingers tended to come in a fixed time sequence—thumb, then index, then middle finger—leading to a topographical map organization. See M. M. Merzenich, 2001, 69.

66 ***the monkey's fingertip had enlarged as the monkey had learned how to touch the disk:*** W. M. Jenkins, M. M. Merzenich, M. T. Ochs, T. Allard, and E. Guíc-Robles. 1990. Functional reorganization of primary somatosensory cortex in adult owl monkeys after behaviorally controlled tactile stimulation. *Journal of Neurophysiology,* 63(1): 82–104.

67 •***The trained neurons fired more quickly:*** M. M. Merzenich, P. Tallal, B. Peterson, S. Miller, and W. M. Jenkins. 1999. Some neurological principles relevant to the origins of—and the cortical

plasticity-based remediation of—developmental language impairments. In J. Grafman and Y. Christen, eds., *Neuronal plasticity: Building a bridge from the laboratory to the clinic.* Berlin: Springer-Verlag, 169–87, especially 172. The team found neurons could process a second signal 15 milliseconds after the first. They also determined that the time chunks in which a brain can process and integrate information range from tens of milliseconds to tenths of seconds. This finding was in response to the question, When we say neurons that fire together wire together, what exactly do we mean by fire "together"? Exactly simultaneously? By reviewing their own work and the work of others, Merzenich and Jenkins determined that "together" turned out to mean that the neurons have to fire within thousandths to tenths of seconds. M. M. Merzenich and W. M. Jenkins. 1995. Cortical plasticity, learning, and learning dysfunction. In B. Julesz and I. Kovács, eds., *Maturational windows and adult cortical plasticity. SFI studies in the sciences of complexity.* Reading, MA: Addison-Wesley, 23:247–64.

68 ***Finally, Merzenich discovered that paying close attention is essential:*** M. P. Kilgard and M. M. Merzenich. 1998. Cortical map reorganization enabled by nucleus basalis activity. *Science,* 279(5357): 1714–18; reviewed in M. M. Merzenich et al., 1999.

70 ***When Tallal originally discovered their problems:*** M. Barinaga. 1996. Giving language skills a boost. *Science,* 271(5245): 27–28.

71 ***The first study results . . . were remarkable:*** P. Tallal, S. L. Miller, G. Bedi, G. Byma, X. Wang, S. S. Nagarajan, C. Schreiner, W. M. Jenkins, and M. M. Merzenich. 1996. Language comprehension in language-learning impaired children improved with acoustically modified speech. *Science,* 271(5245): 81–84.

72 ***The study showed that . . . ability to understand language normalized:*** This study of *Fast ForWord* was a national U.S. field trial. Another study of 452 students had similar findings: S. L. Miller, M. M. Merzenich, P. Tallal, K. DeVivo, K. LaRossa, N. Linn, A. Pycha, B. E. Peterson, and W. M. Jenkins. 1999. *Fast*

ForWord training in children with low reading performance. *Nederlandse Vereniging voor Lopopedie en Foniatrie: 1999 Jaarcongres Auditieve Vaardigheden en Spraak-taal.* [Proceedings of the 1999 Netherlands Annual Speech-Language Association Meeting.]

72 ***new scans showed that their brains had begun to normalize:*** E. Temple, G. K. Deutsch, R. A. Poldrack, S. L. Miller, P. Tallal, M. M. Merzenich, and J. Gabrieli. 2003. Neural deficits in children with dyslexia ameliorated by behavioral remediation: Evidence from functional MRI. *Proceedings of the National Academy of Sciences, USA,* 100(5): 2860–65.

74 ***these same people could detect 75-millisecond* sounds *as well:*** S. S. Nagarajan, D. T. Blake, B. A. Wright, N. Byl, and M. M. Merzenich. 1998. Practice-related improvements in somatosensory interval discrimination are temporally specific but generalize across skin location, hemisphere, and modality. *Journal of Neuroscience,* 18(4): 1559–70.

75 ***One, a language study, showed that* Fast ForWord *quickly moved autistic children . . . to the normal range:*** M. M. Merzenich, G. Saunders, W. M. Jenkins, S. L. Miller, B. E. Peterson, and P. Tallal. 1999. Pervasive developmental disorders: Listening training and language abilities. In S. H. Broman and J. M. Fletcher, eds., *The changing nervous system: Neurobehavioral consequences of early brain disorders.* New York: Oxford University Press, 365–85, especially 377.

75 ***But another pilot study of one hundred autistic children:*** M. Melzer and G. Poglitch. 1998. Functional changes reported after *Fast ForWord* training for 100 children with autistic spectrum disorders. Presentation to the American Speech Language and Hearing Association, November.

80 ***BDNF plays a crucial role in reinforcing plastic changes:*** Z. J. Huang, A. Kirkwood, T. Pizzorusso, V. Porciatti, B. Morales, M. F. Bear, L. Maffei, and S. Tonegawa. 1999. BDNF regulates the maturation of inhibition and the critical period of plasticity in mouse visual cortex. *Cell,* 98:739–55. See also M. Fagiolini and T. K.

Hensch. 2000. Inhibitory threshold for critical-period activation in primary visual cortex. *Nature*, 404(6774): 183–86; E. Castrén, F. Zafra, H. Thoenen, and D. Lindholm. 1992. Light regulates expression of brain-derived neurotrophic factor mRNA in rat visual cortex. *Proceedings of the National Academy of Sciences, USA*, 89(20): 9444–48.

80 ***The fourth and final service that BDNF . . . is to help close down the critical period:*** M. Ridley. 2003. *Nature via nurture: Genes, experience, and what makes us human.* New York: HarperCollins, 166; J. L. Hanover, Z. J. Huang, S. Tonegawa, and M. P. Stryker. 1999. Brain-derived neurotrophic factor overexpression induces precocious critical period in mouse visual cortex. *Journal of Neuroscience*, 19:RC40:1–5.

81 ***If they hear one frequency, the whole auditory cortex starts firing:*** J. L. R. Rubenstein and M. M. Merzenich. 2003. Model of autism: Increased ratio of excitation/inhibition in key neural systems. *Genes, Brain and Behavior*, 2:255–67.

81 •***autistic children have bigger brains:*** Brain scan studies have shown that autistic children have bigger brains than normal. The difference, Merzenich says, is almost entirely caused by the overgrowth of the fatty coat around the nerves that helps conduct signals faster. These differences emerge, he says, "between six and ten months of age," when BDNF is released in large quantities.

82 ***brings the critical period to a premature close:*** L. I. Zhang, S. Bao, and M. M. Merzenich. 2002. Disruption of primary auditory cortex by synchronous auditory inputs during a critical period. *Proceedings of the National Academy of Sciences, USA*, 99(4): 2309–14.

82 • ***The animals are left with undifferentiated brain maps:*** It is not just external noise that can devastate a cortex. Merzenich believes many inherited conditions interfere with the ability of neurons to make strong clear signals that stand out against the background of the brain's other activities, creating the same effect on a brain as white noise. He calls this problem *internal* noise.

82 ***autistic children do indeed process sound in an abnormal way:*** N. Boddaert, P. Belin, N. Chabane, J. Poline, C. Barthélémy, M. Mouren-Simeoni, F. Brunelle, Y. Samson, and M. Zilbovicius. 2003. Perception of complex sounds: Abnormal pattern of cortical activation in autism. *American Journal of Psychiatry,* 160:2057–60.

83 ***Then, after the damage was done, they normalized and redifferentiated the maps:*** S. Bao, E. F. Chang, J. D. Davis, K. T. Gobeske, and M. M. Merzenich. 2003. Progressive degradation and subsequent refinement of acoustic representations in the adult auditory cortex. *Journal of Neuroscience,* 23(34): 10765–75.

83 ***They had found an artificial way to reopen the critical period in adults:*** M. P. Kilgard and M. M. Merzenich. 1998. Cortical map reorganization enabled by nucleus basalis activity. *Science,* 279(5357): 1714–18.

88 ***•Part of the scientific challenge is to find the most efficient way to train the brain:*** To be useful, a brain exercise must "generalize." For instance, say you were trying to train people to improve temporal processing. If you had to train them to get better at recognizing every known time interval (75 milliseconds, 80, 90, and so on), you would need a lifetime of training to improve temporal processing. But Merzenich's team found they only need to train the brain to recognize a few intervals efficiently, and this is sufficient to allow people to recognize many other intervals. In other words, the training generalizes, and the person now has improved his temporal processing for a full range of time intervals.

89 ***first control study:*** H. W. Mahncke, B. B. Connor, J. Appelman, O. N. Ahsanuddin, J. L. Hardy, R. A. Wood, N. M. Joyce, T. Boniske, S. M. Atkins, and M. M. Merzenich. 2006. Memory enhancement in healthy older adults using a brain plasticity–based training program: A randomized, controlled study. *Proceedings of the National Academy of Sciences, USA,* 103(33): 12523–28.

89 ***William Jagust, did "before" and "after" PET . . . scans of people who underwent the training:*** W. Jagust, B. Mormino, C. DeCarli,

J. Kramer, D. Barnes, B. Reed. 2006. Metabolic and cognitive changes with computer-based cognitive therapy for MCI. Poster presentation at the Tenth International Conference on Alzheimer's and Related Disorders, Madrid, Spain, July 15–20.

Chapter 4
Acquiring Tastes and Loves

95 •***Even sexual preference can occasionally change:*** The tendency of some heterosexuals to develop a homosexual attraction when members of the opposite sex are not available is well known (e.g., in prison or in the military), and these attractions tend to be "add-ons." According to Richard C. Friedman, researcher on male homosexuality, when male homosexuals develop a heterosexual attraction, it is almost always an "add-on" attraction, not a replacement (personal communication).

96 •***Yet the human sexual "instinct" seems to have broken free of its core purpose, reproduction, and varies:*** This plasticity is one reason why Freud called sex a "drive" as opposed to an instinct. A drive is a powerful urge that has instinctual roots but is more plastic than most instincts and is more influenced by the mind.

97 •***The brain structure that regulates instinctive behaviors, including sex, called the hypothalamus, is plastic, as is the amygdala, the structure that processes emotion and anxiety:*** The hypothalamus also regulates eating, sleeping, and important hormones. G. I. Hatton. 1997. Function-related plasticity in hypothalamus. *Annual Review of Neuroscience,* 20:375–97; J. LeDoux. 2002. *Synaptic self: How our brains become who we are.* New York: Viking; S. Maren. 2001. Neurobiology of Pavlovian fear conditioning. *Annual Review of Neuroscience,* 24:897–931, especially 914.

97 ***Plasticity exists in the hippocampus:*** B. S. McEwen. 1999. Stress and hippocampal plasticity. *Annual Review of Neuroscience,* 22: 105–22.

97 ***in areas that control our breathing:*** J. L. Feldman, G. S. Mitchell, and E. E. Nattie. 2003. Breathing: Rhythmicity, plasticity, chemosensitivity. *Annual Review of Neuroscience,* 26:239–66.

97 ***in areas that . . . process primitive sensation:*** E. G. Jones. 2000. Cortical and subcortical contributions to activity-dependent plasticity in primate somatosensory cortex. *Annual Review of Neuroscience,* 23:1–37.

97 ***and process pain:*** G. Baranauskas. 2001. Pain-induced plasticity in the spinal cord. In C. A. Shaw and J. C. McEachern, eds., *Toward a theory of neuroplasticity.* Philadelphia: Psychology Press, 373–86.

97 ***It exists in the spinal cord:*** J. W. McDonald, D. Becker, C. L. Sadowsky, J. A. Jane, T. E. Conturo, and L. M. Schultz. 2002. Late recovery following spinal cord injury: Case report and review of the literature. *Journal of Neurosurgery (Spine 2)* 97:252–65; J. R. Wolpaw and A. M. Tennissen. 2001. Activity-dependent spinal cord plasticity in health and disease. *Annual Review of Neuroscience,* 24:807–43.

97 ***•if one brain system changes, those systems connected to it change as well:*** Merzenich has done experiments that show that when change occurs in a sensory processing area—the auditory cortex—it causes change in the frontal lobe, an area involved in planning, to which the auditory cortex is connected. "You can't change the primary auditory cortex," says Merzenich, "without changing what is happening in the frontal cortex. It's an absolute impossibility."

98 ***These more complex melody maps obey the same plastic principles:*** M. M. Merzenich, personal communication; H. Nakahara, L. I. Zhang, and M. Merzenich. 2004. Specialization of primary auditory cortex processing by sound exposure in the "critical period." *Proceedings of the National Academy of Sciences, USA,* 101(18): 7170–74.

98 ***"The sexual instincts," wrote Freud, "are noticeable to us for their plasticity":*** S. Freud. 1932/1933/1964. *New introductory lectures on psycho-analysis.* Translated by J. Stratchey. In *Standard*

edition of the complete psychological works of Sigmund Freud, vol. 22. London: Hogarth Press, 97.

98 ***•Plato . . . argued that human Eros took many forms:*** Plato's Eros is not identical with Freud's libido (or later Eros), but there is some overlap. Platonic Eros is the longing we feel in response to our awareness of our incompleteness as human beings. It is a longing to complete ourselves. One way we try to overcome our incompleteness is by finding another person to love and have sex with. But the speakers in Plato's *Symposium* also emphasize that this same Eros can take many forms, some of which don't appear erotic at first glance, and that erotic longing can have many different kinds of objects.

98 ***a significant body of research now confirms Freud's basic insight that early patterns of relating and attaching to others, if problematic, can get "wired" into our brains:*** A. N. Schore. 1994. *Affect regulation and the origin of the self: The neurobiology of emotional development.* Hillsdale, NJ: Lawrence Erlbaum Associates; A. N. Schore. 2003. *Affect dysregulation and disorders of the self.* New York: W. W. Norton & Co.; A. N. Schore. 2003. *Affect regulation and the repair of the self.* New York: W. W. Norton & Co.

99 ***The idea of the critical period was formulated . . . by embryologists:*** M. C. Dareste. 1891. *Recherches sur la production artificielle des monstruosités.* [Studies of the artificial production of monsters.] Paris: C. Reinwald; C. R. Stockard. 1921. Developmental rate and structural expression: An experimental study of twins, "double monsters," and single deformities and their interaction among embryonic organs during their origin and development. *American Journal of Anatomy,* 28(2): 115–277.

99 ***•They are brief windows of time when new brain systems and maps develop:*** In the first year of life, the average brain goes from weighing 400 grams at birth to 1000 grams at twelve months. We are so dependent on early love and the caregiving of others in part because large areas of our brain don't begin to develop until after we are born. The neurons in the prefrontal cortex, which helps us

regulate our emotions, make connections in the first two years of life, but only with the help of people, which in most cases means the mother, who literally molds her baby's brain.

100 •***Regression can be pleasant and harmless . . . or it can be problematic:*** Sometimes regression is quite unanticipated, and otherwise mature adults become shocked at how "infantile" their behavior can become.

102 •***Pornography, delivered by high-speed Internet connections, satisfies every one of the prerequisites for neuroplastic change:*** In chapter 8, "Imagination," I give the scientific evidence that proves that we can change our brain maps simply by imagining things.

104 ***In the book one boy . . . says, "Anybody got porn?":*** T. Wolfe. 2004. *I Am Charlotte Simmons.* New York: HarperCollins, 92–93.

106 ***Cocaine, almost all other illegal drugs, and even nondrug addictions such as running make . . . dopamine more active:*** E. Nestler. 2001. Molecular basis of long-term plasticity underlying addiction. *Nature Reviews Neuroscience,* 2(2): 119–28.

107 ***When Merzenich used an electrode . . . dopamine release stimulated plastic change:*** S. Bao, V. T. Chan, L. I. Zhang, and M. M. Merzenich. 2003. Suppression of cortical representation through backward conditioning. *Proceedings of the National Academy of Sciences, USA,* 100(3): 1405–8.

107 ***dopamine is also released in sexual excitement:*** T. L. Crenshaw. 1996. *The alchemy of love and lust.* New York: G. P. Putnam's Sons, 135.

107 ***Nondrug addictions . . . lead to . . . permanent changes in the dopamine system:*** E. Nestler. 2003. *Brain plasticity and drug addiction.* Presentation at "Reprogramming the Human Brain" Conference, Center for Brain Health, University of Texas at Dallas, April 11.

108 ***An addict experiences cravings because . . . sensitized:*** K. C. Berridge and T. E. Robinson. 2002. The mind of an addicted brain: Neural sensitization of wanting versus liking. In J. T. Cacioppo,

G. G. Bernston, R. Adolphs, et al., eds., *Foundations in social neuroscience.* Cambridge, MA: MIT Press, 565–72.

108 •***So sensitization leads to increased wanting, though not necessarily liking:*** It is possible to judge whether an animal or a person likes the taste of a food by its facial expressions. Berridge and Robinson have shown, by manipulating dopamine levels while animals eat, that it is possible to make them want more food, even though they don't like it.

108 ***we have two separate pleasure systems in our brains:*** N. Doidge. 1990. Appetitive pleasure states: A biopsychoanalytic model of the pleasure threshold, mental representation, and defense. In R. A. Glick and S. Bone, eds., *Pleasure beyond the pleasure principle.* New Haven: Yale University Press, 138–73.

108 •***The second pleasure system has to do with . . . consummatory pleasure:*** Certain depressed people have trouble experiencing any pleasure at all, and their appetitive and consummatory systems do not function. They can't anticipate having a good time, and should they be dragged out to a meal or some other pleasant activity, they can't enjoy it. But some people who are depressed, while unable to anticipate having fun, will, if dragged out to a meal or social event, find their spirits lifting because, even though the appetitive system is not working properly, the consummatory system is.

109 ***The story of Sean Thomas:*** S. Thomas. 2003. How Internet porn landed me in hospital. *National Post,* June 30, A14. These quotes are from the *National Post* version of an article originally published in the *Spectator,* June 28, 2003, called "Self abuse."

111 ***interest in lesbian sex can express . . . unconscious female identification:*** E. Person. 1986. The omni-available woman and lesbian sex: Two fantasy themes and their relationship to the male developmental experience. In G. I. Fogel, F. M. Lane, and R. S. Liebert, eds., *The psychology of men.* New York: Basic Books, 71–94, especially 90.

112 •***"ugliness becomes beauty":*** Stendhal also described how young girls at the theater fell in love with famously "ugly" actors, such as

Le Kain, who in their performances evoked powerful, pleasurable emotions. By the end of the performance, the girls exclaimed, "Isn't he beautiful!" See Stendhal. 1947. *On love.* Translated by H.B.V. under the direction of C. K. Scott-Moncrieff. New York: Grosset & Dunlap, 44, 46–47.

113 ***In 1950 "pleasure centers" were discovered:*** R. G. Heath. 1972. Pleasure and pain activity in man. *Journal of Nervous and Mental Disease,* 154(1): 13–18.

113 ***our pleasure centers now fire so easily that makes whatever we experience feel great:*** N. Doidge, 1990.

113 ***falling in love also lowers the threshold at which the pleasure centers will fire:*** Ibid.

114 ***"globalization":*** Ibid.

114 ***•when our pleasure centers fire, it is more difficult for . . . pain . . . centers to fire:*** Unfortunately, the tendency of our pleasure and pain centers to inhibit each other also means that a person who is depressed, and who has aversive centers firing, finds it more difficult to enjoy things he normally would.

115 ***"romantic intoxication":*** M. Liebowitz. 1983. *The chemistry of love.* Boston: Little, Brown & Co.

115 ***Recent fMRI . . . scans of lovers:*** A. Bartels and S. Zeki. 2000. The neural basis for romantic love. *NeuroReport,* 11(17): 3829–34; see also H. Fisher. 2004. *Why we love: The nature and chemistry of romantic love.* New York: Henry Holt & Co.

116 ***•When monogamous mates develop a tolerance for each other:*** Tolerance occurs when the brain is inundated with a substance—in this case dopamine—and in response the receptors on the neurons for that substance "down regulate," or decrease in number, so more of the substance is required to get the same effect.

117 ***unlearning . . . is necessary to make room for new memories in our networks:*** E. S. Rosenzweig, C. A. Barnes, and B. L. McNaughton. 2002. Making room for new memories. *Nature Neuroscience,* 5(1): 6–8.

118 ***The work of mourning is piecemeal:*** S. Freud. 1917/1957. *Mourning and melancholia.* Translated by J. Stratchey. In *Standard edition of the complete psychological works of Sigmund Freud,* vol. 14. London: Hogarth Press, 237–58, especially 245.

119 ***in humans oxytocin is released in both sexes during orgasm:*** W. J. Freeman. 1999. *How brains make up their minds.* London: Weidenfeld & Nicolson, 160; J. Panksepp. 1998. *Affective neuroscience: The foundations of human and animal emotions.* New York: Oxford University Press, 231; L. J. Young and Z. Wang. 2004. The neurobiology of pair bonding. *Nature Neuroscience,* 7(10): 1048–54.

119 ***An fMRI study shows that when mothers look at photos of their children, brain regions:*** A. Bartels and S. Zeki. 2004. The neural correlates of maternal and romantic love. *NeuroImage,* 21:1155–66.

119 ***Their oxytocin levels remain low for several years after they have been adopted:*** A. B. Wismer Fries, T. E. Ziegler, J. R. Kurian, S. Jacoris, and S. D. Pollak. 2005. Early experience in humans is associated with changes in neuropeptides critical for regulating social behavior. *Proceedings of the National Academy of Sciences, USA,* 102(47): 17237–40.

119 ***When people sniff oxytocin . . . they are more prone to trust:*** M. Kosfeld, M. Heinrichs, P. J. Zak, U. Fischbacher, and E. Fehr. 2005. Oxytocin increases trust in humans. *Nature,* 435(7042): 673–76.

119 ***•oxytocin . . . makes us commit to our partners and devotes us to our children:*** The ancient Greeks, with simple elegance, described our tendency to develop powerful, not always rational, loving attachments to family and friends, as "love of one's own," and oxytocin seems to be one of several neurochemicals that promote it.

120 ***But if oxytocin is injected . . . she will mother the strange lamb:*** C. S. Carter. 2002. Neuroendocrine perspectives on social attachment and love. In J. T. Cacioppo, G. G. Bernston, R. Adolphs, et al., eds., 853–90, especially 864.

120 ***Freeman suspects that the mother bonds with her first litter using other neurochemicals:*** Personal communication.

120 ***Oxytocin's "ability" to wipe out . . . an amnestic hormone:*** T. R. Insel. 1992. Oxytocin—a neuropeptide for affiliation: Evidence from behavioral, receptor, autoradiographic, and comparative studies. *Psychoneuroendocrinology,* 17(1): 3–35, especially 12; Z. Sarnyai and G. L. Kovács. 1994. Role of oxytocin in the neuroadaptation to drugs of abuse. *Psychoneuroendocrinology,* 19(1): 85–117, especially 86.

120 **•*Freeman proposes that oxytocin melts down existing neuronal connections:*** W. J. Freeman. 1995. *Societies of brains: A study in the neuroscience of love and hate.* Hillsdale, NJ: Lawrence Erlbaum Associates, 122–23; W. J. Freeman, 1999, 160–61.

Freeman points out that hormones that influence behavior, such as estrogen or thyroid, generally need to be released steadily in the body to have their effects. But oxytocin is released only briefly, which strongly suggests that its role is *setting the stage for a new phase,* in which new behaviors replace existing behaviors.

Unlearning may be especially important in mammals because the cycle of reproduction and rearing the young takes so long and requires such a deep bond. For a mother to switch from being totally preoccupied with one litter to caring for the next requires a massive alteration in her goals, intentions, and the neuronal circuits involved.

121 ***"It is the afterplay, not the foreplay, that counts":*** W. J. Freeman, 1995, 122–23.

121 **•*Contrast him with the inveterate bachelor:*** One typical explanation for the rigidity of aging bachelors or bachelorettes, who want to marry but have become too fussy, is that they fail to fall in love because they have become increasingly rigid through living alone. But perhaps they also become increasingly rigid because they fail to fall in love and never get the surge of oxytocin that may facilitate plastic change. In a similar vein, one can ask how much of people's ability to parent well is enhanced by the prior experience of having fallen in love—in a mature way—allowing them to unlearn selfishness and open themselves up to another. If each mature love

experience has the potential to help us unlearn earlier, more selfish intentions and become less self-centered, a mature adult love would be one of the best predictors of the ability to parent well.

122 ***Merzenich has described a number of "brain traps":*** M. M. Merzenich, F. Spengler, N. Byl, X. Wang, and W. Jenkins. 1996. Representational plasticity underlying learning: Contributions to the origins and expressions of neurobehavioral disabilities. In T. Ono, B. L. McNaughton, S. Molochnikoff, E. T. Rolls, and H. Nishijo, eds., *Perception, memory and emotion: Frontiers in neuroscience.* Oxford: Elsevier Science, 45–61, especially 50.

123 •***Nancy Byl . . . teaches people . . . to redifferentiate their finger maps:*** N. N. Byl, S. Nagarajan, and A. L. McKenzie. 2003. Effect of sensory discrimination training on structure and function in patients with focal hand dystonia: A case series. *Archives of Physical Medicine and Rehabilitation,* 84(10): 1505–14. Merzenich has helped Japanese people trying to speak English without an accent get out of their brain traps (see page 122). Knowing that the basis for this problem lies in the absence of a differentiated auditory cortex for certain sounds, Merzenich and his collaborators set out to differentiate them. Using the same kind of approach as *Fast ForWord,* he radically modified the *r* and *l* sounds, so that the difference was *grossly* exaggerated and the Japanese listeners could pick it up. Then the team gradually normalized the sounds, while the subjects were listening. It was essential for the speakers to always pay very close attention throughout the exercises, something they didn't do in normal speech. It took about ten to twenty hours of training to learn to make the distinction. "You can teach anybody to speak an accentless second language as an adult," Merzenich says, "but it requires very intense training."

124 •***perversions:*** The notion of "perversion" implies that our sexual drive is like a river that most naturally flows in a certain channel, until something happens that puts it off course and diverts, or perverts, its direction. People who call themselves "kinky" concede the point, a kink being something with a twist in it.

125 •***Sexual sadism:*** True, some reject the idea that in perversion, aggression gets linked with sexuality. The literary critic Camille Paglia argues that sexuality is by nature aggressive. "My theory," she says, "is that whenever sexual freedom is sought or achieved, sadomasochism will not be far behind." She attacks feminists who believe that sex is all sugar and spice and who argue that it is patriarchal society that makes sex violent. Sex, for Paglia, is about power; society is not the source of sexual violence; sex, the irrepressible natural force, is. If anything, society is the force that inhibits the inherent violence of sex. Paglia is certainly more realistic than those who would deny that perversion is rife with aggression. But in assuming that sex is fundamentally aggressive, and sadomasochistic, she doesn't allow for the plasticity of human sexuality. Just because sex and aggression can unite in a plastic brain, and appear "natural," doesn't mean that that is their only possible expression. We have seen that certain brain chemicals released in sex, such as oxytocin, cause us to be tender to each other. It is no more accurate to say that fully realized sexuality is always violent than to say it is always gentle and sweet. C. Paglia. 1990. *Sexual personae.* New Haven: Yale University Press, 3.

125 ***Robert Stoller . . . did make important discoveries:*** R. J. Stoller. 1991. *Pain and passion: A psychoanalyst explores the world of S & M*. New York: Plenum Press.

125 ***"they had to be confined . . . Hence the perversions":*** Ibid., 25.

129 •***A fetish . . . is an object:*** More precisely, Stoller wrote, "a fetish is a story masquerading as an object."

Chapter 5
Midnight Resurrections

135 ***Stroke is one of the leading causes of disability:*** P. W. Duncan. 2002. Guest editorial. *Journal of Rehabilitation Research and Development,* 39(3): ix–xi.

135 ***Until CI therapy, studies . . . concluded that no existing treatment was effective:*** P. W. Duncan. 1997. Synthesis of intervention trials to improve motor recovery following stroke. *Topics in Stroke Rehabilitation,* 3(4): 1–20; E. Ernst. 1990. A review of stroke rehabilitation and physiotherapy. *Stroke,* 21(7): 1081–85; K. J. Ottenbacher and S. Jannell. 1993. The results of clinical trials in stroke rehabilitation research. *Archives of Neurology,* 50(1): 37–44; J. de Pedro-Cuesta, L. Widen-Holmquist, and P. Bach-y-Rita. 1992. Evaluation of stroke rehabilitation by randomized controlled studies: A review. *Acta Neurologica Scandinavica,* 86:433–39.

137 •***John B. Watson, wrote derisively, "Most of the psychologists talk . . . about the formation of new pathways in the brain":*** The neuroplasticians would show that the arrogant Watson couldn't have been more wrong, and that our thoughts and skills do form new pathways and deepen older ones. J. B. Watson. 1925. *Behaviorism.* New York: W. W. Norton & Co.

139 •***Even . . . voluntary movements . . . require the motor cortex to modify* preexisting *reflexes:*** The idea that everything we do is a reflex had roots that predated Sherrington, and understanding these roots helps one understand why the idea took hold. The German physiologist Ernest Brücke proposed that all brain functioning involved reflex functions. Brücke was wary of the tendency, popular in his day, to describe the nervous system by reference to spiritual or magical but vague "vital forces." Brücke and his followers wanted to describe the nervous system in terms consistent with Newton's laws of action and reaction, and with what was known about electricity. For them, the nervous system, to be a system, had to be mechanistic. The idea of reflex, in which a physical *stimulus* gave rise to an excitation that traveled along a sensory nerve, to a motor nerve, which it excited, giving rise to a *response,* was very appealing to behaviorists, because here was a complex action that didn't involve the mind. For behaviorists, the mind became a passive spectator, and how it influenced or was influenced by the nervous system remained unclear. B. F. Skinner

devoted a major portion of one of his books on behaviorism to the reflexological theory.

139 •***The monkeys . . . started using their deafferented arms:*** Taub eventually discovered that a German, H. Munk, reported performing a deafferentation in 1909 and was able to get the monkey to feed itself if he restrained the good arm and rewarded use of the deafferented arm.

140 •***Ivan Pavlov . . . argued that the brain is plastic:*** He wrote: ". . . our system is self-regulatory in the highest degree,—self-maintaining, repairing, readjusting, and even improving. The chief, strongest, and ever-present impression received from the study of the higher nervous activity by our method, is the extreme plasticity of this activity, its immense possibilities: nothing remains stationary, unyielding; and everything could always be attained, all could be changed for the better, were only the appropriate conditions realized." Cited in D. L. Grimsley and G. Windholz. 2000. The neurophysiological aspects of Pavlov's theory of higher nervous activity: In honor of the 150th anniversary of Pavlov's birth. *Journal of the History of the Neurosciences,* 9(2): 152–163, especially 161. Original passage from I. P. Pavlov. 1932. The reply of a physiologist to psychologists. *Psychological Review,* 39(2): 91–127, 127.

141 ***Spinal shock can last from two to six months:*** G. Uswatte and E. Taub. 1999. Constraint-induced movement therapy: New approaches to outcomes measurement in rehabilitation. In D. T. Stuss, G. Winocur, and I. H. Robertson, eds., *Cognitive neurorehabilitation.* Cambridge: Cambridge University Press, 215–29.

142 ***He then tested whether he could correct learned nonuse several years after it had developed:*** E. Taub. 1977. Movement in nonhuman primates deprived of somatosensory feedback. In J. F. Keogh, ed., *Exercise and sport sciences reviews.* Santa Barbara: Journal Publishing Affiliates, 4:335–74; E. Taub. 1980. Somatosensory deafferentation research with monkeys: Implications for rehabilitation medicine. In L. P. Ince, ed., *Behavioral psychology in rehabilitation*

medicine: Clinical applications. Baltimore: Williams & Wilkins, 371–401.

142 ***Taub believed these discoveries meant that people who had had strokes . . . even years earlier, might be suffering from learned nonuse:*** E. Taub, 1980.

144 ***arson, property destruction . . . acceptable "when they directly alleviate the pain and suffering of an animal":*** K. Bartlett. 1989. The animal-right battle: A jungle of pros and cons. *Seattle Times,* January 15, A2.

144 ***Taub was demonized . . . Nazi Dr. Mengele:*** C. Fraser. 1993. The raid at Silver Spring. *New Yorker,* April 19, 66.

145 ***Taub has always contended that Pacheco's photos were staged:*** E. Taub. 1991. The Silver Spring monkey incident: The untold story. *Coalition for Animals and Animal Research,* Winter/Spring, 4(1): 2–3.

146 ***asked Pacheco if they had been taken . . . to Gainesville, Florida, and he said, "That's a pretty good guess":*** C. Fraser, 1993, 74.

146 ***•By the end of Taub's first trial before a judge, in November 1981, 113 of the 119 charges against him had been dismissed:*** The Department of Agriculture veterinarian who made unannounced visits to the Taub laboratory during the period that Pacheco was there testified that he did not find the unsatisfactory conditions depicted by Pacheco. Taub was not found guilty of cruel or inhumane treatment of the animals but still was fined $3,500 for the remaining charges. It was argued that he should have obtained outside veterinary help for six of his deafferented monkeys instead of treating them himself—though no veterinarian had his expertise with deafferented animals—so six counts remained against him, one for each animal.

Because Taub's convictions in the first trial were for misdemeanors, he was now entitled, by law, to a trial by jury. By the end of that second trial, in June 1982, he was acquitted of five of the six remaining charges, or 118 of 119 original charges. The sole charge remaining was that the lab didn't provide adequate veterinary care for one monkey, Nero, which allegedly caused him to develop a bone infection. Taub has written that there was a pathology report showing that the monkey did *not* have a bone infection. E. Taub, 1991, 6.

146 •***the NIH, which reversed its decision:*** T. Dajer. 1992. Monkeying with the brain. *Discover,* January, 70–71. Few scientists helped Taub, but among them were Neal Miller and Vernon Mountcastle (Merzenich's mentor), who stood up for Taub and helped in his defense.

146 •***When he was finally . . . there were demonstrations:*** A donor with PETA sympathies, who had pledged a million-dollar bequest, said she would withdraw it if Taub was kept. Some Alabama faculty argued that even if he was innocent, he was too controversial.

147 •***80 percent of stroke patients who have lost arm function can improve substantially:*** E. Taub, G. Uswatte, M. Bowman, A. Delgado, C. Bryson, D. Morris, and V. W. Mark. 2005. Use of CI therapy for plegic hands after chronic stroke. Presentation at the Society for Neuroscience, Washington, DC, November 16, 2005. An earlier paper documented a 50 percent improvement rate: G. Uswatte and E. Taub. 1999. Constraint-induced movement therapy: New approaches to outcomes measurement in rehabilitation. In D. T. Stuss, G. Winocur, and I. H. Robertson, eds., *Cognitive neurorehabilitation.* Cambridge: Cambridge University Press, 215–29.

147 ***Many . . . had severe, chronic strokes and showed very large improvements:*** E. Taub, G. Uswatte, D. K. King, D. Morris, J. E. Crago, and A. Chatterjee. 2006. A placebo-controlled trial of constraint-induced movement therapy for upper extremity after stroke. *Stroke,* 37(4): 1045–49. E. Taub, G. Uswatte, and T. Elbert. 2002. New treatments in neurorehabilitation founded on basic research. *Nature Reviews Neuroscience,* 3(3): 228–36.

147 ***Even patients who had had their strokes . . . more than four years before:*** E. Taub, N. E. Miller, T. A. Novack, E. W. Cook, W. C. Fleming, C. S. Nepomuceno, J. S. Connell, and J. E. Crago. 1993.

Technique to improve chronic motor deficit after stroke. *Archives of Physical Medicine and Rehabilitation,* 74(4): 347–54.

149 ***After CI therapy the size of the brain map . . . doubled:*** J. Liepert, W. H. R. Miltner, H. Bauder, M. Sommer, C. Dettmers, E. Taub, and C. Weiller. 1998. Motor cortex plasticity during constraint-induced movement therapy in stroke patients. *Neuroscience Letters,* 250:5–8.

149 ***The second study showed . . . changes . . . in both hemispheres:*** B. Kopp, A. Kunkel, W. Mühlnickel, K. Villringer, E. Taub, and H. Flor. 1999. Plasticity in the motor system related to therapy-induced improvement of movement after stroke. *NeuroReport,* 10(4): 807–10.

149 ***•New neurons have to take over the lost functions, and they may not be quite as effective:*** While plasticity makes recovery possible, competitive plasticity may also be a factor that limits some recoveries in people who get conventional treatment. The brain has neurons that can adapt and take over either lost movement or lost cognitive functions, and that may therefore be used for either function during recovery. University of Toronto researcher Robin Green is studying this phenomenon. Preliminary data—not on patients receiving Taub's treatment, but on patients in an inpatient neurorehabilitation program—show that in some patients who have both movement and cognitive deficits from their strokes there is a trade-off as they improve; the more cognitive improvement they show, the less improvement they make in movement, and vice versa. R. E. A. Green, B. Christensen, B. Melo, G. Monette, M. Bayley, D. Hebert, E. Inness, and W. Mcilroy. 2006. Is there a trade-off between cognitive and motor recovery after traumatic brain injury due to competition for limited neural resources? *Brain and Cognition,* 60(2): 199–201.

154 ***to help stroke patients who have damage to Broca's area:*** F. Pulvermüller, B. Neininger, T. Elbert, B. Mohr, B. Rockstroh, M. A. Koebbel, and E. Taub. 2001. Constraint-induced therapy of chronic aphasia after stroke. *Stroke,* 32(7): 1621–26.

155 ***the CI therapy group had a 30 percent increase in communication:*** Ibid.

156 ***He has begun working with children with cerebral palsy:*** E. Taub, S. Landesman Ramey, S. DeLuca, and K. Echols. 2004. Efficacy of constraint-induced movement therapy for children with cerebral palsy with asymmetric motor impairment. *Pediatrics,* 113(2): 305–12.

161 ***the largest amount of rewiring that had ever been mapped:*** T. P. Pons, P. E. Garraghty, A. K. Ommaya, J. H. Kaas, E. Taub, and M. Mishkin. 1991. Massive cortical reorganization after sensory deafferentation in adult macaques. *Science,* 252(5014): 1857–60.

Chapter 6
Brain Lock Unlocked

164 ***a desperate college student . . . put a gun in his mouth . . . was found . . . cured:*** Associated Press story, February 24, 1988. Cited in J. L. Rapoport. 1989. *The boy who couldn't stop washing.* New York: E. P. Dutton, 8–9.

166 •***The worries can be bizarre—and make no conceivable sense:*** Only in rare cases are people with OCD totally unable to appreciate that their fears are overblown, and sometimes such people have both OCD and a near psychotic, or psychotic, illness.

166 ***A husband . . . thought that there are razor blades attached to his fingernails:*** J. M. Schwartz and S. Begley. 2002. *The mind and the brain: Neuroplasticity and the power of mental force.* New York: ReganBooks/HarperCollins, 19.

167 ***Schwartz describes a man who feared . . . battery acid:*** Ibid., xxvii, 63.

168 ***Schwartz has developed an effective . . . treatment:*** J. M. Schwartz and B. Beyette. 1996. *Brain lock: Free yourself from obsessive-compulsive behavior.* New York: ReganBooks/HarperCollins.

169–170 •***the caudate nucleus . . . allows our thoughts to flow:*** The caudate is right next to a brain area called the putamen that performs a similar function but is the automatic transmission for movement. It knits individual movements into a flowing automatic sequence. When the putamen gets damaged in Huntington's disease, patients can't go automatically from one movement to another. They have to think about every movement they perform, or they literally get stuck. Each movement is as laborious as it was the very first time they learned it. Every movement—brushing, getting out of bed, answering the phone—requires constant, effortful attention. J. J. Ratey and C. Johnson. 1997. *Shadow syndromes.* New York: Pantheon Books, 308–9.

170 ***•OCD . . . can also be caused by infections that swell the caudate:*** National Institutes of Health researchers recently discovered that some children who showed no signs of OCD suddenly developed it overnight after suffering from strep throat. Some became compulsive hand-washers. MRI brain scans showed that their caudates were swollen 24 percent larger than normal. These children had had common group A streptococcal infections, which their bodies' immune systems fought, attacking the infection but also attacking the caudate, developing an autoimmune disease, in which their antibodies attacked their own body along with the invading organism. The usual treatments for an autoimmune disease are drugs that suppress the immune system and washing the antibodies out of the system. With these therapies, the OCD disappeared in these children. A few of the children who got strep throat already had OCD, and they got markedly worse. It was also noted that the swelling of the caudate was proportional to the severity of the OCD.

170 ***unlocking the link between the orbital cortex and the cingulate and normalizing . . . the caudate:*** J. M. Schwartz and S. Begley, 2002, 75.

171 ***pictures of the abnormal OCD brain scan:*** J. M. Schwartz and B. Beyette, 1996.

171 ***"exposure and response prevention" . . . helps about half of OCD patients:*** J. S. Abramowitz. 2006. The psychological treatment of

obsessive-compulsive disorder. *Canadian Journal of Psychiatry,* 51(7): 407–16, especially 411, 415.

171 ***30 percent of patients refused:*** Ibid., 414.

172 ***As Schwartz says, ". . . cognitive distortion is just not an intrinsic part of the disease":*** J. M. Schwartz and S. Begley, 2002, 77.

173 ***"The struggle . . . is* not to give in to the feeling":** J. M. Schwartz and B. Beyette, 1996, 18.

173 ***•any time spent resisting is beneficial:*** If you want to lift a hundred pounds, you don't expect to succeed the first time. You start with a lighter weight and work up little by little. You actually fail to lift a hundred pounds, every day, until the day you succeed. But it is in the days when you are exerting yourself that the growth is occurring.

Chapter 7. Pain

180 ***Phantom limbs . . . give rise to a chronic "phantom pain" in 95 percent of amputees:*** R. Melzack. 1990. Phantom limbs and the concept of a neuromatrix. *Trends in Neuroscience,* 13(3): 88–92; P. Wall. 1999. *Pain: The science of suffering.* London: Weidenfeld & Nicholson.

180 ***"phantom pain" . . . often persists for a lifetime:*** P. Wall, 1999, 10.

180 ***women . . . suffer . . . labor pains* even after *their uteruses have been removed:*** T. L. Dorpat. 1971. Phantom sensations of internal organs. *Comprehensive Psychiatry,* 12:27–35.

180 ***men who still feel ulcer pain* after *the ulcer . . . cut out:*** H. F. Gloyne. 1954. Psychosomatic aspects of pain. *Psychoanalytic Review,* 41:135–59.

180 ***hemorrhoidal pain after their rectums . . . removed:*** P. Ovesen, K. Kroner, J. Ornsholt, and K. Bach. 1991. Phantom-related phenomena after rectal amputation: Prevalence and clinical characteristics. *Pain,* 44:289–91.

180 ***bladders were removed who still . . . need to urinate:*** R. Melzack, 1990; P. Wall, 1999.

180 • ***"acute pain," alerts us to injury:*** Normally pain prevents problems. When we sip a scalding cup of coffee and burn our tongue, we become less likely to swallow and do further damage. Children born with an inability to feel pain, a condition called "congenital analgesia," often die young of what were initially minor ailments. For instance, they do not know to stop walking on a damaged joint and may die of bone infections.

183 ***Ramachandran's finding in the Tom Sorenson case:*** V. S. Ramachandran, D. Rogers-Ramachandran, and M. Stewart. 1992. Perceptual correlates of massive cortical reorganization. *Science,* 258(5085): 1159–60.

183 ***Brain scan studies . . . confirmed a correlation between the amount of plastic change and the . . . pain:*** H. Flor, T. Elbert, S. Knecht, C. Wienbruch, C. Pantev, N. Birbaumer, W. Larbig, and E. Taub. 1995. Phantom-limb pain as a perceptual correlate of cortical reorganization following arm amputation. *Nature,* 375(6531): 482–84.

183 ***he believes, its surviving brain map "hungers" for incoming stimulation:*** V. S. Ramachandran and S. Blakeslee. 1998. *Phantoms in the brain.* New York: William Morrow. Also, personal communication.

183 ***Ramachandran wondered, might a person who is touched, in . . . cross-wiring, feel pain:*** V. S. Ramachandran and S. Blakeslee, 1998, 33.

184 • ***The Penfield brain map shows the genitals next to the feet:*** Martha Farah, of the University of Pennsylvania, has noted that babies curled up in the womb often have their legs crossed and folded up against their genitals. Legs and genitals would thus be jointly stimulated when they touch each other, and then map together, because neurons that fire together wire together.

184–185 ***People are tortured by phantom memories . . . especially if that pain existed at the time of the amputation:*** J. Katz and R. Melzack. 1990. Pain "memories" in phantom limbs: Review and clinical observations. *Pain,* 43:319–36.

185 ***Sometimes a patient can be pain free for decades, and then . . . reactivates:*** W. Noordenbos and P. Wall. 1981. Implications of the failure of nerve resection and graft to cure chronic pain produced by nerve lesions. *Journal of Neurology, Neurosurgery and Psychiatry,* 44:1068–73.

186 ***•imaginary clenching evokes pain because maximum contraction and pain are associated in memory:*** Because the phantom is illusory, the person with clenching pain can't use reality to challenge the memory that associates clenching with pain. So he is locked in the past. Proposed by Ronald Melzack in R. Melzack, 1990.

188 ***half . . . lost their phantom pain:*** V. S. Ramachandran and D. Rogers-Ramachandran. 1996. Synaesthesia in phantom limbs induced with mirrors. *Proceedings of the Royal Society B: Biological Sciences,* 263(1369): 377–86.

188 ***brain scans show that as these patients improve . . . map shrinkage . . . is reversed:*** P. Giraux and A. Sirigu. 2003. Illusory movements of the paralyzed limb restore motor cortex activity. *NeuroImage,* 20:S107–11.

188 ***•and sensory and motor maps normalize:*** Herta Flor of the University of Heidelberg in Germany, inspired by Ramachandran's work, treated amputees with phantom pain using mirror therapy and did fMRI scans to see what was happening in their heads. At first, they showed no activity in their sensory and motor hand maps for the amputated hand. But as the therapy proceeded, their sensory hand maps for the amputation became active again. This study has not yet been published but was reported in *The Economist,* 2006. Science and technology: A hall of mirrors; Phantom limbs and chronic pain. July 22, 380(8487): 88.

188 ***Marilyn Monroe . . . many bodily defects:*** S. Shaw and N. Rosten. 1987. *Marilyn among friends.* London: Bloomsbury, 16.

190 ***the most important article in the history of pain:*** R. Melzack and P. Wall. 1965. Pain mechanisms: A new theory. *Science,* 150(3699): 971–79.

190 •***the brain always controls the pain signals we feel:*** Scientists now think in terms of many pain-responsive regions in the brain, called a "pain matrix," including the thalamus, somatosensory cortex, insula, anterior cingulate cortex, and other regions.

191 ***70 percent of the men who were seriously wounded reported that they were not in pain:*** Study by H. Beecher, cited in P. Wall, 1999.

191 •***the brain closes the "gate," to keep . . . attention riveted on how to get out of harm's way:*** Many people saw the gating phenomenon in 1981, when they saw footage of President Ronald Reagan being shot through the chest with a 9-millimeter bullet, in an assassination attempt. Reagan just stood there feeling nothing. Neither he nor the Secret Service, which slammed him roughly into his car to protect him, knew he had been shot. Reagan said in a CBS documentary, "I had never been shot before, except in the movies. Then you always act as though it hurts. Now I know that does not always happen." Cited ibid., 1999.

191 ***brain scans show that during the placebo effect the brain turns down its own pain-responsive regions:*** T. D. Wager, J. K. Rilling, E. E. Smith, A. Sokolik, K. L. Casey, R. J. Davidson, S. M. Kosslyn, R. M. Rose, and J. D. Cohen. 2004. Placebo-induced changes in fMRI in the anticipation and experience of pain. *Science,* 303(5661): 1162–67.

191 ***the neurons in our pain system are far more plastic:*** R. Melzack, T. J. Coderre, A. L. Vaccarino, and J. Katz. 1999. Pain and neuroplasticity. In J. Grafman and Y. Christen, eds., *Neuronal plasticity: Building a bridge from the laboratory to the clinic.* Berlin: Springer-Verlag, 35–52.

191 •***a chronic injury can make the cells in the pain system fire more easily . . . making a person hypersensitive to pain:*** Hypersensitivity was proposed by J. MacKenzie. 1893. Some points bearing on the association of sensory disorders and visceral diseases. *Brain,* 16:321–54.

191 ***Maps can also enlarge . . . increasing pain sensitivity:*** R. Melzack, T. J. Coderre, A. L. Vaccarino, and J. Katz, 1999, 37.

191 ***pain signals in one map can "spill" into adjacent pain maps, and we may develop "referred pain":*** R. Melzack, T. J. Coderre, A. L. Vaccarino, and J. Katz, 1999, 46.

192 ***"Pain is an opinion on the organism's state":*** V. S. Ramachandran and S. Blakeslee, 1998, 54.

192 ***could he also use the mirror box to make chronic pain in a real limb disappear:*** V. S. Ramachandran. 2003. *The emerging mind: The Reith lectures 2003*. London: Profile Books, 18–20.

193 •***What better way . . . to prevent movement than to make sure the motor command itself triggers pain:*** In the cases Ramachandran described, chronic pain and pathological guarding occurred because the motor command for a movement was wired directly into the pain center, so even the thought of moving caused preemptive guarding and pain. I suspect that something like preemptive guarding and pain occurs when people feel pangs of guilt when they only imagine doing bad things. The motor command for the forbidden wish is wired directly into an anxiety center, so that it triggers anguish, even before it is enacted. This would give guilt the ability to preempt bad actions, not just to make us feel bad after the fact.

193 ***In a study conducted by . . . Patrick Wall:*** C. S. McCabe, R. C. Haigh, E. F. J. Ring, P. W. Halligan, P. D. Wall, and D. R. Black. 2003. A controlled pilot study of the utility of mirror visual feedback in the treatment of complex regional pain syndrome (type 1). *Rheumatology,* 42:97–101. They studied complex regional pain syndrome, or CRPS, which includes a number of syndromes, including reflex sympathetic dystrophy, causalgia, and algodystrophy.

194 ***An Australian scientist, G. L. Moseley:*** G. L. Moseley. 2004. Graded motor imagery is effective for long-standing complex regional pain syndrome: A randomised controlled trial. *Pain,* 108:192–98.

195 ***Postoperative phantom pain can be minimized if surgical patients get . . . local anesthetics . . .* before *the general anesthetic:*** S. Bach, M. F. Noreng, and N. U. Tjéllden. 1988. Phantom limb pain in amputees during the first twelve months following limb

amputation, after preoperative lumbar epidural blockade. *Pain,* 33:297–301; Z. Seltzer, B. Z. Beilen, R. Ginzburg, Y. Paran, and T. Shimko. 1991. The role of injury discharge in the induction of neuropathic pain behavior in rats. *Pain,* 46:327–36; P. M. Dougherty, C. J. Garrison, and S. M. Carlton. 1992. Differential influence of local anesthesia upon two models of experimentally induced peripheral mononeuropathy in rats. *Brain Research,* 570:109–15.

195 •***Pain-killers, administered before surgery, not just afterward . . . prevent plastic change in the brain's pain map:*** R. Melzack, T. J. Coderre, A. L. Vaccarino, and J. Katz, 1999, 35–52, 43–45; Herta Flor has used the same reasoning, to decrease the postoperative pain of patients undergoing amputation, by administering the drug memantine. Following Ramachandran's idea that phantom pain is a memory that has been locked into the system, she uses memantine to block the activity of proteins necessary to form memories. She has found that the drug works if given before, or in the four weeks immediately after, amputations. Reported in *The Economist,* 2006.

195 ***the mirror box is effective on . . . stroke:*** E. L. Altschuler, S. B. Wisdom, L. Stone, C. Foster, D. Galasko, D. M. E. Llewellyn, and V. S. Ramachandran. 1999. Rehabilitation of hemiparesis after stroke with a mirror. *Lancet,* 353(9169): 2035–36.

195 ***mirror therapy was helpful . . . for a Taub-like treatment:*** K. Sathian, A. I. Greenspan, and S. L. Wolf. 2000. Doing it with mirrors: A case study of a novel approach to neurorehabilitation. *Neurorehabilitation and Neural Repair,* 14(1): 73–76.

Chapter 8
Imagination

196 •***A changing magnetic field induces an electric current around it:*** It was Michael Faraday who discovered, in the nineteenth century, that a changing magnetic field induces an electric current around it.

197 ***To determine the function of a specific brain area:*** A. Pascual-Leone, F. Tarazona, J. Keenan, J. M. Tormos, R. Hamilton, and M. D. Catala. 1999. Transcranial magnetic stimulation and neuroplasticity. *Neuropsychologia,* 37:207–17.

197 ***"repetitive TMS":*** A. Pascual-Leone, J. Valls-Sole, E. M. Wassermann, and M. Hallet. 1994. Responses to rapid-rate transcranial magnetic stimulation of the human motor cortex. *Brain,* 117:847–58.

198 ***Pascual-Leone's group was the first to show that rTMS is effective in treating . . . depressed patients:*** A. Pascual-Leone, B. Rubio, F. Pallardo, and M. D. Catala. 1996. Rapid-rate transcranial stimulation of left dorsolateral prefrontal cortex in drug-resistant depression. *Lancet,* 348(9022): 233–37.

198 •***rTMS . . . had fewer side effects:*** Unlike electroconvulsive therapy, or ECT, TMS doesn't require the patient to be anesthetized and doesn't cause a seizure. It also causes fewer short-term cognitive side effects, such as memory problems.

198 ***He studied how people learn . . . Braille:*** A. Pascual-Leone, R. Hamilton, J. M. Tormos, J. P. Keenan, and M. D. Catala. 1999. Neuroplasticity in the adjustment to blindness. In J. Grafman and Y. Christen, eds., *Neuronal plasticity: Building a bridge from the laboratory to the clinic.* New York: Springer-Verlag, 94–108, especially 97.

198 •***When Pascual-Leone used TMS to map the* motor *cortex:*** To map the motor cortex, Pascual-Leone stimulated a part of the cortex, observed which muscle moved, and recorded it. Then he shifted the TMS paddle a centimeter on the subject's head. He observed whether it triggered the same muscle or a different one. To map the size of the *sensory* map, he touched the subject's fingertips and asked if the subject felt it. Then he applied TMS to the subject's brain to see whether he could *block* those sensations. If he could, he knew the area in the brain that he blocked was part of the sensory map. By seeing *how much* transmagnetic stimulation it took to block the person from feeling they were being touched,

he got a sense of how substantial the sensory map was. If he had to turn up the stimulation to a high intensity to block the sensation, he knew there was a lot of cortical map representation for the fingertip. He then moved the TMS paddle around to different positions on the scalp, to determine the map's precise borders. A. Pascual-Leone and F. Torres. 1993. Plasticity of the sensorimotor cortex representation of the reading finger in Braille readers. *Brain,* 116:39–52; A. Pascual-Leone, R. Hamilton, J. M. Tormos, J. P. Keenan, and M. D. Catala, 1999, 94–108.

200 ***•our thoughts can change the material structure of our brains:*** The groundwork for the idea that thoughts can change the physical structure of the brain was proposed five hundred years ago by Thomas Hobbes (1588–1679), then was developed by the philosopher Alexander Bain, Sigmund Freud, and the neuroanatomist Santiago Ramón y Cajal.

Hobbes proposed that our imagination was related to sensation, and that sensation led to physical changes in the brain. T. Hobbes. 1651/1968. *Leviathan.* London: Penguin, 85–88. See also his work *De Corpore.* He argued that when a person is touched, the impact, in the form of movement, travels down the nerves, leading to sensory impressions. The same happens, he argued, when the eye is struck by light—the impact creates "movement" in the nerves. Indeed, this idea that movement extends into the nervous system is still alive in our language when we speak of sense "impressions"—for impressions are usually caused by a moving force applying pressure. Hobbes defined imagination as "nothing but decaying sense." Thus, when we see something, then shut our eyes, we can still imagine it, though more faintly because it is "decaying." He argued that when we "imagine" a fanciful thing like a centaur, we simply combine two images, for a centaur is the image of a man and a horse combined.

Hobbes's idea that the nerves "move" in response to touch, light, sound, and so forth was not a bad guess in an era long before electricity was understood, for he correctly intuited that the nerves convey some kind of physical energy to the brain. (He may have had help from Galileo, whom he visited on a trip to Italy. Hobbes, possibly at Galileo's suggestion, began to apply Galileo's new physical laws of movement to the understanding of the mind and sensation.)

Similarly, Hobbes's assertion that imagination is "nothing but decaying sense" proves to be extremely insightful. PET scans show that imagined visual images are generated by the same visual centers as are real images produced by external stimuli.

Hobbes was a materialist: he thought the nervous system, the brain, and the mind all work on the same principles, so he had no trouble, in principle, understanding how changes in thought might lead to changes in the nerves. His idea was opposed by his contemporary René Descartes, who argued that the mind and brain work by completely different laws. The mind, or the soul as he sometimes called it, has nonmaterial thoughts, and it doesn't obey the same physical laws as the material brain. Our existence consists of this duality, and people who follow Descartes are called "dualists." But Descartes could never credibly explain how the immaterial mind could influence the material brain. For centuries, most scientists followed Descartes, with the result that it seemed impossible to envision the idea that a thought might change the structure of the physical brain.

Two hundred years later, in 1873, philosopher Alexander Bain took Hobbes's idea to the next level and proposed that each time a thought, memory, habit, or train of ideas occurs, there is some "growth in the cell junctions" of the brain. A. Bain. 1873. *Mind and body: The theories of their relation.* London: Henry S. King. Thoughts lead to changes in what would come to be called the "synapses." Then Freud, based on his own neuroscience research, added that "imagination" too led to changes in neuronal connections.

In 1904 Santiago Ramón y Cajal, a Spanish neuroanatomist, speculated that not only physical practice but mental practice leads to changes in these networks. See below and text.

201 ***the "organ of thought is . . . malleable, and perfectible by well-directed mental exercise":*** S. Ramón y Cajal. 1894. The Croonian lecture: La fine structure des centres nerveux. *Proceedings of the Royal Society of London,* 55:444–68, especially 467–68.

201 ***•He also had the intuition that this process would be . . . pronounced . . . in pianists:*** S. Ramón y Cajal wrote, "The work of a pianist . . . is inaccessible for the untrained human, as the acquisition of new abilities requires many years of mental and physical practice. In order to fully understand this complicated phenomenon it is necessary to admit, in addition to the strengthening of pre-established organic pathways, the establishment of new ones, through ramification and progressive growth of dendritic arborizations and nervous terminals . . . Such a development takes place in response to exercise, while it stops and may be reversed in brain spheres that are not cultivated." S. Ramón y Cajal. 1904. *Textura del sistema nervioso del hombre y de los sertebrados.* Cited by A. Pascual-Leone. 2001. The brain that plays music and is changed by it. In R. Zatorre and I. Peretz, eds., *The biological foundations of music.* New York: Annals of the New York Academy of Sciences, 315–29, especially 316.

201 ***The details of the imagining experiment were simple:*** A. Pascual-Leone, N. Dang, L. G. Cohen, J. P. Brasil-Neto, A. Cammarota, and M. Hallett. 1995. Modulation of muscle responses evoked by transcranial magnetic stimulation during the acquisition of new fine motor skills. *Journal of Neurophysiology,* 74(3): 1037–45, especially 1041.

202 ***Glenn Gould relied largely on mental practice:*** B. Monsaingeon. 1983. *Écrits/Glenn Gould, vol. 1, Le dernier puritain.* Paris: Fayard; J. DesCôteaux and H. Leclère. 1995. Learning surgical technical skills. *Canadian Journal of Surgery,* 38(1): 33–38.

203 ***Rüdiger Gamm, a young German . . . human calculator:*** M. Pesenti, L. Zago, F. Crivello, E. Mellet, D. Samson, B. Duroux, X. Seron, B. Mazoyer, and N. Tzourio-Mazoyer. 2001. Mental

calculation in a prodigy is sustained by right prefrontal and medial temporal areas. *Nature Neuroscience,* 4(1): 103–7.

203 ***When people close their eyes and visualize . . . the letter* a:** E. R. Kandel, J. H. Schwartz, and T. M. Jessell, eds. 2000. *Principles of Neural Science,* 4th ed. New York: McGraw-Hill, 394; M. J. Farah, F. Peronnet, L. L. Weisberg, and M. Monheit. 1990. Brain activity underlying visual imagery: Event-related potentials during mental image generation. *Journal of Cognitive Neuroscience,* 1:302–16; S. M. Kosslyn, N. M. Alpert, W. L. Thompson, V. Maljkovic, S. B. Weise, C. F. Chabris, S. E. Hamilton, S. L. Rauch, and F. S. Buonanno. 1993. Visual mental imagery activates topographically organized visual cortex: PET investigations. *Journal of Cognitive Neuroscience,* 5:263–87. Yet the following paper is an exception and does not find evidence for the activation of the primary visual cortex in visual imagery: P. E. Roland and B. Gulyas. 1994. Visual imagery and visual representation. *Trends in Neurosciences,* 17(7): 281–87.

204 ***in action and imagination many of the same parts of the brain are activated:*** K. M. Stephan, G. R. Fink, R. E. Passingham, D. Silbersweig, A. O. Ceballos-Baumann, C. D. Frith, and R. S. J. Frackowiak. 1995. Functional anatomy of mental representation of upper extremity movements in healthy subjects. *Journal of Neurophysiology,* 73(1): 373–86.

204 ***Those who only* imagined *. . . increased their muscle strength by 22 percent*:** G. Yue and K. J. Cole. 1992. Strength increases from the motor program: Comparison of training with maximal voluntary and imagined muscle contractions. *Journal of Neurophysiology,* 67(5): 1114–23.

205 ***These [thought translation] machines were developed in a few simple steps:*** J. K. Chapin. 2004. Using multi-neuron population recordings for neural prosthetics. *Nature Neuroscience,* 7(5): 452–55.

205 ***Miguel Nicolelis and John Chapin began a behavioral experiment, with the goal of learning to read an animal's thoughts:***

M. A. L. Nicolelis and J. K. Chapin. 2002. Controlling robots with the mind. *Scientific American,* October, 47–53.

206 ***The team has since taught . . . monkeys to use only their thoughts to move a robotic arm:*** J. M. Carmena, M. A. Lebedev, R. E. Crist, J. E. O'Doherty, D. M. Santucci, D. F. Dimitrov, P. G. Patil, C. S. Henriquez, and M. A. L. Nicolelis. 2003. Learning to control a brain-machine interface for reaching and grasping by primates. *PLOS Biology,* 1(2): 193–208.

207 •***After four days of practice he was able to move a computer cursor . . . using his thoughts:*** L. R. Hochberg, M. D. Serruya, G. M. Friehs, J. A. Mukand, M. Saleh, A. H. Caplan, A. Branner, D. Chen, R. D. Penn, and J. P. Donoghue. 2006. Neuronal ensemble control of prosthetic devices by a human with tetraplegia. *Nature,* 442(7099): 164–71; A. Pollack. 2006. Paralyzed man uses thoughts to move cursor. *New York Times,* July 13, front page. This breakthrough followed work Donoghue had done with Mijail D. Serruya, which involved teaching rhesus monkeys to move cursors on computers with their thoughts, using only six neurons. M. D. Serruya, N. G. Hatsopoulos, L. Paninski, M. R. Fellows, and J. P. Donoghue, 2002. Brain-machine interface: Instant neural control of a movement signal. *Nature,* 416(6877): 141–42.

207 ***Some scientists hope to develop a technology less invasive than microelectrodes:*** A. Kübler, B. Kotchoubey, T. Hinterberger, N. Ghanayim, J. Perelmouter, M. Schauer, C. Fritsch, E. Taub, and N. Birbaumer. 1999. The thought translation device: A neurophysiological approach to communication in total motor paralysis. *Experimental Brain Research,* 124:223–32; N. Birbaumer, N. Ghanayim, T. Hinterberger, I. Iversen, B. Kotchoubey, A. Kübler, J. Perelmouter, E. Taub, and H. Flor. 1999. A spelling device for the paralyzed. *Nature,* 398(6725): 297–98.

207 ***Most people who are right-handed find that their "mental left hand" is slower:*** J. Decety and F. Michel. 1989. Comparative analysis of actual and mental movement times in two graphic tasks. *Brain and Cognition,* 11:87–97; J. Decety. 1996. Do imagined and

executed actions share the same neural substrate? *Cognitive Brain Research,* 3:87–93; J. Decety. 1999. The perception of action: Its putative effect on neural plasticity. In J. Grafman and Y. Christen, eds., 109–30.

207–208 ***patients with stroke or Parkinson's disease . . . took longer to imagine moving the affected limb:*** Reviewed in M. Jeannerod and J. Decety, 1995. Mental motor imagery: A window into the representational stages of action. *Current Opinion in Neurobiology,* 5:727–32.

208 ***•Both mental imagery and actions are . . . products of the same motor program:*** Decety has also shown that when people imagine walking with a heavy load, their autonomic nervous system—breathing and heart rate—is activated.

211 ***Pascual-Leone, working with Roy Hamilton . . . propose a theory:*** A. Pascual-Leone and R. Hamilton. 2001. The metamodal organization of the brain. In C. Casanova and M. Ptito, eds., *Progress in Brain Research,* Vol. 134. San Diego, CA: Elsevier Science, 427–45.

212 ***•Someone . . . memorizing Homer's* Iliad, *might blindfold himself to recruit operators:*** Such manipulation of senses and the brain is not so uncommon. The anthropologist Edmund Carpenter, who worked with Marshall McLuhan (discussed in appendix 1), observed that "every culture has a sensory profile, and native cultures, for example, to maximize sound will minimize sight. So the dancer is often blinded, deliberately. Or, you may find, they will deliberately turn sound into a textile thing, so they will plug their ears when they sing. If you begin to examine cultures I think you will find all peoples do this. We go into an art gallery and the sign says 'Do Not Touch.' A concert goer closes his eyes. To maximize [reading] in a library, it says, 'Silence.'" From the film *McLuhan's Wake.* 2002. Written by David Sobelman; directed by Kevin McMahon. National Film Board of Canada, section Voices, audio interview, with Edmund Carpenter.

214 ***•the immaterial, ghostlike soul Descartes placed within:*** There are those who argue that Descartes may not have believed his proposal that the rational soul is not a physical thing and that he voiced it so as not to offend the Catholic Church. The Church considered the soul a *supernatural* phenomenon, which could not be physical because it was immortal and survived death and the physical, material body.

Descartes was part of the movement that sought to revolutionize humanity by using modern science to explain all living things, a project that brought him into direct conflict with the Church of the time, which had its own explanations for nature, life, the body, the brain, and the mind. Descartes had reasons to be careful: Galileo was shown the instruments of torture by the Inquisition when his theories and observations about the physical world seemed to challenge Church teaching. When Descartes found this out, he chose to suppress many of his own writings. In later years Descartes often kept only one step ahead of various persecutors, who alleged he was an atheist. In the last thirteen years of his life he resided at twenty-four different addresses.

Descartes dropped hints that he did not always write exactly what he believed and took political realities into account. He wrote, "I have composed my philosophy in such a way as not to shock anyone, and so that it can be received everywhere." R. Descartes. 1596–1659. *Oeuvres.* C. Adam and P. Tannery, eds. 1910. Paris: L. Cerf, 5:159. His chosen epigraph for his tombstone was from Ovid: "Bene qui latuit, bene vixit," or "He who hid well, lived well." Also see A. R. Damasio. 1994. *Descartes' error: Emotion, reason and the human brain.* New York: G. P. Putnam's Sons.

214 ***Any plasticity . . . existed in the mind, with its changing thoughts, not in the brain:*** C. Clemente. 1976. Changes in afferent connections following brain injury. In G. M. Austin, ed., *Contemporary aspects of cerebrovascular disease.* Dallas, TX: Professional Information Library, 60–93.

214 •*While we have yet to understand exactly* **how** ***thoughts actually change brain structure:*** Jeffrey Schwartz, who invented the brain lock treatment, has proposed a theory that uses quantum mechanics to attempt to explain how mental activities might alter the neural structure. I lack the competence to assess it. In J. M. Schwartz and S. Begley. 2002. *The mind and the brain: Neuroplasticity and the power of mental force.* New York: ReganBooks/HarperCollins.

Chapter 9
Turning Our Ghosts into Ancestors

218 ***Kandel was the first to show that as we learn, our individual neurons alter their structure:*** E. R. Kandel. 2003. The molecular biology of memory storage: A dialog between genes and synapses. In H. Jörnvall, ed., *Nobel Lectures, Physiology or Medicine, 1996–2000.* Singapore: World Scientific Publishing Co., 402. Also http:/nobelprize.org/nobel_prizes/medicine/laureates/2000/kandel-lecture.html.

218 ***Kandel's hope was to "trap" a learned response in the smallest possible group of neurons:*** E. R. Kandel. 2006. *In search of memory: The emergence of a new science of mind.* New York: W. W. Norton & Co., 166.

219 ***This was the first proof that learning led to neuroplastic strengthening:*** E. R. Kandel. 1983. From metapsychology to molecular biology: Explorations into the nature of anxiety. *American Journal of Psychiatry,* 140(10): 1277–93, especially 1285.

219 ***When the snails developed learned fear, the presynaptic neurons released more:*** Ibid.; E. R. Kandel, 2003, 405.

219 •***snails could be taught to recognize a stimulus as harmless:*** Learning to recognize a stimulus as harmless is called "habituation" and is a form of learning that we all do when we learn to tune out background noise.

219 •***Finally Kandel was able to show that snails can also learn to associate two different events and that their nervous systems***

change: What Kandel demonstrated was the neural analog of classical Pavlovian conditioning. This demonstration was crucial to him. Aristotle, the British empiricist philosophers, and Freud had all argued that learning and memory are the result of the mind's associating the events, ideas, and stimuli we experience. Pavlov, who founded behaviorism, discovered classical conditioning, a form of learning in which an animal or person is taught to associate two stimuli. A typical example would be to expose an animal to a benign stimulus, such as the sound of a bell, followed immediately by an unpleasant one, such as a shock, and repeat this a number of times, so that the animal soon begins to respond to the bell alone with fear.

220 •***the changes in the neurons lasted as long as three weeks:*** E. R. Kandel, J. H. Schwartz, and T. M. Jessel. 2000. *Principles of neural science,* 4th ed. New York: McGraw-Hill, 1250. In terms of training effects, they also found that if a snail was given a mild stimulus forty times in a row, the resulting habituation of the gill reflex would last a day. But if ten stimuli were given every day for four days, the effect would last for weeks. So appropriate spacing of learning is a key factor in developing long-term memory. E. R. Kandel, 2006, 193.

220 ***the individual* molecules *. . . involved in forming long-term memories:*** E. R. Kandel, J. H. Schwartz, and T. M. Jessel, 2000, 1254.

220 ***for short-term memories to become long-term, a new protein had to be made:*** E. R. Kandel, 2006, 241.

220 •***a single neuron . . . might go from having 1,300 to 2,700 synaptic connections:*** This was work by Craig Baily and Mary Chen. If the same cell developed a long-term memory for habituation, it would go from having 1,300 connections to 850, of which only about 100 would be active. Ibid., 214.

221 •***Kandel argues that when psychotherapy changes people, "it presumably does so through learning, by producing changes in gene expression":*** E. R. Kandel. 1998. A new intellectual framework for psychiatry. *American Journal of Psychiatry,* 155(4): 457–69,

especially 460. Along similar lines, neuroscientist Joseph LeDoux has argued that psychiatric disorders might be thought of as malconnection syndromes that occur between synapses of various regions and functions, and that "if the self can be disassembled by experiences that alter connections, presumably it can be reassembled by experiences that establish, change, or renew connections." J. LeDoux. 2002. *The synaptic self: How our brains become who we are.* New York: Viking, 307.

221 ***the talking cure works by "talking to neurons":*** S. C. Vaughan. 1997. *The talking cure: The science behind psychotherapy.* New York: Grosset/Putnam.

221 ***"I remember Kristallnacht even today, more than sixty years later, almost as if it were yesterday":*** E. R. Kandel. 2001. Autobiography. In T. Frängsmyr, ed., *Les Prix Nobel: The Nobel Prizes 2000.* Stockholm: The Nobel Foundation. Also on the Internet at http://nobelprize.org/nobel_prizes/medicine/laureates/2000/kandel-autobio.html.

222 ***psychoanalysis . . . "outlined by far the most coherent, interesting and nuanced view of the human mind":*** E. R. Kandel, 2000, Autobiography.

222 ***how a country . . . could become "so radically dissociated":*** Ibid.

222 ***•Freud began . . . as a laboratory neuroscientist, but because he was too impoverished to continue:*** Despite his brilliance, Freud did not progress through the ranks at the University of Vienna, in part because of his ideas, in part because he was a Jew. He became a lecturer in 1885, and it took seventeen years until he was made a professor. The average span between those appointments was eight years. In the meantime he had a family to support. P. Gay. 1988. *Freud: A life for our time.* New York: W. W. Norton & Co., 138–39.

223 ***wrote a book titled* On Aphasia:** S. Freud. 1891. *On aphasia: A critical study.* New York: International Universities Press.

223 ***In 1895 Freud completed the "Project for a Scientific Psychology":*** S. Freud. 1895/1954. Project for a scientific psychology. Translated

by J. Strachey. In *Standard edition of the complete psychological works of Sigmund Freud,* vol. 1. London: Hogarth Press.

223 ***•still admired for its sophistication:*** Admired by Karl Pribram and Nobel Prize winner Gerald Edelman, among others.

223 ***•The first plastic concept Freud developed is the law that neurons that fire together wire together:*** It is no coincidence that Freud developed plastic concepts after rejecting the simplified localizationism of his day. Having argued that the brain constructs new functional systems connecting neurons spread throughout the brain, in novel ways, as it learns new tasks, he needed to think through how this might play out on a neuronal level, and how it might affect memory and other mental functions. In essence, he developed a more dynamic view of the brain, one that inspired the work of Luria and the birth of neuropsychology. S. Freud, 1891; O. Sacks. 1998. The other road: Freud as neurologist. In M. S. Roth, ed., *Freud: Conflict and culture.* New York: Alfred A. Knopf, 221–34. The "Project" wasn't published until 1954, six years before Kandel began trying to show that learning leads to changes in synapses. (For background on the "Project," see P. Amacher. 1965. *Freud's neurological education and its influence on psychoanalytic theory.* New York: International Universities Press, 57–59; S. Freud, 1895/1954, 319, 338; K. H. Pribram and M. M. Gill. 1976. *Freud's "Project" re-assessed: Preface to contemporary cognitive theory and neuropsychology.* New York: Basic Books, 62–66, 80.) Kandel also knew of Santiago Ramón y Cajal's 1894 proposal that mental activity might strengthen connections between neurons or lead to the formation of new connections. Cajal wrote, "Mental exercise facilitates a greater development of the protoplasmic apparatus and of the nervous collaterals in the parts of the brain in use. In this way, pre-existing connections between groups of cells could be reinforced by multiplication of the terminal branches . . . But the pre-existing connections could also be reinforced by the formation of new collaterals and . . . expansions." S. Ramón y Cajal. 1894. The Croonian lecture: La fine structure des

centres nerveux. *Proceedings of the Royal Society of London,* 55:444–68, especially 466.

223– •***all our mental associations . . . are expressions of links formed***
224 ***in our memory networks:*** The relation of memory networks to neuronal networks in associations is implicit and is spelled out in more detail in M. F. Reiser. 1984. Mind, brain, body: Toward a convergence of psychoanalysis and neurobiology. New York: Basic Books, 67.

224 •***His law of association by simultaneity implicitly links changes in neuronal networks with changes in our memory networks:*** For instance, in the "Project," after discussing contact barriers, or synapses, Freud goes on to discuss memory and writes, "A main characteristic of nervous tissue is memory: that is a capacity for being permanently altered by single occurrences." S. Freud, 1895/1954, 299; K. H. Pribram and M. M. Gill, 1976, 64–68.

224 •***Freud's second plastic idea was . . . the related idea of sexual plasticity:*** Freud wrote, "The sexual instincts are noticeable to us for their plasticity, their capacity for altering their aims, their replaceability, which admits of one instinctual satisfaction being replaced by another, and their readiness for being deferred." S. Freud. 1932/1933/1964. New introductory lectures on psycho-analysis. Translated by J. Strachey. In *Standard edition of the complete psychological works of Sigmund Freud,* vol. 22. London: Hogarth Press, 97.

224 ***What happens during these critical periods has an inordinate effect on our ability to love and relate:*** A. N. Schore. 1994. *Affect regulation and the origin of the self: The neurobiology of emotional development.* Hillsdale, NJ: Lawrence Erlbaum Associates; A. N. Schore. 2003. *Affect dysregulation and disorders of the self.* New York: W. W. Norton & Co.; A. N. Schore. 2003. *Affect regulation and the repair of the self.* New York: W. W. Norton & Co.

224 ***In 1896 Freud wrote that . . . memory traces are subjected to "...a retranscription":*** J. M. Masson, trans. and ed. 1985. *The complete letters of Sigmund Freud to Wilhelm Fliess.* Cambridge, MA: Harvard University Press, 207.

224 ***"analogous in every way to the process by which a nation constructs legends":*** S. Freud. 1909. Notes upon a case of obsessional neurosis. In *Standard edition of the complete psychological works,* vol. 10, 206.

224–225 ***To be changed . . . memories had to be conscious . . . as neuroscientists have since shown:*** F. Levin. 2003. Psyche and brain: The biology of talking cures. Madison, CT: International Universities Press.

226 ***The right hemisphere has just completed a growth spurt:*** A. N. Schore, 1994.

226 ***The right hemisphere . . . allows us to . . . read facial expressions, and it connects us:*** A. N. Schore. 2005. A neuropsychoanalytic viewpoint: Commentary on a paper by Steven H. Knoblauch. *Psychoanalytic Dialogues,* 15(6): 829–54.

226 ***It also processes the musical component of speech:*** J. S. Sieratzki and B. Woll. 1996. Why do mothers cradle babies on their left? *Lancet,* 347(9017): 1746–48.

226 ***our right hemisphere dominates the brain for the first three years:*** A. N. Schore. 2005. Back to basics: Attachment, affect regulation, and the developing right brain: Linking developmental neuroscience to pediatrics. *Pediatrics in Review,* 26(6): 204–17.

226 ***Brain scans show that . . . the mother . . . communicates nonverbally with her right hemisphere to reach her infant's right:*** A. N. Schore. 2005. A neuropsychoanalytic viewpoint.

226 ***a key area of the right frontal lobe . . . will allow infants both to maintain human attachments and to regulate their emotions:*** A. N. Schore, 1994.

226 **•right orbitofrontal system:** The full name is "the right orbital area of the prefrontal cortex."

227 ***If others cannot help him . . . he learns to "autoregulate":*** A. N. Schore, 2005. Personal communication.

228 ***René Spitz studied infants:*** R. Spitz. 1965. *The first year of life: A psychoanalytic study of normal and deviant development of object relations.* New York: International Universities Press.

228 ***"During the first 2–3 years . . . relies primarily on its procedural memory systems":*** E. R. Kandel. 1999. Biology and the future of psychoanalysis: A new intellectual framework for psychiatry revisited. *American Journal of Psychiatry,* 156(4): 505–24.

229 ***•It helps us to organize our memories by time and place:*** The hippocampus is also involved in spatial organization, and perhaps this is why it helps provide a context for our explicit memories, which help us to recall them. But this is speculation. A recent issue of the journal *Hippocampus* has several articles exploring this question. See J. R. Manns and H. Eichenbaum. 2006. Evolution of declarative memory. *Hippocampus,* 16:795–808.

232–233 ***•Mr. L. began to . . . reexperience . . . memories of searching for his mother that had been frozen in time:*** The idea that an image from the traumatic past can be frozen in the mind and remain unchanged since the time of the trauma is not unlike what happens to patients who have their wounded limbs put in casts and then develop frozen phantom limbs after amputation, as we saw in chapter 7, "Pain." Because the parent is no longer present, the child cannot use the parent as feedback to help modify his mental image of him. The image of a parent lost in earliest childhood can haunt a child the way a phantom limb does and can be experienced as a felt presence that makes unpredictable distressing intrusions.

233 ***•positive bonds appear to facilitate neuroplastic change . . . dissolving existing neuronal networks:*** Recent studies, inspired in part by Kandel's work, by Karim Nader of McGill University, show that when memories are activated, they enter a labile state, when they can be altered. In fact, before evoked memories go back into storage, they must be reconsolidated and new proteins must be made. This may be another reason why remembering traumas, or repeating transferences in psychotherapy, can lead to psychic change: memories must be reactivated to have their neuronal connections altered, so they can be retranscribed and changed. K. Nader, G. E. Schafe, and J. E. LeDoux. 2000. Fear

memories require protein synthesis in the amygdala for reconsolidation after retrieval. *Nature,* 406(6797): 722–26; J. Debiec, J. E. LeDoux, and K. Nader. 2002. Cellular and systems reconsolidation in the hippocampus. *Neuron,* 36(3): 527–38.

233 ***"There is no longer any doubt . . . that psychotherapy can result in detectable changes in the brain":*** A. Etkin, C. Pittenger, H. J. Polan, and E. R. Kandel. 2005. Toward a neurobiology of psychotherapy: Basic science and clinical applications. *Journal of Neuropsychiatry and Clinical Neurosciences,* 17:145–58.

233 ***When patients . . . have flashbacks . . . the flow of blood to the prefrontal and frontal lobes . . . decreases:*** S. L. Rauch, B. A. van der Kolk, R. E. Fisler, N. M. Alpert, S. P. Orr, C. R. Savage, A. J. Fischman, M. A. Jenike, and R. K. Pitman. 1996. A symptom provocation study of PTSD using PET and script-driven imagery. *Archives of General Psychiatry,* 53(5): 380–87.

233 ***"The aim of the talking cure . . . to extend the . . . influence of the prefrontal lobes":*** M. Solms and O. Turnbull. 2002. *The brain and the inner world.* New York: Other Press, 287.

233 ***•A study of depressed patients treated with interpersonal psychotherapy:*** Dr. Myrna Weissman, who developed interpersonal psychotherapy, did so by reviewing the risk factors for depression and was also influenced by the work of two psychoanalysts, John Bowlby and Harry Stack Sullivan, who focused on how relationships and loss affect the psyche (personal communication). This study of Interpersonal Psychotherapy and change is in A. L. Brody, S. Saxena, P. Stoessel, L. A. Gillies, L. A. Fairbanks, S. Alborzian, M. E. Phelps, S. C. Huang, H. M. Wu, M. L. Ho, M. K. Ho, S. C. Au, K. Maidment, and L. R. Baxter, 2001. Regional brain metabolic changes in patients with major depression treated with either paroxetine or interpersonal therapy: Preliminary findings. *Archives of General Psychiatry,* 58(7): 631–40. Another study of depressed patients showed that cognitive-behavior therapy—a form of treatment that corrects the exaggerated forms of negative thinking in depression—also worked by normalizing the prefrontal

lobes. K. Goldapple, Z. Segal, C. Garson, M. Lau, P. Bieling, S. Kennedy, and H. Mayberg. 2004. Modulation of cortical-limbic pathways in major depression. *Archives of General Psychiatry,* 61(1): 34–41.

234 ***A more recent fMRI brain scan study of anxious patients with panic . . . reduced following psychoanalytic psychotherapy:*** M. E. Beutel. 2006. Functional neuroimaging and psychoanalytic psychotherapy—Can it contribute to our understanding of processes of change? Presentation, Arnold Pfeffer Center for Neuro-Psychoanalysis at the New York Psychoanalytic Institute, Neuro-Psychoanalysis Lecture Series. October 7.

237 • ***"I believe this was a memory of my mother's wake":*** Some might question whether Mr. L.'s memory of his mother's wake was a "true" memory or merely a wish he could remember. If it was merely a wishful fantasy, it was one he had been incapable of having when he started analysis. But even if a fantasy, it was hardly wishful thinking—it was an extremely painful experience for him and certainly was not a magical denial of reality, because he verified that he was present at the wake. As we shall see in this chapter (and the following notes), research now shows that some children at twenty-six months are capable of some explicit memories.

Major life traumas, as the Israeli psychoanalyst and psychiatrist Yoram Yovell, who worked in Kandel's lab, points out, can have a dual impact on the hippocampus as it forms memories. The glucocorticoids that are released lead to patchy memories. But the adrenaline and noradrenaline released by stressful events can cause the hippocampus to form "flashbulb memories," which are enhanced, vivid, explicit memories. That is probably why people who have experienced traumas have both hypervivid memories for some aspects of the trauma and patchy memories for other aspects of it. The sight of his mother dead might well have produced a flashbulb memory in Mr. L.

Ultimately, Mr. L.'s own prudent statement says it best: the image of the open coffin came into his mind "tagged" as a memory, but he prefaced his report of it with a cautionary "I believe." See Y. Yovell. 2000. From hysteria to posttraumatic stress disorder. *Journal of Neuro-Psychoanalysis,* 2:171–81; L. Cahill, B. Prins, M. Weber, and J. L. McGaugh. 1994. β-Adrenergic activation and memory for emotional events. *Nature,* 371(6499): 702–4.

237 ***new studies show that infants in the first and second years can store . . . facts and events, including traumatic ones:*** P. J. Bauer. 2005. Developments in declarative memory: Decreasing susceptibility to storage failure over the second year of life. *Psychological Science,* 16(1): 41–47; P. J. Bauer and S. S. Wewerka. 1995. One- to two-year-olds' recall of events: The more expressed, the more impressed. *Journal of Experimental Child Psychology,* 59:475–96; T. J. Gaensbauer. 2002. Representations of trauma in infancy: Clinical and theoretical implications for the understanding of early memory. *Infant Mental Health Journal,* 23(3): 259–77; L. C. Terr. 2003. "Wild child": How three principles of healing organized 12 years of psychotherapy. *Journal of the American Academy of Child and Adolescent Psychiatry,* 42(12): 1401–9; T. J. Gaensbauer. 2005. "Wild child" and declarative memory. *Journal of the Academy of Child and Adolescent Psychiatry,* 44(7): 627–28.

237 •***While the explicit memory system is not robust in the first few years, research . . . shows it exists:*** We have underestimated the development of the explicit memory system for facts and events in infants because we usually test the explicit memory system by asking people questions, which are answered with words. Obviously preverbal infants cannot tell us whether they consciously recollect a particular event. But recently researchers have found ways to test infants by getting them to kick when they recognize the repetition of events, and they can remember them. C. Rovee-Collier. 1997. Dissociations in infant memory: Rethinking the development of implicit and explicit memory. *Psychological Review,* 104(3): 467–98; C. Rovee-Collier. 1999. The development of infant memory. *Current Directions in Psychological Science,* 8(3): 80–85.

238 ***Infants can remember events from the first few years of life if . . . reminded:*** C. Rovee-Collier, 1999.

238 ***children can remember events that occurred before they could talk and . . . put those memories into words:*** T. J. Gaensbauer, 2002, 265.

238 •***his core dream:*** Indeed, Mr. L.'s core dream, "I am searching for something lost, I know not what, maybe part of me . . . and I will know it when I find it," articulated perfectly that he had a problem with his memory and recall. He knew he could not, on his own, recall what was lost but also that should it be put in front of him, he would recognize it, recognition being an even more basic form of remembering than recall. In this sense, his dream's prediction was accurate, for when he finally found what he was looking for, he did recognize it, in a way that shook him to his core.

238 •***This kind of progressive dream series shows the mind and brain . . . must unlearn:*** Nobel laureate Francis Crick and Graeme Mitchison proposed that a kind of "reverse learning" occurs in dreaming, because the dreaming brain has, as one of its tasks, unlearning various spurious images that we have learned in the course of developing perceptual memories. F. Crick and G. Mitchison. 1983. The function of dream sleep. *Nature,* 304(5922): 111–14. See also G. Christos. 2003. *Memory and dreams: The creative mind.* New Brunswick, NJ: Rutgers University Press. In their model "we dream in order to forget." It makes sense that if the dreaming brain is trying to classify events and images, it will find some to be important and worth remembering and many more that are worth forgetting. This theory is best at accounting for why we forget our dreams. But it is weak in explaining why so much can be learned from dreams, or the post-traumatic, recurring dreams that Mr. L. had and couldn't get out of his head.

239 •***The newest brain scans show that when we dream:*** Dreams are often higgledy-piggledy and hard to understand because certain "higher" mental functions are not operating in the way they do

when we are awake. Allen Braun, a researcher at the National Institutes of Health in Bethesda, Maryland, has used positron emission tomography (PET) scans to measure brain activity in dreaming subjects. He has shown that the region known as the limbic system—which processes emotion; sexual, survival, and aggressive instincts; and interpersonal attachments—shows high activity. The ventral tegmental area that is associated with pleasure seeking (which we discussed with respect to the pleasure systems in chapter 4, "Acquiring Tastes and Loves") is also activated. But the prefrontal cortex, the area responsible for achieving goals and discipline, postponing gratification and controlling our impulses, shows lower activity.

With the emotional-instinctual processing areas of the brain turned on, and the part of the brain that controls our impulses relatively inhibited, it is no wonder that wishes and impulses that we normally restrain or are even unaware of are more likely to be expressed in dreams as Freud, and Plato before him, noted.

But why are our dreams hallucinations, in which we experience things that aren't happening as real? When we are awake, we first take the world in through our senses. For vision, input comes through the eyes. Then the primary visual zone in the brain receives direct input from the retina. Next, the secondary visual zone processes colors and movement and recognizes objects. Finally, a tertiary zone further down the line of perceptual processing (in the occipito-tempero-parietal junction) brings together these visual sense perceptions and relates them to other sensory modalities. Thus events that we have concretely perceived are related to each other, and once that happens, more abstract thought and meaning can emerge.

Freud argued that in hallucinations and in dreaming, the mind "regresses." By this he meant it processes images in backward or reverse order. We begin not with perceptions of the outside world and then form abstract ideas about them, but with our own abstract ideas, which become represented in a concrete, often visual way, as though they were perceptions happening in the world.

Allen Braun has shown with brain scans of dreamers that the parts of the brain that are first to receive incoming visual input—the primary visual areas—shut down. But the secondary visual zones that integrate the different kinds of visual input (e.g., color, movement) into objects are active. So what we experience in dreams is images that come not from the external world but from within us and that are experienced as hallucinations. This is consistent with the assertion that in dreaming, perception is processed in a backward direction.

A proper dream interpretation starts from the hallucinatory dream perceptions, which seem to be bizarre and unconnected to each other, and traces them back to the more abstract dream thoughts that produced them.

Studies by neuropsychoanalyst Mark Solms of patients who have had strokes shed much light on dreams. Working with these patients, Solms has shown that dreams don't consist just of confusing visual images but of thinking. He worked with patients with damage to an area of the brain that is necessary for producing visual images. In waking life, these patients suffer from a well-known neurological syndrome called "irreminiscence" and cannot form whole visual images in their heads. One woman who'd had a stroke in this area couldn't recognize the faces of her family but could recognize their voices. In her dreams, Solms found, she'd hear voices but had no images; in other words, she had nonvisual dreaming.

Another patient with a similar deficit, which came about after a brain tumor was removed, reported dreaming, "My mother and another lady were holding me down." When Solms asked him how he knew it, since he had no visual images, he answered, "I just knew it," and reported clearly feeling held down. He said that since his operation his dreams were "thinking dreams." In other words, behind the visual imagery of dreams, a kind of thinking occurs.

Now what of patients with damage to those tertiary zones of the brain that form abstract thoughts? According to Freud, that part of the brain actually generates dreams. Solms has found that when those tertiary zones that generate abstract thought are damaged, dreaming stops. Clearly, this region is crucial to generating dreams.

Solms theorizes that dreams are typically hard to understand because in dreams abstract ideas are visually represented. How might this play out? Clinically, one often finds that an abstract idea such as "I am special and don't have to follow the rules other people have to" might be represented visually by "I'm flying." The abstract idea that "deep down, I fear my ambition is out of control" might be represented by a dream of Mussolini's body after he was executed. K. Kaplan-Solms and M. Solms. 2002. *Clinical studies in neuro-psychoanalysis.* New York: Karnac; M. Solms and O. Turnbull, 2002, 209–10.

239 ***studies show that sleep helps us consolidate learning and memory and effects plastic change:*** R. Stickgold, J. A. Hobson, R. Fosse, and M. Fosse. 2001. Sleep, learning, and dreams: Off-line memory reprocessing. *Science,* 294(5544): 1052–57.

239 ***When we learn . . . better at it . . . if we have a good night's sleep:*** Ibid.

239 ***sleep enhances neuroplasticity during the critical period:*** M. G. Frank, N. P. Issa, and M. P. Stryker. 2001. Sleep enhances plasticity in the developing visual cortex. *Neuron,* 30(1): 275–87.

239 ***the effects of REM sleep . . . on their brain structure:*** G. A. Marks, J. P. Shaffrey, A. Oksenberg, S. G. Speciale, and H. P. Roffwarg. 1995. A functional role for REM sleep in brain maturation. *Behavioral Brain Research,* 69:1–11.

239 ***REM . . . to retain emotional memories:*** U. Wagner, S. Gais, and J. Born. 2001. Emotional memory formation is enhanced across sleep intervals with high amounts of rapid eye movement. *Learning and Memory,* 8:112–19.

239–240 ***REM . . . for allowing the hippocampus to turn short-term memories of the day before into long-term:*** During our dreams the hippocampus works by interacting with the cortex, to make long-term memories.

When we have a perceptual experience while awake, we register it in our cortex. The look of your friend turns on cells in your visual cortex, the sound of his voice triggers neurons in your auditory cortex, and when the two of you hug, sensory and motor areas light up. Your limbic system, which deals with emotion, is also triggered. All these different areas send streams of signals at once, and you recognize that this is your friend. These signals are sent simultaneously to the hippocampus, where they are briefly stored, and are "bound" together. (That is why, when you remember a conversation with your friend, you automatically also see his face.) If seeing the friend is an important event, the hippocampus turns it from a short-term memory into a long-term explicit memory. But that memory is not stored in the hippocampus. Rather, it is sent back to the parts of the cortex where it came from and is stored in the original cortical networks that first produced its various sights, sounds, and so on. So the memory is widely distributed throughout your brain.

Scientists can measure the brain waves given off by the hippocampus and the cortex, when they are active. By looking at the timing of when these various areas fire during sleep, they have come up with an intriguing proposal. During REM sleep our cortex downloads its signals into the hippocampus. During non-REM sleep the hippocampus, after having worked these short-term memories over, uploads them back into the cortex, where they will remain as long-term memories. It may be that during our dreams we are at times consciously experiencing the downloading of many bits of experience from the various parts of our cortex that are being fired. R. Stickgold, J. A. Hobson, R. Fosse, and M. Fosse, 2001.

These recent findings were anticipated in a remarkable study in the 1970s by Dr. Stanley Palombo, who had a patient in psychoanalysis just after the patient's father died. As part of Dr.

Palombo's study, the patient spent nights between psychoanalytic sessions at a sleep lab and was awakened at the end of each REM sleep cycle, and his dreams were recorded. Palombo discovered that over the course of each night the patient's dreams worked over new experiences he had had during the day, and he progressively matched them with his past experiences, determining which of his memories they were to be linked to and hence stored with. S. R. Palombo. 1978. *Dreaming and memory: A new information-processing model.* New York: Basic Books.

240 •***early stresses predispose . . . motherless animals to . . . illness for the rest of their lives:*** Psychologist Seymour Levine found that rat pups separated from their mothers protest immediately, giving off high-intensity cries, and search for their mothers until they show signs of despair. Their heart rate and body temperature drop and they become less alert, like the children Spitz observed, who seemed turned off and unreachable, with a faraway look in their eyes. Levine then discovered that their brains triggered a "stress response," releasing high quantities of glucocorticoid, the "stress hormone." These stress hormones are good for the body for short periods, because they mobilize it to deal with emergencies by increasing heart rate and sending blood to the muscles. But if they are released repeatedly, they lead to stress-related illnesses and wear the body out prematurely.

Recent research by Michael Meaney, Paul Plotsky, and others showed that if pups were separated from their mothers for periods of three to six hours each day for two weeks, the mothers soon ignored their pups, and these pups showed an increased release of the glucocorticoid stress hormones *that lasted into adulthood.* Early trauma can have lifelong effects, and its victims are more easily stressed thereafter.

Pups removed from their mother only briefly during the first two weeks of life made the usual cries, which summoned their mothers, who licked them more than normal, groomed them more, and carried them around more than pups that had not been removed. The effect of this maternal response was *to reduce, for the rest of the animal's life,* its tendency to secrete glucocorticoids and to develop stress-related illness and experience fear. Such is the power of good mothering in the critical period of attachment. This lifelong benefit may be related to plasticity because the pups got this close maternal attention during the critical period of development for their brains' stress response systems. S. Levine. 1957. Infantile experience and resistance to physiological stress. *Science,* 126(3270): 405; S. Levine. 1962. Plasma-free corticosteroid response to electric shock in rats stimulated in infancy. *Science,* 135(3506): 795–96; S. Levine, G. C. Haltmeyer, G. G. Karas, and V. H. Denenberg. 1967. Physiological and behavioral effects of infantile stimulation. *Physiology and Behavior,* 2:55–59; D. Liu, J. Diorio, B. Tannenbaum, C. Caldji, D. Francis, A. Freedman, S. Sharma, D. Pearson, P. M. Plotsky, and M. J. Meaney. 1997. Maternal care, hippocampal glucocorticoid receptors, and hypothalamic-pituitary-adrenal responses to stress. *Science,* 277(5332): 1659–62, especially 1661; P. M. Plotsky and M. J. Meaney. 1993. Early, postnatal experience alters hypothalamic corticotropin-releasing factor (CRF) mRNA, median eminence CRF content and stress-induced release in adult rats. *Molecular Brain Research,* 18:195–200.

240 ***When they undergo long separations, the gene . . . gets turned on:*** P. M. Plotsky and M. J. Meaney, 1993; C. B. Nemeroff. 1996. The corticotropin-releasing factor (CRF) hypothesis of depression: New findings and new directions. *Molecular Psychiatry,* 1:336–42; M. J. Meaney, D. H. Aitken, S. Bhatnagar, and R. M. Sapolsky. 1991. Postnatal handling attenuates certain neuroendocrine, anatomical and cognitive dysfunctions associated with aging in female rats. *Neurobiology of Aging,* 12:31–38.

240–241 ***adult survivors of childhood abuse also show . . . glucocorcoid supersensitivity:*** C. Heim, D. J. Newport, R. Bonsall, A. H. Miller, and C. B. Nemeroff. 2001. Altered pituitary-adrenal axis responses to provocative challenge tests in adult survivors of childhood abuse. *American Journal of Psychiatry,* 158(4): 575–81.

241 ***Depression, high stress, and childhood trauma all . . . kill cells:*** R. M. Sapolsky. 1996. Why stress is bad for your brain. *Science,* 273(5276): 749–50; B. L. Jacobs, H. van Praag, and F. H. Gage. 2000. Depression and the birth and death of brain cells. *American Scientist,* 88(4): 340–46.

241 ***The longer people are depressed, the smaller their hippocampus:*** B. L. Jacobs, H. van Praag, and F. H. Gage, 2000.

241 ***The hippocampus of . . . adults who suffered . . . childhood trauma is 18 percent smaller:*** M. Vythilingam, C. Heim, J. Newport, A. H. Miller, E. Anderson, R. Bronen, M. Brummer, L. Staib, E. Vermetten, D. S. Charney, C. B. Nemeroff, and J. D. Bremner. 2002. Childhood trauma associated with smaller hippocampal volume in women with major depression. *American Journal of Psychiatry,* 159(12): 2072–80.

241 ***•If the stress . . . is too prolonged, the damage is permanent.*** According to Kandel, "Stress early in life produced by separation of the infant from its mother produces a reaction in the infant that is stored primarily by the procedural memory system, the only well-differentiated memory system that infant has early in its life, but this action of the procedural memory systems leads to a cycle of changes that ultimately damages the hippocampus and thereby results in a persistent change in declarative [i.e., explicit] memory." E. R. Kandel. 1999. Biology and the future of psychoanalysis: A new intellectual framework for psychiatry revisited. *American Journal of Psychiatry,* 156(4): 505–24, especially 515. See also L. R. Squire and E. R. Kandel. 1999. *Memory: From molecules to memory.* New York: Scientific American Library; B. S. McEwen and R. M. Sapolsky. 1995. Stress and cognitive function. *Current Opinion in Neurobiology,* 5:205–16.

241 ***•As people recover . . . their hippocampi can grow:*** B. L. Jacobs, H. van Praag, and F. H. Gage, 2000. This paper cites a report by Premal Shah and colleagues, at the Royal Edinburgh Hospital,

showing that hippocampal volume is smaller in chronically depressed patients but not in those who have recovered.

241 ***Rats given Prozac . . . increase in . . . cells:*** Ibid.

242 ***a "depletion of the plasticity" . . . in many older people:*** S. Freud. 1937/1964. Analysis terminable and interminable. In *Standard edition of the complete psychological works,* vol. 23, 241–42.

242 ***"some people . . . retain this mental plasticity":*** S. Freud. 1918/1955. An infantile neurosis. In *Standard edition of the complete psychological works,* vol. 17, 116.

Chapter 10. Rejuvenation

249 ***"In adult [brain] centers . . . Everything may die, nothing may be regenerated":*** S. Ramón y Cajal. 1913, 1914/1991. *Cajal's degeneration and regeneration of the nervous system.* J. DeFelipe and E. G. Jones, eds. Translated by R. M. May. New York: Oxford University Press, 750.

250 ***discovered these cells in 1998, in the hippocampus:*** P. S. Eriksson, E. Perfilieva, T. Björk-Eriksson, A. Alborn, C. Nordborg, D. A. Peterson, and F. H. Gage. 1998. Neurogenesis in the adult human hippocampus. *Nature Medicine,* 4(11): 1313–17.

250 ***"neurogenesis," and it goes on until the day that we die:*** H. van Praag, A. F. Schinder, B. R. Christie, N. Toni, T. D. Palmer, and F. H. Gage. 2002. Functional neurogenesis in the adult hippocampus. *Nature,* 415(6875): 1030–34; H. Song, C. F. Stevens, and F. H. Gage. 2002. Neural stem cells from adult hippocampus develop essential properties of functional CNS neurons. *Nature Neuroscience,* 5(5): 438–45.

250 ***•in 1965 . . . discovered them in rats:*** Finding neuronal stem cells in rats was an important finding because rats (and mice) share over 90 percent of their DNA with humans.

251 ***forty thousand new neurons:*** G. Kempermann, H. G. Kuhn, and F. H. Gage. 1997. More hippocampal neurons in adult mice living in an enriched environment. *Nature,* 386(6624): 493–95.

251 ***older mice raised in the enriched environment . . . a fivefold increase:*** G. Kempermann, D. Gast, and F. H. Gage. 2002. Neuroplasticity in old age: Sustained fivefold induction of hippocampal neurogenesis by long-term environmental enrichment. *Annals of Neurology,* 52:135–43.

252 ***After a month on the wheel, the mice had doubled the number:*** H. van Praag, G. Kempermann, and F. H. Gage. 1999. Running increases cell proliferation and neurogenesis in the adult mouse dentate gyrus. *Nature Neuroscience,* 2(3): 266–70.

253 ***as we age, we . . . perform cognitive activities in different lobes:*** M. V. Springer, A. R. McIntosh, G. Wincour, and C. L. Grady. 2005. The relation between brain activity during memory tasks and years of education in young and older adults. *Neuropsychology,* 19(2): 181–92.

254 ***one theory is that as we age . . . the other hemisphere compensates:*** R. Cabeza. 2002. Hemispheric asymmetry reduction in older adults: The HAROLD model. *Psychology and Aging,* 17(1): 85–100.

254 ***The more education . . . the less . . . Alzheimer's:*** R. S. Wilson, C. F. Mendes de Leon, L. L. Barnes, J. A. Schneider, J. L. Bienias, D. A. Evans, and D. A. Bennett. 2002. Participation in cognitively stimulating activities and risk of incident Alzheimer disease. *JAMA,* 287(6): 742–48.

254 ***musical instrument, playing board games, reading, and dancing—are associated with a lower risk:*** J. Verghese, R. B. Lipton, M. J. Katz, C. B. Hall, C. A. Derby, G. Kuslansky, A. F. Ambrose, M. Sliwinski, and H. Buschke. 2003. Leisure activities and the risk of dementia in the elderly. *New England Journal of Medicine,* 348(25): 2508–16.

254–255 •***It is possible that people with very early onset but undetectable Alzheimer's:*** The idea that Alzheimer's might begin in early adulthood and not be detectable for years comes from a famous study of nuns that found that those who developed Alzheimer's had used much simpler language in their twenties.

255 •***the common practices we should all be pursuing:*** I leave aside the question of supplements to the diet, which is not my subject, except to say that the old idea of eating fish, or fish oils with omega fatty acids, seems wise. But there are many other potential supplements. M. C. Morris, D. A. Evans, C. C. Tangney, J. L. Bienias, and R. S. Wilson. 2005. Fish consumption and cognitive decline with age in a large community study. *Archives of Neurology*, 62(12): 1849–53.

255 ***exercise stimulates . . . BDNF:*** S. Vaynman and F. Gomez-Pinilla. 2005. License to run: Exercise impacts functional plasticity in the intact and injured central nervous system by using neurotrophins. *Neurorehabilitation and Neural Repair*, 19(4): 283–95.

256 ***being social, which also preserves brain health:*** J. Verghese et al., 2003.

256 ***meditative aspect, which has been proven very effective:*** A. Lutz, L. L. Greischar, N. B. Rawlings, M. Ricard, and R. J. Davidson. 2004. Long-term meditators self-induce high-amplitude gamma synchrony during mental practice. *Proceedings of the National Academy of Sciences, USA*, 101(46): 16369–73.

256 ***Harvard Study of Adult Development:*** G. E. Vaillant. 2002. *Aging well: Surprising guideposts to a happier life from the landmark Harvard study of adult development.* Boston: Little, Brown, & Co.

257 ***people in their sixties and seventies . . . as productive as . . . in their twenties:*** H. C. Lehman. 1953. *Age and achievement.* Princeton, NJ: Princeton University Press; D. K. Simonton. 1990. Does creativity decline in the later years? Definition, data, and theory. In M. Permutter, ed., *Late life potential.* Washington, DC: Gerontological Society of America, 83–112, especially 103.

257 ***Casals . . . "Because I am making progress":*** Cited in G. E. Vaillant, 2002, 214. From H. Heimpel. 1981. Schlusswort. In M. Planck, ed., *Hermann Heimpel zum 80. Geburtstag.* Institut für Geschichte. Göttingen: Hubert, 41–47.

Chapter 11
More than the Sum of Her Parts

273 •***He suggested that Renata begin . . . thinking exercises:*** Grafman used the Preview, Question, Read, Study Test Method to help Renata improve her thinking and reading abilities.

274 •***Vietnam Head Injury Study:*** Most of the Vietnam veterans Grafman studied suffered *penetrating* head injuries—bullets, shrapnel, and flying metal had pierced their skulls and brains. The victim of a penetrating injury often does not lose consciousness, so about half the soldiers with these injuries walked into the surgical triage unit on their own and told the doctors they needed help.

274 ***IQ was . . . predictor of how well he would recover . . . brain functions:*** J. Grafman, B. S. Jonas, A. Martin, A. M. Salazar, H. Weingartner, C. Ludlow, M. A. Smutok, and S. C. Vance. 1988. Intellectual function following penetrating head injury in Vietnam veterans. *Brain,* 111:169–84.

276 ***Grafman has identified four kinds of plasticity:*** J. Grafman and I. Litvan. 1999. Evidence for four forms of neuroplasticity. In J. Grafman and Y. Christen, eds., *Neuronal plasticity: Building a bridge from the laboratory to the clinic.* Berlin: Springer-Verlag, 131–39; J. Grafman. 2000. Conceptualizing functional neuroplasticity. *Journal of Communication Disorders,* 33(4): 345–56.

276 ***"mirror region takeover" . . . grew out of work . . . with a boy I shall call Paul:*** H. S. Levin, J. Scheller, T. Rickard, J. Grafman, K. Martinkowski, M. Winslow, and S. Mirvis. 1996. Dyscalculia and dyslexia after right hemisphere injury in infancy. *Archives of Neurology,* 53(1): 88–96.

278 •***Migration of a mental function:*** Children with damage to their *right* nonverbal hemisphere (like Paul) don't do nearly so well in reorganizing their left hemispheres to take over the lost functions as Michelle did reorganizing her right hemisphere to take over her lost functions. This may be because key language

functions often develop earlier than nonverbal functions, and so when those "right-hemisphere" nonverbal functions seek to migrate to the left, they find that the left hemisphere is already committed to language.

280 ***Betty Edwards's popular book:*** B. Edwards. 1999. *The new drawing on the right side of the brain.* New York: Jeremy P. Tarcher/-Putnam, xi.

281 ***•In Grafman's view, Michelle's superior registration of events:*** Normally, the left prefrontal lobe registers a sequence of events. Grafman theorizes that after the right prefrontal lobe extracts the theme or meaning of those events, the same right prefrontal lobe probably inhibits the memory of those events in the left, because there is no need to keep all these details in their pure, vivid form. The ability to remember the day before, and what was important about it, Grafman says, "is a compromise between the details and the meaning." In Michelle, there is less of a compromise, because she doesn't have a separate hemisphere to inhibit the event registration. Hence the vividness of events persists.

Appendix 1
The Culturally Modified Brain

288 ***As Merzenich puts it, "Our brains are vastly different, in fine detail:*** Interview in S. Olsen. 2005. Are we getting smarter or dumber? CNet News.com. http://news.com.com/Are+we+getting+smarter+or+dumber/2008-1008_3-5875404.html.

289 ***•sunlight . . . "refracted":*** Refraction occurs because light bends as it passes from a substance of one density to another. The human eye is a terrestrial eye, evolved to accommodate light as it passes into it from the air, not from water.

289 ***Anna Gislén . . . studied the Sea Gypsies':*** A. Gislén, M. Dacke, R. H. H. Kröger, M. Abrahamsson, D. Nilsson, and E. J. Warrant. 2003. Superior underwater vision in a human population of Sea Gypsies. *Current Biology,* 13:833–36.

289 ***•pupil adjustment has been thought to be a fixed, innate reflex:*** The brain and the sympathetic and parasympathetic branches of the nervous system adjust pupil size.

289 ***the Sixth Paganini Etude . . . eighteen hundred notes per minute:*** T. F. Münte, E. Altenmüller, and L. Jäncke. 2002. The musician's brain as a model of neuroplasticity. *Nature Reviews Neuroscience,* 3(6): 473–78.

289 ***the more these musicians practice, the larger the brain maps:*** T. Elbert, C. Pantev, C. Wienbruch, B. Rockstroh, and E. Taub. 1995. Increased cortical representation of the fingers of the left hand in string players. *Science,* 270(5234): 305–7.

290 ***in trumpeters . . . maps that respond to "brassy" sounds enlarge:*** C. Pantev, L. E. Roberts, M. Schulz, A. Engelien, and B. Ross. 2001. Timbre-specific enhancement of auditory cortical representations in musicians. *NeuroReport,* 12(1): 169–74.

290 ***Imaging also shows that musicians . . . playing before the age of seven have larger brain areas:*** T. F. Münte, E. Altenmüller, and L. Jäncke, 2002.

290 ***As Vasari writes,* ". . . Michelangelo . . . could only read and look at designs in that posture":** G. Vasari. 1550/1963. *The lives of the painters, sculptors and architects,* vol. 4. New York: Everyman's Library, Dutton, 126.

290 ***•inversion glasses . . . turn the world upside down . . . perceptual centers "flip," so that they . . . read books held upside down:*** There are countless other examples of the brain adapting to unusual situations. Plasticity researcher Ian Robertson notes that NASA has found that after a flight, it takes four to eight days for astronauts to regain their balance, which Robertson argues is likely a plastic effect; in the condition of weightlessness the sense of balance does not tell them where their bodies are in space, so they must rely on their eyes. Thus weightlessness leads to two brain alterations. The balance system, which gets no input, is stepped down (in a case of use it or lose it), and the eyes, given massed practice, are stepped up to inform the astronaut where he is in space.

290 ***London taxi drivers show . . . larger . . . hippocampus:*** E. A. Maguire, D. G. Gadian, I. S. Johnsrude, C. D. Good, J. Ashburner, R. S. J. Frackowiak, and C. D. Frith. 2000. Navigation-related structural change in the hippocampi of taxi drivers. *Proceedings of the National Academy of Sciences, USA,* 97(8): 4398–4403.

290 ***meditators . . . have a thicker insula:*** S. W. Lazar, C. E. Kerr, R. H. Wasserman, J. R. Gray, D. N. Greve, M. T. Treadway, M. McGarvey, B. T. Quinn, J. A. Dusek, H. Benson, S. L. Rauch, C. I. Moore, and B. Fischl. 2005. Meditation experience is associated with increased cortical thickness. *NeuroReport,* 16(17): 1893–97.

291 •***plasticity, also part of our genetic heritage:*** We are just beginning to understand the genetics of neuroplasticity. Frederick Gage and his team, which proved that mice reared in enriched environments develop new neurons and larger hippocampi, have also discovered that one of the most powerful predictors of whether a mouse will be able to make new neurons is genetically determined.

292 •***"cognitive fluidity" . . . probably has its basis in brain plasticity:*** Cognitive fluidity, according to cognitive archaeologist Steven Mithen, may explain one of the great mysteries of human prehistory, namely the sudden explosion of human culture.

Homo sapiens first walked the earth about 100,000 years ago, and for the next 50,000 years, based on archaeological evidence, human culture was static and hardly more complex than that of the other prehuman species that had preceded us for almost a million years. Archaeological remains from this protracted period of cultural monotony pose several riddles. First, human beings used only stone or wood to make tools and not bone, ivory, or antlers, which were also available. Second, while these humans had invented a general-purpose ax, they never developed an ax, or any tool, for specific purposes. All spear points were the same size and made in the same way. Third, no tools were ever made of several components, such as the Inuit harpoon that has hard stone points, ivory shafts, thongs for retrieval, and inflated seal skins so they float after being thrown. And finally, there were no signs of art, decoration, or religion.

Then fifty thousand years ago, all of a sudden, with no fundamental change in our brain size or genetic makeup, this all changed, and art, religion, and complex technologies developed. Boats were invented that took human beings over the sea to Australia; cave drawings emerged; imaginative bone and ivory carvings of hybrid creatures made of animal and human shapes, and bead and pendant decorations for the human body, became common. They began burying their dead in pits, and human corpses now had the carcasses of animals beside them—"grave goods" of food supplies for the afterlife—the first evidence of religion. And for the first time, tools were designed for specific purposes, and spearheads were tailored for different-size prey and took into account the thickness of their prey's skin and their habitat.

Mithen argues that the period of cultural monotony occurred because Homo sapiens had three separate intelligence modules, which each worked independently. The first module was a natural history intelligence, shared with many animals, which allowed humans to understand the habits of game animals, weather, and geography; how tracks in the ground and feces of a certain kind predicted finding an animal ahead; or that birds' leaving predicted the coming of winter. The second module was technical intelligence, understanding how to manipulate objects, such as stones, and turn them into blades. The third module was social intelligence, also shared with other animals, which allowed humans to interact with and read the emotions of other humans and to understand dominance and submission hierarchies, courtship rituals, and how to nurture the young.

Mithen theorizes that cultural monotony existed because the three intelligence modules were separate in the mind. Thus early humans never carved bone or ivory, because bone was an animal product, and they had a mental barrier between technical intelligence and animal intelligence, so they couldn't think of using animals for tools. They didn't have specific kinds of tools for different purposes, or complex tools, because creating them would have required integrating natural history intelligence (the thickness of hides, the size of animals, different habits) with technical intelligence. A barrier must have existed between social and technical intelligence, since no beads, pendants, or other bodily decorations (which designate a person's social affiliation, religion, and status, much as wedding rings, crucifixes, and diamonds do in the West) have been found.

Fifty thousand years ago these barriers broke down. Complex tools useful for different purposes emerged. Art showed the mixing of all three kinds of intelligence, as in the case of a statue of a lion-man found in southern Germany. This carved object (technical intelligence) depicted the body of a man (social intelligence), combined with the head of a lion and the tusk of a mammoth (natural history intelligence). In France, ivory beads were carved to mimic seashells, a mingling of natural history and technical intelligence, and new tools were found with animals carved on them. Primitive religion, sometimes called "totemism," developed, which merges a human social group's identity with a totem animal—suddenly giving the natural world a social meaning.

Mithen argues that all this creativity, in the absence of a change in brain size, came about because "cognitive fluidity" permitted the breakdown of barriers among the three intelligence modules and allowed the mind to reorganize. And what might have allowed these modules to link up?

I would argue that the plasticity of the brain could have caused these three different neuronal groups or modules to link up and is the neural analog of cognitive fluidity. Why did the modules not link up sooner? Because plasticity is always a double-edged sword and can lead to rigidity as well as flexibility; if these modules evolved in animals and primates for specialized purposes, they would tend to continue to be used for their original purpose—the way a sled that makes tracks on its first run tends to stay in them. But that would not mean that the intelligence modules could never mix, only that they were predisposed to remain separate—until it was discovered, perhaps by accident, that mixing them gave Homo sapiens a distinct advantage. See S. Mithen. 1996. *The prehistory of the mind: The cognitive origins of art, history and science.* London: Thames & Hudson.

292 ***An fMRI study . . . we recognize cars and trucks . . . faces:*** I. Gauthier, P. Skudlarski, J. C. Gore, and A. W. Anderson. 2000. Expertise for cars and birds recruits brain areas involved in face recognition. *Nature Neuroscience,* 3(2): 191–97.

293 ***According to Merzenich, "Our brains are different from those of all humans before us":*** Interview in S. Olsen, 2005.

293–294 ***Robert Sapolsky points out . . . A regulatory gene . . . differs between humans and chimps:*** R. Sapolsky. 2006. The 2% difference. *Discover,* April, 27(4): 42–45.

294 ***Edelman writes, "If we considered the number of possible neural circuits":*** G. M. Edelman and G. Tononi. 2000. *A universe of consciousness: How matter becomes imagination.* New York: Basic Books, 38.

295 ***as Gerald Edelman argues, "smaller parts . . . new functions":*** G. Edelman. 2002. A message from the founder and director. *BrainMatters.* San Diego: Neurosciences Institute, Fall, 1.

295–296 ***Neville . . . found that deaf people intensify their peripheral vision:*** H. J. Neville and D. Lawson. 1987. Attention to central and peripheral visual space in a movement detection task: An event-related potential and behavioral study. II. Congenitally deaf adults. *Brain Research,* 405(2): 268–83.

299 ***•culture shock is brain shock:*** Learning a new culture as an adult requires that one use new parts of the brain, at least for language. Brain scans show that people who learn one language and then, after a time lag, learn another store the languages *in separate areas.* When bilingual people have strokes, they sometimes lose the ability to speak one language but not the other. Such people

have distinct neuronal networks for their two languages, and perhaps for other aspects of their two cultures. But brain scans also show that children raised learning two languages simultaneously during the critical period develop an auditory cortex that represents both languages together. This is why Merzenich advocates learning as many *different* language sounds as possible in early childhood: such children develop a single, large cortical library of sounds and have an easier time learning languages later in life. For brain scan studies, see S. P. Springer and G. Deutsch. 1998. *Left brain, right brain: Perspectives from cognitive science,* 5th ed. New York: W. H. Freeman & Co., 267.

300 ***Merlin Donald . . . argued in 2000 that culture changes our functional cognitive architecture:*** M. Donald. 2000. The central role of culture in cognitive evolution: A reflection on the myth of the "isolated mind." In L. Nucci, ed., *Culture, thought and development.* Mahwah, NJ: Lawrence Erlbaum Associates, 19–38.

300 ***"where people in one culture differ . . . it can't be because they have different cognitive processes":*** R. E. Nisbett. 2003. *The geography of thought: How Asians and Westerners think differently . . . and why.* New York: Free Press, xii–xiv.

301 ***the peoples of the East . . . and . . . the West . . . perceive in different ways:*** R. E. Nisbett, K. Peng, I. Choi, and A. Norenzayan. 2001. Culture and systems of thought: Holistic versus analytic cognition. *Psychological Review,* 291–310.

301 ***• Westerners approach the world "analytically":*** The word "analyze," which comes from ancient Greek, means "break up into pieces," and to analyze a problem means to break it into parts. The analytic habit of mind affected how the Greeks saw the world. Greek scientists were the first to argue that matter was formed of discrete objects called atoms; Greek physicians learned by dissection, cutting the body into pieces, and developed surgery to remove malfunctioning parts; logic, being typically Greek in origin, solves a problem by *isolating* a portion of it—the structure of the argument—from its original context.

301 ***•Easterners tend to approach the world . . . by looking at "the whole":*** Instead of seeing matter as discrete atoms, the Chinese saw it as continuous interpenetrating substances. They were more interested in understanding an object's context than in focusing on it in isolation. Chinese scientists were interested in fields of forces and how things influence each other; they had early insights into magnetism and acoustic resonance and discovered, long before Westerners, that the moon moves the tides. In medicine, the Chinese—after practicing dissection and surgery for some time—abandoned them and pioneered holistic medicine, preferring to look at the body as a single system.

301 ***•The left hemisphere . . . more sequential and analytical . . . the right hemisphere . . . simultaneous and holistic:*** The left hemisphere is more involved in processing abstract verbal analytical thought (and some believe logic) and in perceiving things *sequentially*. Right-hemisphere thinking is more holistic and perceives things all at once, or *simultaneously,* and hence is often called more synthetic, intuitive, or Gestalt-like. (S. P. Springer and G. Deutsch. 1998. *Left brain, right brain: Perspectives from cognitive science,* 5th ed. New York: W. H. Freeman & Co., 292.) But even if Western civilization favors the left hemisphere, and the East the right, there still must be a mechanism by which this occurs. There's good reason to believe that this mechanism is based on plasticity, and not just genetics, because when people try to change civilizations, their perception alters.

301 ***•he believed that we all perceive and reason in the same way:*** R. E. Nisbett, 2003. Nisbett, a specialist in understanding reason, initially believed that reasoning, like perception, was universal, innate, and hardwired in the brain. So certain was he that reason was hardwired that he believed it could not be taught and set out to prove it couldn't. In his experiments he attempted to teach people rules for reasoning, to use in their everyday lives. To his surprise, his experiments showed the opposite: reasoning could be taught. This was an important finding because

education, particularly in America, had moved away from teaching abstract rules for reasoning, in part because of a disbelief in plasticity. Critiquing the classical curriculum, which extended back to Plato, the greatest American psychologist of his day, William James, mocked the study of abstract reasoning rules because it implied that it was possible to exercise some nonexistent "muscles of the mind." Cited in R. E. Nisbett, ed. 1993. *Rules for Reasoning*. Hillsdale, NJ: Lawrence Erlbaum Associates, 10. In Plato's *Republic*, studying mathematics is described as a "gymnastic" practice, a form of mental exercise. Plato. 1968. *The Republic of Plato*. Translated by A. Bloom. New York: Basic Books, 526b, p. 205.

302 •***when people change cultures, they learn to perceive in a new way:*** Shinobu Kitayama, using the kinds of perceptual experiments that Nisbett developed, showed that Americans who lived in Japan for a few months started to perform like the Japanese on the perceptual tests. Japanese who lived in America for a few years became indistinguishable from Americans. These time frames are what one might expect for a plastic alteration in perceptual learning circuits. Holistic or analytic ways of perceiving are never formally taught to immigrants, of course, but immersion in a civilization causes perceptual learning, because the environment—its language, tastes, aesthetics, philosophy, approach to science, and everyday life—continually reiterates the basic perceptual premises of that civilization, so that the visitors cannot avoid having their brains undergo massed practice. Philip Zelazo, at the University of Toronto, is currently involved in comparing the effects of culture on the development of attention and frontal lobe functions in China and the West; he has found that one's culture has an impact on cognitive development and believes it probably affects neural development as well.

302 ***children of Asian-American immigrants perceive:*** R. E. Nisbett, 2003, *The geography of thought*.

303 ***people raised in a bicultural situation actually alternate:*** Ibid.

303 ***the stationary eye is virtually incapable of perceiving:*** A. Luria. 1973. *The working brain: An introduction to neuropsychology.* London: Penguin, 100.

303 ***Both our sensory* and *our motor cortices are always involved in perceiving:*** Ibid.; A. Noë. 2004. *Action in perception.* Cambridge, MA: MIT Press.

303 ***"higher" levels of perception affect . . . neuroplastic change in the "lower," sensory parts:*** M. Fahle and T. Poggio. 2002. *Perceptual learning.* Cambridge, MA: A Bradford Book, MIT Press, xiii, 273; W. Li, V. Piëch, and C. D. Gilbert. 2004. Perceptual learning and top-down influences in primary visual cortex. *Nature Neuroscience,* 7(6): 651–57.

304 ***He replied . . . "They don't know how to look":*** B. Simon. Sea Gypsies see signs in the waves. March 20, 2005. www.cbsnews.com/stories/2005/03/18/60minutes/main681558.shtml.

304 ***Bruce Wexler . . . in his book:*** B. E. Wexler. 2006. *Brain and culture: Neurobiology, ideology, and social change.* Cambridge, MA: MIT Press.

305 ***For instance, North Korea, the . . . totalitarian regime:*** P. Goodspeed. 2005. Adoration 101. *National Post,* November 7; P. Goodspeed. 2005. Mysterious kingdom: North Korea remains an enigma to the outside world. *National Post,* November 5.

306 ***brainwashing . . . "unlearn" their preexisting mental structures:*** W. J. Freeman. 1995. *Societies of brains: A study in the neuroscience of love and hate.* Hillsdale, NJ: Lawrence Erlbaum Associates; W. J. Freeman. 1999. *How brains make up their minds.* London: Weidenfeld & Nicolson; R. J. Lifton. 1961. *Thought reform and the psychology of totalism.* New York: W. W. Norton & Co.; W. Sargant. 1957/1997. *Battle for the mind: A physiology of conversion and brain-washing.* Cambridge, MA: Malor Books.

306 ***"The Internet is":*** Michael Merzenich interviewed in S. Olsen. 2005. Are we getting smarter or dumber? CNet News.com. http://news.com.com/Are+we+getting+smarter+or+dumber/2008-1008_3-5875404.html.

306 ***believed that "the mind exists and develops entirely in the head":*** M. Donald, 2000, 21.

307 ***early exposure to television . . . correlates with problems paying attention and controlling impulses:*** D. A. Christakis, F. J. Zimmerman, D. L. DiGiuseppe, and C. A. McCarty. 2004. Early television exposure and subsequent attentional problems in children. *Pediatrics,* 113(4): 708–13.

307 ***This study, as psychologist Joel T. Nigg argues, did not perfectly control:*** Joel T. Nigg. 2006. *What causes ADHD?* New York: Guilford Press.

307 ***Forty-three percent of U.S. children two years or younger watch television daily:*** V. J. Rideout, E. A. Vandewater, and E. A. Wartella. 2003. *Zero to six: Electronic media in the lives of infants, toddlers, and preschoolers.* Publication no. 3378. Menlo Park, CA: Kaiser Family Foundation, 14.

307 ***a quarter have TVs in their bedrooms:*** J. M. Healy. 2004. Early television exposure and subsequent attention problems in children. *Pediatrics,* 113(4): 917–18; V. J. Rideout, E. A. Vandewater, and E. A. Wartella, 2003, 7, 17.

307 ***Healy . . . in her book* Endangered Minds:** J. M. Healy. 1990. *Endangered minds: Why our children don't think.* New York: Simon & Schuster.

307–308 ***attention deficit traits, which are not genetic:*** E. M. Hallowell. 2005. Overloaded circuits: Why smart people underperform. *Harvard Business Review,* January, 1–9.

308 ***brain exercises to treat attention deficit disorder:*** R. G. O'Connell, M. A. Bellgrove, P. M. Dockree, and I. H. Robertson. 2005. Effects of self alert training (SAT) on sustained attention performance in adult ADHD. Cognitive Neuroscience Society, Conference, April, poster.

308 ***"The medium is the message":*** M. McLuhan, 1964/1994; W. T. Gordon, ed. *Understanding media: The extensions of man, critical edition.* Corte Madera, CA: Ginkgo Press, 19.

308 ***brain scan study to test whether the medium is indeed the message:*** E. B. Michael, T. A. Keller, P. A. Carpenter, and M. A. Just. 2001. fMRI investigation of sentence comprehension by eye and by ear: Modality fingerprints on cognitive processes. *Human Brain Mapping,* 13:239–52; M. Just. 2001. The medium and the message: Eyes and ears understand differently. *EurekAlert,* August 14, www.eurekalert.org/pub_releases/2001-08/cmu-tma081401.php.

309 ***"The ratio among our senses is altered":*** E. McLuhan and F. Zingrone, eds. 1995. *Essential McLuhan.* Toronto: Anansi, 119–20.

309 ***dopamine . . . released in the brain during . . . games:*** M. J. Koepp, R. N. Gunn, A. D. Lawrence, V. J. Cunningham, A. Dagher, T. Jones, D. J. Brooks, C. J. Bench, and P. M. Grasby. 1998. Evidence for striatal dopamine release during a video game. *Nature,* 393(6682): 266–68.

309 ***•Television . . . high-speed transitions:*** The show *24* has many more characters and plots and subplots than similar shows from twenty years before. A single forty-four-minute episode had twenty-one distinct characters, each with a clearly defined story. S. Johnson. 2005. Watching TV makes you smarter. *New York Times,* April 24.

309 ***television . . . activating . . . the "orienting response":*** R. Kubey and M. Csikszentmihalyi. 2002. Television addiction is no mere metaphor. *Scientific American,* February, 23.

311 ***"Now man is beginning to wear his brain outside":*** M. McLuhan. 1995. *Playboy* interview. In E. McLuhan and F. Zingrone, eds., 264–65.

311 ***"Today . . . we have extended our central nervous system":*** M. McLuhan, 1964/1994.

Appendix 2: Plasticity and the Idea of Progress

313 ***•Rousseau . . . argued that nature was alive and had a history:*** Rousseau was inspired by the naturalist Buffon, who discovered that the earth was much older than people had thought, and that

its rocks contained fossils of animals that once existed, but no longer did, confirming that even the bodies of animals, once thought to be immutable, could change. A new science called natural history emerged in Rousseau's time that saw all living things as having a history.

One reason that Rousseau might have been so open to the idea of natural history and plasticity was his immersion in the ancient Greek classics. As we have seen (in the third note to chapter 1), the Greeks viewed nature as a vast *living organism.* Because all nature was alive, it is not likely that they would have been opposed to the idea of plasticity in principle. As we have seen, Socrates, in the *Republic,* argued that a person could train his mind the way gymnasts trained their muscles.

After the discoveries of Galileo, the second great idea of nature emerged, that of *nature as mechanism,* which drained the brain of life and tended to be opposed to the idea of plasticity, almost in principle.

The third, grander idea of nature, inspired by Buffon, Rousseau, and others, restored life to nature, depicting it as an evolving *historical process* that is changing over time, and brought back much of the vitality that was inherent in the ancient Greek view of it. See R. G. Collingwood. 1945. *The idea of nature.* Oxford: Oxford University Press; R. S. Westfall. 1977. *The construction of modern science: Mechanisms and mechanics.* Cambridge: Cambridge University Press, 90.

313 ***our nervous systems are not like machines, he said, but are alive and able to change:*** J. J. Rousseau. 1762/1979. *Emile, or on education.* Translated by A. Bloom. New York: Basic Books, 272–82, especially 280.

313 ***the "organization of the brain" was affected by our experience, and that we need to "exercise" our senses and mental abilities:*** Ibid., 132; also 38, 48, 52, 138.

314 ***•brought the French word* perfectibilité *into vogue:*** He also saw it as a mixed blessing and wrote, "Why is man alone liable to

become an imbecile? Is it not that he thus returns to his primitive state and that, whereas the Beast, which has acquired nothing and also has nothing to lose, always keeps its instinct, man, losing through old age or other accidents all that his *perfectibility* had made him acquire, thus relapses lower than the Beast itself? It would be sad for us to be forced to agree that this distinctive and almost unlimited faculty is the source of all of man's miseries; that it is the faculty which, by dint of time, draws him out of that original condition in which he would spend calm and innocent days; that it is the faculty which, over the centuries, causes his enlightenment and his errors, his vices and his virtues to arise, and eventually makes him his own and Nature's tyrant." J. J. Rousseau. 1755/1990. *The first and second discourses, together with the replies to critics and essay on the origin of languages.* Translated and edited by V. Gourevitch. New York: Harper Torchbooks, 149, 339.

314 ***the balance between our senses and our imagination can be disturbed by the wrong kinds of experience:*** J. J. Rousseau, 1762/1979, 80–81; J. J. Rousseau, 1755/1990, 149, 158, 168; L. M. MacLean, 2002. The free animal: Free will and perfectibility in Rousseau's *Discourse on Inequality.* Ph.D. thesis, University of Toronto, 34–40.

314 •***Charles Bonnet (1720–1793):*** Bonnet made important discoveries about a form of reproduction in which unfertilized eggs reproduce themselves, by themselves, without sperm. He was especially interested in regeneration and studied how animals, such as crabs, could re-create lost limbs after they were bitten off. Of course, when a crab's claw regenerates, so does the nervous tissue within that claw; thus Bonnet had an interest in adult nervous tissue growth. Of interest is that Bonnet, like Rousseau, was also Swiss, also of Geneva; he became Rousseau's ardent enemy, attacked Rousseau's political writings in print, and worked to ban them.

314–315 ***Bonnet . . . wrote to . . . Malacarne . . . proposing that neural tissue might respond to exercise as do muscles:*** M. J. Renner and

M. R. Rosenzweig. 1987. *Enriched and impoverished environments: Effects on brain and behavior.* New York: Springer-Verlag, 1–2; C. Bonnet. 1779–1783. *Oeuvres d'histoire naturelle et de philosophie.* Neuchâtel: S. Fauche.

315 ***Malacarne's work:*** M. J. Renner and M. R. Rosenzweig, 1987; M. Malacarne. 1793. *Journal de physique,* vol. 43: 73, cited in M. R. Rosenzweig. 1996. Aspects of the search for neural mechanisms of memory. *Annual Review of Psychology,* 47:1–32, especially 4; G. Malacarne. 1819. *Memorie storiche intorno alla vita ed alle opere di Michele Vincenzo Giacinto Malacarne.* Padua: Tipografia del Seminario, 88.

316 ***Rousseau used the term "perfectibility" in an ironic sense:*** R. L. Velkley. 1989. *Freedom and the end of reason: On the moral foundation of Kant's critical philosophy.* Chicago: University of Chicago Press, 53.

316 ***"the perfectibility of man is truly indefinite":*** A.-N. de Condorcet. 1795/1955. *Sketch for a historical picture of the progress of the human mind.* Translated by J. Barraclough. London: Weidenfeld & Nicolson, 4.

316 ***Jefferson . . . introduced to Condorcet by Benjamin Franklin:*** V. L. Muller. 1985. *The idea of perfectibility.* Lanham, MD: University Press of America.

316 ***"I believe also, with Condorcet . . . that his mind is perfectible":*** T. Jefferson. 1799. To William G. Munford, 18 June. In B. B. Oberg, ed., 2004. *The papers of Thomas Jefferson,* vol. 31: *1 February 1799 to 31 May 1800.* Princeton: Princeton University Press, 126–30.

316 ***Tocqueville . . . remarked that Americans . . . seemed to believe in the "indefinite perfectibility of man":*** A. de Tocqueville. 1835/1840/2000. *Democracy in America.* Translated by H. C. Mansfield and D. Winthrop. Chicago: University of Chicago Press, 426.

317 ***Thomas Sowell has observed, "While the use of the word 'perfectibility' . . . 'the human being is highly plastic material'":*** T. Sowell. 1987. *A conflict of visions.* New York: William Morrow, 26.

Index